NECROBOTICS FOR HEALTHCARE APPLICATIONS AND MANAGEMENT

NECROBOTICS FOR HEALTHCARE APPLICATIONS AND MANAGEMENT

Edited by

HEMACHANDRAN K.
Department of Artificial Intelligence & Business Analytics, Woxsen University, Hyderabad, India

RAUL V. RODRIGUEZ
Woxsen University, Hyderabad, India

MURUGAPPAN M.
Kuwait College of Science and Technology, Kuwait

SIVARAMAKRISHNAN R.
Guidehouse Digital LLC, West McLean, VA, United States

RAJESH KUMAR K.V.
Department of Artificial Intelligence & Business Analytics, Woxsen University, Hyderabad, India

ELSEVIER

ACADEMIC PRESS
An imprint of Elsevier

Academic Press is an imprint of Elsevier
125 London Wall, London EC2Y 5AS, United Kingdom
525 B Street, Suite 1650, San Diego, CA 92101, United States
50 Hampshire Street, 5th Floor, Cambridge, MA 02139, United States

ISBN 978-0-443-24832-0

For information on all Academic Press publications
visit our website at https://www.elsevier.com/books-and-journals

Publisher: Mara Conner
Acquisitions Editor: Sonnini Yura
Editorial Project Manager: Tessa Katryn
Production Project Manager: Kamesh R
Cover Designer: Greg Harris

Typeset by STRAIVE, India

Contents

Contributors

Harshini Alapati
Woxsen University, School of Business, Hyderabad, India

N.V. Aniruddha Madhav
Woxsen University, School of Business, Hyderabad, India

Kande Archana
Malla Reddy College of Engineering, Hyderabad, Telangana, India

Maram Ashok
Department of Computer Science and Engineering, Malla Reddy College of Engineering, Hyderabad; Malla Reddy College of Engineering, Secunderabad, Telangana, India

Shrilekha Banerjee
National University of Juridical Science, Law, Kolkata, West Bengal, India

Avnish Bora
Jodhpur Institute of Technology, Jodhpur, India

P. DeepanRamkumar
Vellore Institute of Technology, Amaravathi, India

Ritu Gautam
Sharda University, Greater Noida, India

Puranjoy Ghosh
KIIT (Deemed to be University), School of Law, Bhubaneswar, Odisha, India

Shyam Joshi
AI Research Centre, Woxsen University, Hyderabad, India

Antony Kovilpillai Judeson
Alliance University, CSE, Bengaluru, India

Dhanya Madhuri Kandari
Woxsen University, School of Business, Hyderabad, India

Kamal Karthik Karri
Woxsen University, School of Business, Hyderabad, India

Christian Kaunert
Dublin City University, Dublin, Ireland

Basu Kaustav
Woxsen University, School of Technology, Hyderabad, India

Priyanka Ketha
Woxsen University, School of Business, Hyderabad, India

Aadil Mir
Woxsen University, Department of Analytics, Hyderabad, India

Murugappan M.
Kuwait College of Science and Technology, Intelligent Signal Processing (ISP) Research Lab, Department of Electronics and Communication Engineering, Kuwait, Kuwait

Jadav Nehru
Woxsen University, School of Business, Hyderabad, India

Radhakrishanan Niranchana
Alliance University, Department of Computer Science & Engineering, Bengaluru, India

Vaishnavi Paluru
Woxsen University, School of Business, Hyderabad; Woxsen University, School of Business, Nellore, India

Gajbhiye Pranjali
Woxsen University, AI Research Centre, SOB, Hyderabad, India

V Kamakshi Prasad
Jawaharlal Nehru Technological University, Hyderabad, Telangana, India

Partha Pratim
Department of Computer Applications, Sikkim University, Gangtok, India

Ray
Department of Computer Applications, Sikkim University, Gangtok, India

Harsha Vardhan Reddy
SRM Institute of Science and Technology, College of Engineering and Technology, Biomedical Engineering Department, Chennai, Tamil Nadu, India

Roshan A. Reji
Woxsen University, School of Business, Hyderabad, India

P.R. Sai Soumya
Woxsen University, School of Business, Hyderabad, India

Vandanapu Shashank
Woxsen University, MBA General, Hyderabad, India

Anusha Siddadapu Venkata
Woxsen School of Business, MBA Business Analytics, Sadashivpet, India

Bhupinder Singh
Sharda School of Law, Sharda University, Greater Noida, India

Babu Gole Srihari
Geethanjali College of Engineering and Technology, Department of CSE—DS, Hyderabad, India

Gulam Mahabub Subhani
Koneru Lakshmaiah Education Foundation; Department of Computer Science and Engineering, Malla Reddy College of Engineering, Hyderabad, Telangana, India

Sani Sudeshna
Woxsen University, AI Research Centre, SOB, Hyderabad, India

Ch Sumalakshmi
Department of Computer Science and Engineering, Koneru Lakshmaih Education Foundation, Hyderabad, Telangana, India

Snekhalatha Umapathy
SRM Institute of Science and Technology, Biomedical Engineering, Kattankulathur, Tamil Nadu, India

CHAPTER 1

Introduction to necrobotics

Partha Pratim and Ray
Department of Computer Applications, Sikkim University, Gangtok, India

1 Introduction

Necrobotics, a term first introduced to the scientific community in July 2022 by researchers at Rice University's Preston Innovation Lab (Yap et al., 2022), delineates the innovative practice of repurposing deceased organisms for robotic applications. This pioneering concept, vividly demonstrated through the transformation of dead spiders into functional robotic grippers, represents a significant stride in the interdisciplinary melding of biology and engineering. By utilizing the inherent organic hydraulic systems of spiders, the researchers crafted efficient, simple-to-produce grippers capable of manipulating small and lightweight objects. This approach not only showcases a novel utilization of biological materials but also positions necrobotics as a potential game-changer in the realm of robotic design. The fascinating journey from conceptualization in a lab in Houston, Texas, to practical demonstrations underscores necrobotics' role in redefining the traditional paradigms of robotic capabilities and efficiency. Through the ingenuity of converting a spider's natural hydraulic mechanism into a pneumatic one, this field offers a tantalizing glimpse into the future of biohybrid systems, blending the precision and adaptability of robotics with the intricate complexity of biological structures.

With necrobotics, biological materials, scrupulously harvested from deceased organisms, are seamlessly woven into the fabric of robotic devices, thereby birthing systems where the organic and the artificial not only coexist but synergistically thrive (Williams, 2022; Ackerman, 2022). This delicate dance between life and machine gives rise to entities that simultaneously draw from the remarkable faculties inherent to biological organisms and the pinpoint accuracy and programmability characteristic of robotic systems. In essence, necrobotics is the beacon of relentless technological progression, with its roots deeply embedded in a fascination with the biological wonders crafted meticulously by nature over millennia and a resolute dedication to the finesse of engineering (Kambouris et al., 2023; Vincent, 2022). The discourse presented herein navigates through the labyrinthine alleys of necrobotics, offering readers a granular view into its multifaceted dimensions. We embark on a comprehensive exploration, dissecting the meticulous processes involved in the selection and preparation of biological materials, the intricate design and control mechanisms woven into these novel systems, and the ethical considerations foundational to the responsible and sustainable

Necrobotics for Healthcare Applications and Management
https://doi.org/10.1016/B978-0-443-24832-0.00001-5

development of this field (*Scientists Turned Dead Spiders into Robots*, 2022; Smith et al., 2023). The text further illuminates the challenges intertwined with preserving the integrity of biological components, crafting interfaces that seamlessly marry biology and electronics, and ensuring the operational reliability and consistency that are paramount for the effective functionality of these pioneering systems. Moreover, the scholarly exposition unfurls the canvas of necrobotics' potential applications, painting a vivid picture of its wide-ranging impact across diverse sectors, including healthcare, environmental conservation, the arts, entertainment, and space exploration. In this academic odyssey, readers will delve into the depths of a field that transcends the mere creation of machines, embodying a forward-thinking paradigm that shapes the future trajectory of technology where the lines demarcating the organic and synthetic become indiscernible (Mestre et al., 2023; Raman et al., 2016).

As we navigate this comprehensive narrative, the text promises to shed light on a realm where innovation is unfettered by traditional constraints, presenting a vista where potential is boundless and teeming with promise (Zolfagharian et al., 2017; Klein, 2016). In our hands lies a gateway to understanding the realm of necrobotics, a field pulsating with transformative energy and laden with the seeds of change that hold the promise of a future radiant with technological marvels and discoveries.

The key contributions of this article are multifaceted, as mentioned below.

- The paper provides a detailed background on the underling processes involved in the selection and preparation of biotic materials, illuminating the pathway for translating the theoretical underpinnings of necrobotics into practical, feasible applications.
- It delineates the prospective design and fabrication strategies for necrobotic devices, offering insights into how these systems can be leveraged effectively for diverse applications.
- The text establishes foundational guidelines for envisaged control mechanisms integral to the functioning of necrobotics systems.
- It outlines prospective use cases for necrobotics, highlighting its potential impact and relevance across various domains.
- The paper identifies and discusses key challenges that need to be addressed to propel the field of necrobotics forward, providing a roadmap for future research endeavors.
- Finally, it sketches future directions for the field, offering a lens through which we might envision and navigate the path ahead in this exciting and rapidly evolving domain of innovation.

2 Necrobotic system design

The careful selection and meticulous preparation of biotic materials are fundamental in the creation of effective and reliable necrobotic devices. This process begins with a clear understanding of the device's objectives and required functionalities, guiding the identification of appropriate biological materials that align with these goals. The mechanical and structural needs of the device, including attributes like tensile strength, compressibility, and elasticity, are considered, ensuring the chosen materials can endure operational stresses without premature failure (Rousis et al., 2012). The selected biotic materials, whether muscle tissues for movement, bone structures for support, or sensory organs for environmental perception, are then subjected to a comprehensive preparation process. This involves careful cleaning to remove extraneous matter, preservation to halt decay, modification to optimize compatibility and performance within robotic systems, and stringent quality control protocols to ensure reliability and consistency in the final necrobotic device. Equally critical is the adherence to ethical standards and legal norms in sourcing and handling these materials, emphasizing respect for donors, animal welfare, cultural sensitivities, and sustainability. Furthermore, the storage conditions for these prepared materials require vigilant monitoring and management to preserve their quality, integrity, and functionality, entailing controlled environments regarding temperature, humidity, cleanliness, and security (Pasquero and Poletto, 2022; Saric et al., 2018). Through this intricate process of selection, preparation, and storage, necrobotic devices are endowed with biotic materials that not only meet the immediate technical specifications but are also adaptable and scalable for future improvements and modifications.

2.1 Study design requirements

To initiate a comprehensive study of necrobotics, a nuanced grasp of diverse domains is indispensable. Necrobotics is a deeply interdisciplinary field, bridging the foundational principles of biology, engineering, and computer science to facilitate the innovative development of robotic systems. These systems distinctively incorporate biological materials, creating a synergy that results in advanced and responsive robotic mechanisms. Outlines the essential requirements necessary for embarking on the design and research of a study in the dynamic realm of necrobotics.

2.1.1 Biological knowledge

An understanding of anatomy and physiology is nonnegotiable, providing crucial insights into the Fig. 1.1 functions and structures of various biological elements, laying the groundwork for their effective integration into robotic systems. Furthermore, proficiency in cell biology and biochemistry is mandatory. This foundational knowledge encompasses the detailed study of cellular structures, functions, and the myriad biochemical processes occurring within living organisms. Such knowledge is indispensable for the effective manipulation and preservation of biological materials in applications within the field of necrobotics. Additionally, engaging deeply with the ethical considerations in biology is of paramount importance, ensuring practices within necrobotics are not only responsible but also respectfully acknowledge the value and sanctity of biological materials utilized.

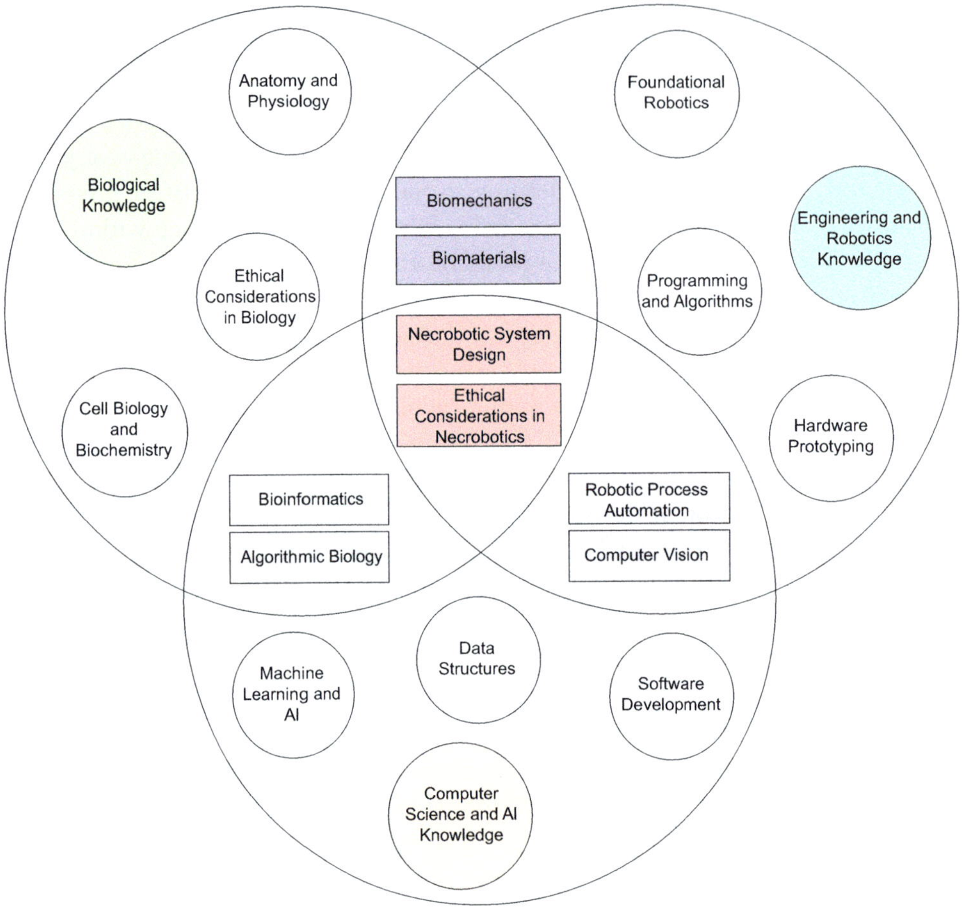

Fig. 1.1 ***Necrobotic study requirements.*** Essential requirements necessary for embarking on the design and research of a study in the dynamic realm of necrobotics. *(No permission required.)*

2.1.2 Engineering and robotics knowledge

A solid grounding in foundational robotics is imperative, offering essential insights into the fundamental principles of robotics. This includes a deep understanding of control systems and mechanical design, thereby supporting the seamless development and integration of robotic components with biological materials. Furthermore, acquiring proficiency in programming and algorithms is vital. This skill enables the crafting of sophisticated control systems and the effective processing of data, both of which are integral to applications within necrobotics. Moreover, developing practical skills in hardware prototyping is equally important, requiring hands-on experience with various tools and components, including but not limited to, microcontrollers, sensors, and actuators.

2.1.3 Computer science and AI knowledge

Deep dives into machine learning and AI are crucial. An in-depth understanding of the various AI and machine learning algorithms, with special attention to their applications in robotics and data processing within the context of necrobotics, is necessary. A grasp on software development is also critical, with a focus on developing proficiency in creating efficient software solutions. This proficiency should be accompanied by an understanding of system architecture and robust design principles, both crucial for applications within necrobotics. Furthermore, achieving mastery over data structures and algorithms is imperative for the development of efficient, responsive, and advanced software and control systems integral to necrobotics.

2.1.4 Interdisciplinary knowledge

Bio-robotics knowledge involves the exploration of biomechanics, engaging with the mechanical aspects of living organisms, and understanding how these principles can be applied to robot design to simulate or mimic biological entities. Understanding biomaterials, focusing on their unique properties and applications within robotic systems, is also crucial. Robotic computing knowledge necessitates learning about robotic process automation (RPA), understanding its principles, and efficiently automating rule-based tasks with algorithms. A deep understanding of computer vision is also required, a field crucial for interpreting and understanding visual information and enhancing robotic perception and navigation. Computational Biology knowledge includes Bioinformatics, using tools from computer science to analyze complex biological data, and Algorithmic Biology, developing algorithms for analyzing and modeling biological processes. For necrobotics specialization, an integrated skill set combining biology, engineering, and computer science is crucial for the effective design of necrobotic systems. Equally important is understanding the ethical considerations and challenges in necrobotics, crucial for responsible development and deployment in this field.

2.2 Material identification

2.2.1 Objective analysis

The first step in the creation of a necrobotic device is a clear definition of its objective and desired functionality (Zhengwei Li and Saif, 2021; Ramshankar et al., 2023). The purpose of the device will guide the selection process for the appropriate biotic material. The functionalities of the necrobotic device could be diverse, ranging from precision gripping to complex locomotion, sensitive sensing capabilities, or a combination of these. Understanding mechanical and structural requirements: It is crucial to thoroughly understand the mechanical and structural prerequisites needed to achieve the device's objectives. For instance, if the necrobotic device is intended for precision gripping, the selected biotic material should possess characteristics like flexibility, strength, and perhaps even a certain degree of tactile sensitivity. Materials need to be analyzed for their mechanical properties, such as tensile strength, compressibility, and elasticity, to ensure they can withstand the stresses encountered during operation without premature wear or failure.

The specific functionalities that the device needs to perform must be outlined and analyzed. If the device is expected to move or actuate, the biotic material should have inherent or modifiable structures that can support these actions, like muscle tissues for contraction and expansion or joint-like structures for articulated movement. For sensing applications, the material should exhibit a high degree of sensitivity to environmental stimuli, like changes in temperature, pressure, or chemical composition.

After defining the objective and understanding the mechanical and structural needs, a meticulous selection process begins to identify deceased organisms or biological materials that align best with the outlined requirements. This process may involve researching and consulting scientific literature, collaborating with biologists and materials scientists, and possibly conducting preliminary tests on various candidate materials to gauge their suitability for the intended application. The selected material should not only meet the current needs but should also offer some degree of adaptability and scalability for future improvements and modifications to the device.

2.2.2 Organism characteristics and types of biological materials

Understanding the characteristics of different organisms is pivotal when selecting biological materials for necrobotic applications (Poletto and Pasquero, 2023; Gul et al., 2018). The selected organism should align closely with the intended application of the robotic system, considering various factors such as size, strength, flexibility, and complexity.

- **Size:** The size of the selected organisms should match the scale of the desired necrobotic device. For applications requiring precision and finesse, smaller organisms or parts of larger organisms might be selected, while applications needing robust strength may require materials from larger organisms.
- **Strength:** Depending on the application, the biological material needs to exhibit the necessary strength. If the necrobotic device is expected to bear loads or exert force, the biological material should have the inherent strength or be modifiable to withstand these stresses without deforming or breaking.
- **Flexibility:** Flexibility is a crucial characteristic, especially for applications involving movement. The selected biological materials should allow for a range of motion that supports the tasks the necrobotic device will perform, ensuring that it can navigate and interact with its environment effectively.
- **Complexity:** The complexity of the biological material also plays a role. Simple structures may be easier to work with and integrate, but complex structures might offer more advanced functionalities.

2.2.3 Specific types of biological materials

- **Muscle tissues:** For instance, consider the use of a scorpion's complex muscular system. Scorpions possess highly responsive and powerful muscles that allow for precise and swift movements, vital for their survival in the wild. In necrobotics, these meticulously harvested and preserved scorpion muscle tissues could be integrated into robotic systems, facilitating fast and accurate maneuvers with a degree of fluidity and precision hard to achieve with traditional robotic actuators. The biochemical treatment and electrical stimulation of these tissues would allow engineers to harness their natural contractile and expansive properties, providing robots with a distinctive advantage in applications requiring rapid and accurate responses.
- **Bone structures:** Similarly, spider-derived materials offer fascinating possibilities. Spiders have an exoskeleton that provides a rigid frame, offering support and protection while being lightweight. The unique properties of spider exoskeletons, predominantly made of chitin, can be exploited in necrobotics to create lightweight yet durable robotic frames. Furthermore, spiders possess specialized sensory organs that allow them to perceive vibrations and movements with extraordinary sensitivity. By extracting and integrating these sensory components into robotic systems, engineers could design robots capable of detecting the faintest of vibrations or movements, which is crucial in applications like search and rescue operations or sensitive environmental monitoring tasks. Such integration would not only augment the robot's sensory perception but also ensure a robust and harmonious interface between the biological and electronic elements, enhancing the overall functionality of the developed necrobotic systems.
- **Sensory organs:** The potential applications and innovations that can emerge from the integration of these carefully selected and treated animal-derived biological materials into robotic systems are boundless. Through the diligent and ethical sourcing and processing of such materials, the field of necrobotics is poised to unlock unprecedented technological advancements, bridging the gap between the organic and inorganic, and paving the way for a future where robots more seamlessly interact and engage with their environment in ways previously thought unimaginable.

2.2.4 Prospective necrobotic materials

Various deceased organisms provide unique structures and functionalities that can be valuable for necrobotic applications. Selecting appropriate animal candidates should be based on specific application requirements, ethical considerations, and the potential impact on public perception and acceptance of the resulting necrobotic devices. Additionally, legal frameworks and cultural sensitivities regarding the use of deceased organisms must be diligently observed and respected. Beyond spiders, various animals exhibit anatomical and physiological characteristics beneficial for necrobotic applications.

This section illuminates alternative deceased animal materials with potential applications in necrobotics, exploring their unique attributes and possible uses. Table 1.1 shows the comparison of prospective dead bodies of animals for use as necrobotic materials.

- **Spider:** Spiders possess a unique hydraulic system within their limbs, facilitating precise and efficient movement. The pressurized hemolymph (spider blood) acts within the limbs as a hydraulic mechanism, allowing the spider to exert controlled movements and grips, making it an ideal candidate for necrobotic material.
- **Scorpions (Order: Scorpiones):** Scorpions possess segmented bodies, pincers, and a venomous stinger. Their exoskeletons are sturdy and lightweight, providing structural integrity suitable for mechanized applications. Due to their pincer-like chelae and stinger, dead scorpions can serve as efficient gripping or injecting devices. Additionally, their complex sensory systems might offer insights into the development of advanced sensing technologies.
- **Crabs (Infraorder: Brachyura):** Crabs boast hard exoskeletons and articulated limbs, providing robustness and flexibility. They are known for their strong pincers, which can apply significant force. The deceased bodies of crabs can function as resilient structures for gripping and holding devices. Their exoskeleton can also serve as protective casings for delicate internal machinery in robotic applications.
- **Cockroaches (Order: Blattodea):** Cockroaches exhibit remarkable resilience and adaptive capacities. Their dead bodies are flexible and robust, capable of traversing through challenging terrains. Dead cockroaches can be utilized for creating small, resilient, and flexible robotic structures designed for exploration or reconnaissance in constrained or hazardous environments.
- **Octopuses (Order: Octopoda):** Octopuses have highly flexible, muscular, and dexterous arms, with each arm capable of independent movement. They also possess suction cups for effective grasping. Deceased octopuses provide materials suitable for creating soft robotics capable of intricate movements and manipulations, valuable in medical devices and underwater exploration units.
- **Bats (Order: Chiroptera):** Bats have lightweight bones and membranous wings, offering opportunities for constructing flight mechanisms. Their echolocation abilities may inspire advanced sensing technologies. Dead bats can serve as prototypes for designing flying robotic units or drones. Studying their echolocation system might also aid in the development of superior navigation and detection systems for robots.
- **Snakes (Suborder: Serpentes):** Snakes offer elongated, flexible, and muscular bodies. Their unique locomotion mechanisms can inspire innovative robotic movement designs. The deceased bodies of snakes can be employed as templates for creating robots that need to navigate through tight or irregular spaces, useful in search and rescue operations or medical applications.

Table 1.1 Comparison of prospective dead bodies of animals for use as necrobotic materials.

Organism	Physical characteristics	Processing method	Applications and possibilities	Obligations and considerations
Spider order: Araneae	Lightweight exoskeleton, hydraulic limb movement	Pressurized air injection into limbs, electrical control through glued electrodes	Grippers for small objects; navigators in tight spaces	Ethical concerns over use; potential hygiene issues; limited durability due to decomposition
Scorpions Order: Scorpiones	Sturdy exoskeleton, venomous stinger, segmented body	Removal of venom; manipulation of chelae and tail through mechanical or electrical means	Gripping and injecting devices; sensory technology development	Must ensure safe removal of venom; ethical considerations
Crabs infraorder: Brachyura	Hard exoskeleton, strong pincers	Dismantling and reassembly of limbs and pincers; possible integration with motors or actuators	Gripping and holding devices; protective casings for internal machinery	Ethical concerns; potential challenges in processing and assembling
Cockroaches order: Blattodea	Resilient and adaptive bodies, ability to traverse various terrains	Dissection and utilization of the exoskeleton and limbs, with possible motor integration	Robots for exploration and reconnaissance in constrained environments	Ethical considerations; public perception and acceptance issues
Octopuses order: Octopoda	Flexible and muscular arms with suction cups	Utilization of arms and suction cups, possible embedding of sensors or actuators	Soft robotics for medical devices and underwater units	Ethical concerns; careful processing to maintain arm functionality
Bats order: Chiroptera	Lightweight bones, membranous wings, echolocation system	Wings and bones utilization for flight mechanisms, study, and implementation of echolocation systems	Flying drones; advanced navigation and detection systems	Ethical and legal considerations regarding use; potential public pushback
Snakes suborder: Serpentes	Elongated and flexible bodies	Dissection and utilization of flexible segments, possible integration with actuators	Robots for navigating tight spaces in search and rescue or medical applications	Ethical considerations; public perception issues

No permission required.

2.3 Ethical considerations and material sourcing

The deliberate navigation through the ethical milieu is indispensable when engaging in the selection and procurement of biological materials earmarked for necrobotics. Implementing biotic elements, regardless of their provenance, engenders a labyrinthine ethical tableau necessitating conscientious deliberation and navigation, underpinning the sanctity and public endorsement of necrobotic technologies. Navigating the ethical landscape is vital in the process of selecting and sourcing biological materials for necrobotics. The incorporation of biotic materials, irrespective of their origin, introduces a complex set of ethical considerations that must be diligently addressed to uphold the integrity of the field and ensure public trust and acceptance of the technology.

- **Ethical sourcing:** For animal-derived materials, sourcing should be done with utmost respect for animal life and welfare. If the materials are obtained postmortem, the death of the animal should not be a direct result of material sourcing. Instead, materials should be sourced from animals that have died of natural causes or have been ethically euthanized due to unrelated reasons. Procurement from animal entities mandates an unequivocal commitment to animal welfare. Material sourcing should not instigate or expedite the demise of an animal; conversely, it should be contingent on natural or ethically justified euthanasia incidents. The treatment of animals, both pre- and post-mortem, should be approached with a reverence for life, ensuring that no unnecessary suffering or harm is inflicted at any juncture of the sourcing process.
- **Respecting cultural and legal norms:** Different cultures have varying beliefs and norms regarding the handling of deceased organisms. Adherence to the mosaic of cultural perspectives on death and the deceased is imperative. Acknowledging and navigating these viewpoints with sensitivity and respect minimizes the risk of cultural alienation or disrespect, fostering a more inclusive and accepted technological landscape. It is imperative to understand and respect these cultural sensitivities, navigating them carefully to avoid offending or alienating communities. Similarly, compliance with legal standards and regulations regarding the use of biological materials from deceased organisms is nonnegotiable. Ensuring alignment with the extant legal framework governing the procurement and utilization of postmortem biological materials is nonnegotiable. The legal tapestry can exhibit substantial variance across different jurisdictions, necessitating a nuanced understanding and adherence to the local, state, and international laws and regulations pertinent to necrobotics. Researchers and practitioners in the field of necrobotics must be aware of and adhere to these legal requirements, which may vary significantly from jurisdiction to jurisdiction.
- **Sustainable practices:** Sustainability in sourcing is another crucial factor. The practice of necrobotics should not encourage wastefulness or the unnecessary use of biological materials. Necrobotic practices should epitomize sustainability by eschewing wasteful tendencies. Prioritizing materials that are typically discarded or are

by-products of other processes not only encapsulates sustainability but also mitigates ethical dilemmas tethered to the intentional termination of life for material acquisition. Endeavoring for efficiency in the use of biological materials underscores a commitment to sustainability, ensuring that each procured material is utilized to its maximum potential, thereby minimizing waste and promoting responsible consumption and production within the field. Where possible, the field should prioritize the use of by-products or materials that would otherwise be discarded. This approach not only aligns with the principles of sustainability but also minimizes the ethical concerns associated with sourcing, as it does not incentivize or necessitate the death of organisms solely for material acquisition.

- **Documentation and transparency:** Maintaining transparent records of sourcing practices is also essential. Transparency builds trust with both the public and regulatory bodies, ensuring that the field can continue to operate and develop without facing undue scrutiny or backlash. Transparent and meticulous documentation of sourcing practices and protocols functions as a bulwark against public skepticism and regulatory backlash. Such transparency facilitates a constructive dialogue with the public while fostering a culture of trust and accountability within and outside the field. Proper documentation of sourcing practices, including the origin of materials, consent obtained, and any other relevant information, should be standard practice within the field. Adopting practices that champion transparency, including voluntary disclosure of sourcing origins, consent acquisition protocols, and other relevant data, should be ingrained as a cornerstone within the operational ethos of necrobotics.

2.4 Material preparation

2.4.1 Cleaning process

The initial stage of preparing biological material for integration into necrobotic devices involves a meticulous cleaning process. This process is crucial as it purges the selected biological materials of any superfluous biological matter while safeguarding the integrity of the structures that are vital for the intended functionality of the necrobotic devices.

- **Removing extraneous biological matter:** When preparing biological materials, it is essential to remove all nonessential tissues, fluids, and other biological matter that do not contribute to the desired function of the final necrobotic device. This step might involve washing, scraping, or otherwise cleaning the material to remove blood, fat, and other tissues that are not needed. The aim is to isolate and preserve the biological structures that will be utilized—like muscle tissues, bone structures, or sensory organs—while discarding the rest.
- **Use of cleaning agents:** Depending on the type of biological material being prepared, different cleaning agents may be employed. These agents must be selected carefully to avoid damaging the vital structures of the biological material. Enzymatic cleaners might be used to break down unwanted tissues gently, while antiseptic

solutions could be used to minimize the risk of contamination from microorganisms. In every case, the chosen cleaning agents should be compatible with both the biological material and the artificial components with which they will be integrated later.
- **Handling with care:** Throughout the cleaning process, the biological materials should be handled with extreme care to prevent damage to the delicate structures that will be incorporated into the necrobotic device. This might involve using specialized tools and equipment that allow for precise manipulation of the material without causing harm. The work environment should also be kept sterile to prevent the introduction of contaminants that could jeopardize the functionality and longevity of the biological materials in the final device.
- **Final inspection:** After the cleaning process, a thorough inspection is necessary to ensure that all unwanted biological matter has been successfully removed and that the structures to be used are intact and undamaged. This step often requires the use of microscopes or other magnification tools to examine the material closely and verify that it has been prepared to the required standard.

2.4.2 Preservation and material preservation methods

Once the biological material has undergone the essential cleaning process, the next critical step in its preparation for integration into necrobotic devices is preservation. Preservation techniques are applied to prevent decay and degradation of the biological material over time, ensuring its durability and stability throughout the lifespan of the necrobotic device. Several preservation methods are commonly employed:
- **Chemical preservation:** Chemical preservation involves treating the biological material with specific chemicals to halt decay and decomposition. One common method is formaldehyde fixation, which is widely used due to its ability to fix the tissue and cellular structures in place. Formaldehyde crosslinks proteins and nucleic acids, effectively preserving the material's structural integrity. Careful consideration is given to the choice and concentration of chemicals to minimize any adverse effects on the material's natural properties. The goal is to preserve the biological material while causing as little alteration as possible.
- **Cryopreservation:** Cryopreservation is a method used to preserve biological materials at extremely low temperatures, typically below freezing. This technique involves freezing the material to inhibit biochemical reactions and microbial growth. Cryopreservation is particularly useful for long-term storage, as it can extend the lifespan of biological materials without significant degradation. It is crucial to establish controlled freezing and thawing protocols to prevent damage to the material's structure during the process.
- **Lyophilization (freeze-drying):** Lyophilization, commonly known as freeze-drying, is another preservation method employed in necrobotics. This process removes moisture from the biological material while preserving its structure. Freeze-drying

involves freezing the material and then subjecting it to a vacuum environment, causing the frozen water to sublimate directly from ice to vapor. This removes moisture, reducing the risk of decay and microbial growth. Freeze-drying is advantageous because it maintains the material's integrity, making it suitable for long-term storage and integration into necrobotic systems.

2.4.3 Modification and enhancement

In the realm of necrobotics, it is often necessary to enhance and modify biological materials to optimize their functionality and compatibility within robotic systems. These modifications are crucial to ensuring that the biological components seamlessly integrate with electronic elements while preserving their inherent properties. Here are the key aspects of modification and enhancement in necrobotics.

- **Coating and sealing:** One common method for enhancing the durability and stability of biological materials is the application of protective coatings and seals. These protective layers act as a barrier against environmental factors, preventing moisture, contaminants, and other potentially damaging agents from affecting the biological material. Coating and sealing techniques are carefully selected to ensure they do not compromise the natural properties of the material. For example, coatings may be chosen to be biocompatible and nonreactive to maintain the material's functionality and structural integrity.
- **Biochemical modification:** Biochemical modification involves altering the biochemical composition of biological materials to optimize their responsiveness and functionality in various environments. This modification can include adjustments to the material's metabolic processes or cellular components. The goal is to enhance the material's ability to interact with electronic components, adapt to changing conditions, or exhibit specific behaviors required for the necrobotic application. Care is taken to ensure that these modifications do not adversely affect the material's overall stability and suitability for integration.
- **Interface development:** Developing interfaces that facilitate seamless interaction between electronic and biological components is a fundamental aspect of modification and enhancement in necrobotics. These interfaces serve as connection points where signals, energy transfer, or data exchange occur between the biological material and electronic systems. Crafting effective interfaces is essential to ensuring that the two disparate components work harmoniously. The design of these interfaces is meticulous, taking into account factors such as mechanical compatibility, electrical conductivity, and the prevention of biological contamination or damage.

The modification and enhancement of biological materials in necrobotics are tailored to the specific needs of each robotic application. Engineers and researchers carefully balance the preservation of the material's natural properties with the necessary modifications to optimize performance. This iterative process of modification, enhancement, and testing contributes to the development of robust and reliable necrobotic systems that can operate effectively in diverse environments and applications.

2.4.4 Quality control and standardization

Quality control and standardization are critical aspects of the material preparation phase in necrobotics. These processes ensure that the biological materials used meet specific standards and specifications, leading to the production of reliable and consistent necrobotic devices. Here is a closer look at quality control and standardization in this context (Seki et al., 2016; Blackiston et al., 2023; Yu et al., 2018).

- **Quality control protocols:** Implementing robust quality control protocols is essential to assess the quality and suitability of biological materials for use in necrobotics. These protocols involve a series of tests and evaluations designed to measure various physical properties of the material. For example, the strength, flexibility, and structural integrity of the material are assessed to ensure it can withstand the mechanical demands of the intended robotic application. Additionally, tests may be conducted to evaluate the material's resistance to environmental factors, such as temperature, humidity, and chemical exposure.
- **Standardization procedures:** Developing standardized procedures for the selection, preparation, and integration of biological materials is crucial for achieving consistency and replicability in necrobotic devices. Standardization helps ensure that each unit produced adheres to the same set of specifications and performance criteria. These procedures encompass guidelines for sourcing materials, preservation methods, modification techniques, and quality control assessments. By following standardized protocols, researchers and engineers can streamline the development process and reduce the likelihood of variations between individual necrobotic devices.
- **Material testing and validation:** Material testing is a core component of quality control in necrobotics. It involves subjecting the biological materials to a battery of tests to validate their suitability for integration into robotic systems. These tests may include tensile strength tests to measure the material's resistance to stretching or pulling forces, flexural tests to assess its bending properties, and impact tests to evaluate its durability when exposed to sudden forces or shocks. Through comprehensive material testing, any weaknesses or inconsistencies in the biological material can be identified and addressed, leading to higher-quality necrobotic devices.
- **Documentation and traceability:** To maintain accountability and traceability, meticulous documentation of the material selection and preparation process is essential. Detailed records of the sourcing, preservation, modification, and quality control procedures should be maintained. This documentation ensures that the entire development process can be audited and that any issues can be traced back to their source, facilitating continuous improvement and refinement of the necrobotic technology.

2.4.5 Storage

Storage in the necrobotics realm is a crucial facet, predominantly during the material preparation phase, ensuring that the biotic materials integral to the development of necrobotic systems are preserved and maintained effectively. The delicacy of these biotic materials necessitates a stringent adherence to specific storage conditions, which are paramount in safeguarding their inherent quality, integrity, and functionality over time.

- **Temperature control:** A pivotal element in this storage process is the careful control of temperature. Given the exquisite sensitivity of biological materials to temperature variations, establishing and maintaining a controlled temperature environment is imperative. Each type of biological material may necessitate a unique temperature range for storage, with some requiring refrigeration or freezing, while others might need to be kept at ambient temperatures. The meticulous regulation of temperature is aimed at impeding or slowing down the inherent degradation processes of these materials, thereby prolonging their viability and usefulness.
- **Humidity management:** Equally significant is the management of humidity within the storage spaces. The level of humidity can profoundly influence the state and longevity of biological materials. An environment with excessive humidity can precipitate issues like the proliferation of microorganisms, decay, or the formation of mold, all of which compromise the integrity of the materials. On the opposite spectrum, an overly dry environment can result in the materials becoming brittle and susceptible to degradation. Hence, achieving and sustaining an optimal level of humidity is a nonnegotiable requirement in the storage process.
- **Environmental control:** Moreover, broader environmental control measures should be vigilantly implemented. The environments designated for storage must be meticulously kept free from a spectrum of potential threats, including contaminants, pollutants, pests, and rodents. Ensuring that the storage spaces are not only clean but also well-ventilated is essential to forestall any form of contamination or deterioration that could adversely affect the biological materials.
- **Packaging and sealing:** The materials also demand specific packaging and sealing protocols. The choice of containers and packaging materials is crucial, as they need to offer protection against external environmental factors. Techniques like hermetic sealing or vacuum packaging might be employed to engender an airtight environment, which is fundamental in minimizing exposure to oxygen, moisture, and other potential contaminants.
- **Documentation and tracking:** Maintaining comprehensive documentation and engaging in precise tracking of the storage conditions is also indispensable. Detailed records, encompassing data on temperature, humidity, duration of storage, and the overall condition of the biological materials, must be diligently kept. This practice is essential for traceability and quality control purposes, ensuring that the materials, when retrieved for use, are still in prime condition.

- **Regular inspection and monitoring:** Furthermore, implementing a regimen of regular inspections and monitoring is vital. Through consistent visual inspections, and utilizing sensors where applicable, any signs of degradation or other related issues can be swiftly identified and addressed. This proactive approach ensures that any deviations from the optimal storage conditions are promptly corrected, maintaining the integrity of the stored materials.
- **Security and access control:** Lastly, considering the sensitive nature of these biological materials, implementing robust security and access control measures within the storage facilities is nonnegotiable. This involves restricting access to the storage areas to a select group of authorized personnel, thereby preventing any unauthorized access, tampering, or mishandling of the precious biological materials destined for incorporation into necrobotic systems.

3 Design and fabrication of necrobotics

This section presents a comprehensive overview of the design and fabrication process of necrobotics, detailing a meticulous approach from initial inspiration to tangible product. Beginning with bio-inspired design, the process involves thoughtful observation and analysis of deceased organisms' inherent functionalities and mechanisms, with designers drawing inspiration from these unique structures and sensory capabilities. The section outlines the importance of not only mimicking but also enhancing and adapting these biological structures for various applications, ensuring that the necrobotic systems are not mere replications but optimized versions of the biological originals. It delves into prototyping, emphasizing the transition from conceptual designs to functional devices and the vital role of rapid prototyping technologies and iterative development (Garcia et al., 2022; Gomi, 2003). The text also discusses the intricacies of integrating biological materials with artificial components, highlighting the importance of material compatibility, custom interfaces, and the precision assembly process in developing effective hybrid systems. The section concludes with an overview of testing and evaluation, underscoring the importance of functional testing, rigorous protocols, and iterative development for continual improvement and refinement of necrobotic systems. The narrative demonstrates a complex, careful, and precise process, resulting in innovative devices that seamlessly blend biology and robotics, ready to address challenges in various fields and applications (Sharkey, 2006; Kim et al., 2018).

Recently, the researchers utilized the deceased spiders, specifically wolf spiders, as they could lift weights heavier than themselves, showcasing potential for tasks like object manipulation (Yap et al., 2022). Interestingly, the team observed that smaller spiders could lift relatively heavier loads, providing valuable insights for future research. The experiments showed that the spider-based grippers, created by injecting small air amounts, could successfully perform various tasks, with the limbs enduring up to 1000 open-close cycles before showing signs of wear. The wear is believed to be due to joint dehydration, a challenge that the team thinks can be mitigated with polymeric coatings. This necrobotic approach offers various practical applications, like sorting small objects and assembling microelectronics. Moreover, these bio-grippers can blend into natural environments due to their inherent camouflage, making them ideal for tasks like capturing insects in the wild. The biodegradable nature of the spiders ensures that this technology will not add significant waste, presenting an eco-friendly alternative to traditional robotic components. Fig. 1.2 shows a depiction of necrobotic setup.

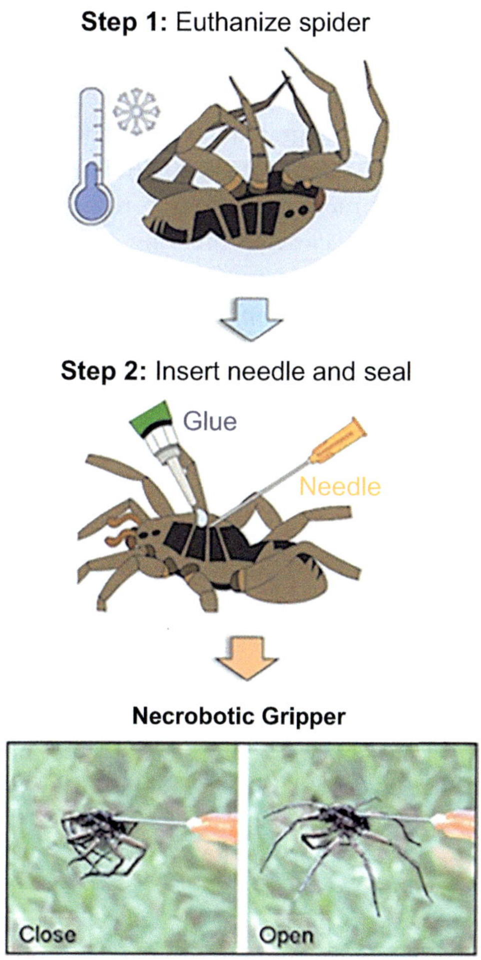

Fig. 1.2 ***Deceased spider as necrobotic gripper.*** An image depicts the technique by which mechanical engineers at Rice University transform deceased spiders into necrobotic grippers, devices capable of picking up items via hydraulic pressure activation. The process, taking place in the Preston Innovation Laboratory, is a noteworthy contribution to the burgeoning field of necrobotics (Williams, 2022). *(Preston Innovation Laboratory.)*

3.1 Bio-inspired design

3.1.1 Drawing inspiration

When approaching the design and fabrication of necrobotic systems, it is imperative to start with a deep and thoughtful observation of the selected deceased organism's inherent functionalities and mechanisms. The chosen organism, serving as the biological material source, possesses a set of unique structural dynamics, movement patterns, and sensing capabilities that have evolved over time, providing invaluable insights for the design process. To effectively draw inspiration, designers and engineers should conduct a thorough analysis of the organism's physiological and biomechanical properties. This involves studying the intricacies of its skeletal structure, muscle arrangement, sensory organs, and any other relevant anatomical features. Understanding these elements is crucial, as it aids in identifying which biological components can be utilized and how they can be integrated into the robotic system to achieve desired functionalities (Niu et al., 2021; Cheng et al., 2019). The process also entails appreciating the organism's movement mechanics, from its navigation techniques to its interaction with the environment. By closely observing and mimicking these movement patterns, engineers can create robotic systems that not only replicate but possibly enhance the natural movements found in the selected biological material. This biomimicry approach often results in robotic devices that operate more efficiently and adaptively in various environments, as they embody the tested and refined strategies employed by living organisms. Furthermore, acknowledging the organism's sensory capabilities provides a roadmap for developing robots that can interact with their surroundings in sophisticated ways. For instance, if the organism has exceptional auditory or visual sensors, engineers might integrate similar sensory technology into the robotic design, allowing the device to perceive and respond to environmental stimuli in a manner analogous to the original organism (Jiang et al., 2014; Faruque et al., 2023). Drawing inspiration from these observations, the bio-inspired design process aims to develop necrobotic systems that seamlessly blend the advantages of biological materials with the capabilities of modern robotics. Through this meticulous approach, designers can innovate robotic devices that are not only functionally effective but also biologically coherent, providing a unique convergence of the living and the artificial in the realm of robotics.

3.1.2 Mimicry and adaptation

In the pursuit of crafting advanced necrobotic systems, mimicry and adaptation of biological structures are paramount. The design process begins with mimicking the selected biological structures, but it does not end there. A careful balance between imitation and innovation is necessary, ensuring that the final design is not a mere replication but an enhanced version of the biological original. Mimicry involves closely observing and understanding the biological structures' functionalities and integrating them into the design. For instance, the specific arrangement of muscle fibers, the hardness of bones, or the sensitivity of sensory organs are meticulously studied and mirrored in the robotic system. However, this process is not about blind replication. It is vital to recognize and respect the intricacy and efficiency of these natural designs while working toward augmenting them for enhanced performance. Adaptation, on the other hand, is about making informed modifications to the mimicked biological structures to suit the specific application for which the robotic system is intended. While biological structures offer a fantastic starting point, they often need to be adapted to perform optimally in various artificial scenarios (Monta et al., 1998; Huang and Rust, 2021). This adaptation process might involve strengthening particular structural components, resizing or reshaping parts for better efficiency, or even integrating additional features not present in the original biological material. For instance, if the necrobotic device is intended for underwater operation, adaptations should ensure that the biological materials can withstand prolonged exposure to water without degradation while maintaining buoyancy and maneuverability. Similarly, if the application requires precision and accuracy, like in medical devices, the design must incorporate features that enhance stability and control. Customization is another critical aspect of the adaptation process. Given that each application has unique operational needs and environmental challenges, the design must be tailored to meet these specific requirements. Engineers might need to make modifications to the biological materials or combine them with synthetic components in novel ways to achieve the desired functionalities (Ensslin, 2017; Murphy, 2019). This might entail employing advanced materials, implementing innovative fabrication techniques, or developing unique control systems. Ultimately, the process of mimicry and adaptation in necrobotics design seeks to create devices that not only emulate the remarkable capabilities found in biology but also surpass them through thoughtful design and engineering innovations. By doing so, it is possible to realize necrobotic systems that offer unprecedented levels of performance, efficiency, and versatility, ready to tackle the diverse challenges posed by real-world applications.

3.1.3 Prototyping

The prototyping stage is crucial as it breathes life into the bio-inspired designs, transitioning them from abstract concepts to tangible, functional devices (Ianniello et al., 2022; Rasakatla et al., 2022). By leveraging rapid prototyping technologies and adopting an iterative development approach, engineers can expedite the realization of innovative and effective necrobotic systems that truly encapsulate the synergy between biology and robotics (Samani and Saadatian, 2012; Delda et al., 2021).

- **Initial prototype creation:** Prototyping is an integral step within the design and fabrication process for necrobotics, serving as the bridge between conceptual design and functional device. The process begins with the production of initial prototypes that embody the bio-inspired designs envisioned in earlier stages (Mattar, 2013; Ramanujam et al., 2022). These prototypes act as tangible representations that facilitate in-depth analysis and real-world testing, thereby allowing for a more comprehensive evaluation of the proposed system's functionality and performance. Creating these prototypes necessitates advanced technological tools and methodologies to ensure precision and fidelity to the bio-inspired design elements. Rapid prototyping technologies, including 3D printing and CNC machining, play a pivotal role in this stage. These technologies facilitate the quick and accurate transformation of digital designs into physical components, offering an efficient means to explore, test, and iterate design ideas.

- **Rapid prototyping technologies:** 3D printing offers unparalleled flexibility in creating components with complex geometries and intricate details, which is often required to replicate the nuanced structures found within biological organisms. It allows for the production of custom parts that can be adjusted easily between iterations, fostering a dynamic and responsive development process. Furthermore, various materials can be utilized in 3D printing, ranging from plastics and metals to biocompatible materials, providing a broad selection to match the mechanical and aesthetic requirements of the prototype. Conversely, CNC machining provides high precision in crafting components from a wide range of materials, including metals, plastics, and composites. This technology is especially valuable when the prototype requires parts with tight tolerances and high rigidity, or when materials unsuitable for 3D printing need to be used.
- **Evaluation and iteration**: Once initial prototypes are developed, they undergo rigorous testing and evaluation. Engineers and designers scrutinize their functionality, durability, and performance under various conditions (Hakim Khalili et al., 2023; Zhou et al., 2016). This phase is critical for identifying any discrepancies between the theoretical design and the practical application, highlighting areas for improvement and refinement. Feedback from the testing phase informs subsequent iterations of the prototype, with each cycle aimed at refining and optimizing the design. Through a series of iterations, the prototype gradually evolves, with its design being fine-tuned to meet the desired specifications and requirements. This iterative process continues until a final prototype that accurately encapsulates the bio-inspired design while delivering reliable and effective performance is realized.

3.2 Integration

3.2.1 Material compatibility

Material compatibility is fundamental to the successful integration within necrobotic systems. Through careful selection, modification, and design of materials and interfaces, engineers can foster a harmonious interaction between the biological and artificial components, laying the groundwork for a system that is not only functional but also stable and reliable over time. This meticulous approach to compatibility ensures that the intrinsic properties of each material are preserved and utilized effectively, culminating in a necrobotic device that truly embodies the convergence of life and machine (Meng et al., 2015; Yang et al., 2023). The synergy between biological materials and artificial components is at the heart of necrobotics. Achieving harmonious integration is imperative for constructing a functional and efficient system. Material compatibility is a foundational consideration in this integration process (Duncan, 2019; Won et al., 2020).

- **Biochemical compatibility:** Firstly, biochemical compatibility is vital to prevent adverse reactions between the biological and artificial materials. Biological materials are inherently complex, sensitive, and responsive to their biochemical environment. To ensure a stable and safe interaction, artificial components must be chosen and treated in a way that avoids disrupting the biochemical balance of the biological materials. This might involve selecting materials that are inert or nonreactive with the biological tissues, or applying coatings that act as a biochemical buffer at the interface.
- **Electrical compatibility**: Electrical compatibility is another critical aspect of integration. The electrical properties of biological materials can be quite distinct from those of artificial materials, especially considering the ionic conduction mechanism prevalent in biological tissues versus the electronic conduction in metals and semiconductors (Iberite et al., 2022). It is crucial to develop interfaces that facilitate efficient electrical communication between these different conduction mechanisms. Strategies may include the use of materials that can translate between ionic and electronic signals, or designing interfaces that minimize signal loss and distortion during this translation process.
- **Mechanical compatibility:** Lastly, mechanical compatibility should be meticulously addressed. The mechanical properties of biological tissues, such as elasticity, tensile strength, and compressibility, often differ significantly from those of artificial materials (Boblan et al., 2007; Bhaumik, 2018). When integrated into a system, these materials should move and interact cohesively without generating stress or strain at the interface. Engineers may need to select or develop artificial materials with mechanical properties that closely match or complement those of the biological counterpart. Moreover, the interface design should also facilitate a smooth transition between the different materials, possibly through graded or composite interfaces that gradually change from biological to artificial characteristics.

3.2.2 Hybrid system development

Hybrid system development is an intricate dance of balancing and integrating disparate elements into a cohesive whole. This careful assembly process in necrobotics aims to create devices that are not only innovative and powerful but also reliable and efficient, opening new possibilities in various fields and applications (Zolfagharian et al., 2016). Through the meticulous design and integration of biological and artificial components, necrobotics can bring forth a new class of devices that truly blend the living and the mechanical in unprecedented ways.

- **Integrating biotic material and artificial components:** The central task in developing hybrid systems within necrobotics is achieving an effective integration of biological materials with artificial elements such as sensors, actuators, and microcontrollers. This process requires a meticulous approach where the distinct characteristics of both material types are carefully considered and respected. Biological materials bring unique and often irreplicable properties to the system, such as specific actuation mechanisms or sensory functions. These qualities should be preserved and leveraged during the integration process, providing the system with capabilities that purely artificial systems might struggle to replicate. Simultaneously, artificial components offer reliability, programmability, and functionality that are essential for the operation and control of the hybrid system (Xiao et al., 2019).
- **Designing custom interfaces:** Crafting custom interfaces is imperative for facilitating seamless communication between biological and artificial elements. These interfaces should be designed to minimize any loss or distortion of signals transmitted between the components, ensuring that the system operates smoothly and responsively (Mousa et al., 2020). Custom interfaces also play a crucial role in physically joining the different materials in a way that is secure yet nondamaging to the fragile biological tissues. This might involve developing connectors, mounts, or scaffolds that can accommodate the shape and mechanical properties of the biological material while providing attachment points or interfaces for the artificial components. These interfaces might also need to incorporate elements that protect the biological materials from the environmental conditions or mechanical stresses encountered during operation.
- **Enhancing system performance**: The end goal of this intricate integration process is to develop a hybrid system where biological and artificial components work harmoniously together, thereby enhancing the overall performance and capabilities of the necrobotic device. The added complexity brought by the integration should be justified by significant improvements in the system's function. For instance, the biological materials might offer nuanced and delicate movement or sensory capabilities that artificial systems find challenging to mimic accurately. At the same time, the artificial components provide the control, precision, and repeatability required for reliable operation (Boch et al., 2023; Jiang et al., 2022). Together, these materials should create a system that combines the best of both worlds: the sophisticated capabilities of biological tissues and the reliability and control of modern electronics and mechanics.

3.2.3 Precision assembly

- **Precision assembly techniques and tools**: Precision assembly in the context of necrobotics is indispensable given the delicate nature of biological materials and the need for secure connections with artificial components. Techniques employed should be sensitive to the integrity of the biological structures, ensuring that they are neither damaged nor altered detrimentally during the assembly process. Employing specialized tools and equipment that can handle both types of materials with the requisite delicacy and precision is crucial. These might include fine-tuned robotic arms, precision grippers, or other specialized assembly devices that can maneuver and place components with high accuracy and minimal force (Wang and Chortos, 2022). For biological materials, this might involve using tools that can accommodate their softness and irregularity without causing harm. The assembly process should also consider the biological material's vulnerability to environmental conditions, employing controlled atmospheres or other protective measures as necessary to preserve their function and structure.
- **Quality control and testing procedures:** Implementing a robust quality control and testing regimen is essential throughout the assembly process. This rigorous approach ensures that the integration of biological and artificial components is executed flawlessly, and it allows for the immediate identification and rectification of issues or defects that might emerge during integration. Quality control procedures should involve both visual inspections and functional tests to verify that components are assembled correctly and that they operate as intended (Raisbeck, 2022). This might involve checking the physical integrity of the biological materials postassembly, ensuring that connections are secure, and verifying that the interface between the biological and artificial components functions correctly. Functional tests are crucial to ascertain that the assembled device operates according to specifications and expectations. These tests should be designed to assess both the biological and artificial components' performance, ensuring that they work synergistically (Bhattacharjee et al., 2016). During this phase, the device might be subjected to various operational scenarios to evaluate its reliability and responsiveness under different conditions.
- **Importance of precision in assembly**: Precision in assembly is not merely a technical requirement but is fundamental to the success of necrobotic devices. Given the hybrid nature of these systems, any misalignment or improper connection between components can lead to functional failures or reduced performance. The precise assembly ensures that the unique capabilities of the biological materials are effectively harnessed and that the artificial components can adequately support and control these functions (Pagandiaz, 2023). In conclusion, precision assembly, characterized by careful techniques, specialized tools, and stringent quality control, is pivotal in the creation of functional and reliable necrobotic devices. Each step of the assembly

process must be approached with an understanding and respect for the materials involved, ensuring that the final device is not only a marvel of engineering but also a testament to the seamless integration of the biological and artificial.

3.3 Testing and evaluation

3.3.1 *Functional testing*

- **Comprehensive functional evaluation**: Functional testing is integral to the development process of necrobotic systems as it serves as the primary method for evaluating the system's functionality and overall performance. This phase involves running the system through a series of tests designed to assess each function it is intended to perform (Silvis-Cividjian, 2017). Comprehensive testing should scrutinize all aspects of the system's operation, from the responsiveness and efficacy of the biotic materials to the reliability and precision of the integrated artificial components. During functional testing, engineers and developers should identify and document areas that may need improvement or optimization. These areas might include responsiveness, accuracy, power consumption, and durability of the system. Recognizing these aspects early in the development process allows for timely modifications and enhancements, thereby ensuring that the final product meets or exceeds the predefined performance criteria.
- **Varied environmental and operational testing:** It is imperative to evaluate the necrobotic system's performance under various environmental conditions and operational scenarios to guarantee its reliability and robustness in real-world applications. Testing environments should simulate the conditions under which the device is expected to operate, including variations in temperature, humidity, pressure, and light (Hu, 2023; Torresen, 2018). Different operational scenarios should also be considered, mimicking the diverse challenges and tasks the system will encounter during its usage. For instance, if a necrobotic system is intended for use in medical surgeries, it should be tested under sterile conditions and subjected to simulations of surgical procedures to assess its precision and stability. If the system is designed for outdoor environmental monitoring, it should be evaluated under various weather conditions and exposed to the different elements it would typically encounter.
- **Importance of rigorous testing protocols:** Rigorous and systematic testing protocols are vital in unveiling any hidden issues or potential failures that might not be apparent under normal or ideal conditions. Functional testing must, therefore, be both exhaustive and meticulous, covering the full spectrum of the system's intended functions and operational environment (Wang et al., 2023). The insights gathered from this intensive testing phase are instrumental in refining the system, making necessary adjustments, and ultimately enhancing its performance and reliability for end-users. Through a combination of comprehensive functional evaluations and varied environmental and operational testing, developers can ensure the creation of a necrobotic system that is not only innovative but also dependable, resilient, and aptly

suited for its intended applications. The goal is to foster confidence in the technology's capabilities and reliability among potential users and stakeholders, paving the way for its successful integration into various fields and industries.

3.3.2 Iterative development

Process of continuous improvement

Iterative development is a cornerstone of necrobotics development, serving as a continuous cycle of improvement and refinement to optimize the system's performance and functionality. After the initial round of functional testing, developers should meticulously analyze the results to identify any shortcomings, inefficiencies, or failures within the system. This analysis should encompass not only the performance of individual components but also the efficiency of the system as a whole, including the integration and interaction between the biotic and abiotic elements (Fang et al., 2022; Must et al., 2015). Through this analysis, engineers can pinpoint specific areas where modifications are necessary, whether it involves adjusting the control algorithms, enhancing the integration interfaces, or even changing the type of biological or artificial materials used. Every round of modifications should be followed by subsequent testing to evaluate the impact of the changes on the system's overall performance and reliability.

- **Importance of a feedback loop:** An effective iterative development process relies heavily on a robust feedback loop between the testing and development phases. Developers need to use the insights gained from each testing phase as a guide for making informed and strategic adjustments to the system. This approach ensures that the modifications are not only addressing the identified issues but are also contributing positively to the system's overall functionality and efficiency. The iterative process is cyclical, with each cycle bringing the system closer to its optimal form. It is crucial for developers to remain open to making continuous improvements, understanding that each cycle offers valuable insights and lessons that are vital for refining the technology (Huang et al., 2018). This willingness to engage in ongoing learning and improvement is fundamental to achieving a final product that meets the desired specifications and standards for performance and reliability.
- **Finalizing development:** After several cycles of testing and refinement, the necrobotic system should exhibit a level of performance and reliability that aligns with the project's objectives and expectations. However, it is important to acknowledge that the iterative development process is never truly complete (Nandakumar et al., 2022). Even after the system is deployed, there should be mechanisms in place for collecting feedback from end-users and monitoring the system's performance in real-world conditions. This ongoing feedback collection allows for continual improvements and adaptations to be made, ensuring the necrobotic technology remains effective and relevant in an ever-changing environment and can respond adequately to the emerging needs and challenges of the fields in which it is applied.

4 Control mechanisms of necrobotics

This section presents essential elements regarding the control mechanism of necrobotics, highlighting crucial aspects in actuation control. It details the development and optimization of biotic algorithms, emphasizing the importance of mimicking natural actuation processes efficiently and understanding unique characteristics of biological materials (Hamet and Tremblay, 2017). The section underscores the necessity for optimization for rapid response times and robustness, along with the incorporation of learning mechanisms for long-term success. Furthermore, it explores integrated control, user interface design, and the importance of sensory feedback for responsiveness and adaptability. The section also elaborates on adaptive control, focusing on environmental awareness and self-calibration mechanisms to ensure optimum performance and reliability over time, and discusses safety and reliability measures, including sophisticated error handling and redundancy in control mechanisms and sensors, all crucial for bolstering reliability and fault tolerance in various application scenarios.

4.1 Actuation control

4.1.1 Algorithm development

- **Biotic algorithms:** Developing control algorithms for necrobotic systems begins with an in-depth understanding of the selected deceased organism's natural actuation mechanisms. These could range from hydraulic or muscular systems in the case of insects or small mammals, to more intricate and subtle mechanisms present in other organisms. It is crucial to construct algorithms that not only accurately mimic but also efficiently trigger these natural actuation processes, resulting in movements and operations that appear fluid, natural, and highly efficient. Crafting such algorithms requires a combination of biological understanding and technical proficiency (Auerbach et al., 2019). The goal is to develop algorithms that embody the intricacies and subtleties of the organism's inherent movement mechanisms while allowing for electronic and mechanical control. This duality ensures a seamless integration of the biological actuation with the artificial components of the system, promoting efficiency and functionality in the end application.
- **Optimization:** Optimizing these control algorithms is an essential phase in the development process. This optimization process aims to refine the algorithms to achieve rapid response times, pinpoint accuracy, and enhanced energy efficiency in the actuation control. Every organism's biological material possesses unique characteristics and constraints; hence, the algorithms must be tailored and fine-tuned to account for these specific traits. Optimization also extends to ensuring the algorithms are robust and resilient, capable of handling various operating conditions and potential disturbances without compromising on performance (Yin and Müller, 2021). This fine-tuning requires iterative testing and refinement, often incorporating feedback from real-world applications to continually improve and adapt the algorithm to changing demands and environments.

- **Adaptation learning:** Incorporating learning mechanisms into the control algorithms is vital for the long-term success and efficiency of necrobotic systems. Adaptation learning mechanisms enable the robotic system to self-improve and refine its actuation control progressively. This aspect is particularly important as the biotic materials within the system may undergo changes over time due to wear, aging, or external environmental influences (Joyee et al., 2022). These learning algorithms must be sophisticated enough to discern subtle changes in the biological material's responses and properties. By continuously learning from these shifts, the necrobotic system can dynamically adjust its control parameters to maintain optimal performance. This adaptive capability not only prolongs the system's operational lifespan but also ensures it remains effective and precise in its functions throughout its use.

4.1.2 Control system

- **Integrated control:** The Integrated Control is a pivotal component, meticulously designed to amalgamate seamlessly with the biotic actuation mechanisms inherent in the selected organism's material. This integration necessitates a deep understanding of the nuanced interplay between the biological and electronic elements to craft a control system that facilitates smooth and precise manipulation of the robotic device's movements and functions. The control system must possess the finesse to not only execute accurate and refined movements but also incorporate safeguard mechanisms that protect the delicate biological components from potential damage due to abrupt or excessive motions (Zhang et al., 2018). The development process of this system requires a collaborative effort between experts in biology, robotics, and control engineering to ensure a harmonious marriage between the diverse set of technologies and materials involved. This integrated approach guarantees that the control system can reliably translate electronic commands into biologically-based actions, providing a level of control fidelity and responsiveness that mirrors the inherent capabilities of the selected biological materials.
- **User interface:** Creating an intuitive and user-friendly interface is equally imperative. The interface is the gateway through which operators interact with the necrobotic device, and as such, it must be thoughtfully designed to provide a smooth and intuitive user experience. The design should allow even individuals without an extensive technical background to easily control and manipulate the robotic device. This interface must provide real-time feedback on the device's status and movements, offering operators an immediate and clear understanding of how the device is responding to their inputs (Kim et al., 2021). Controls should be laid out in a manner that is logical and straightforward, with functionalities that closely correspond to the device's actual movements and operations. For example, if the device is equipped with a gripping mechanism, the interface should allow for precise control over the opening, closing, and pressure exerted by the grip. Moreover, considering that the device incorporates biological materials, the interface should also provide crucial

information regarding the status of these components, alerting the user to any signs of degradation or malfunction in real-time. This feature is crucial for maintaining the integrity and functionality of the necrobotic device over time.

4.1.3 Use of AI algorithms in control systems

AI plays a pivotal role in enhancing the functionality and autonomy of necrobotic systems, offering sophisticated algorithms for data processing, decision-making, and control. Here is an exploration of AI algorithms pertinent to various applications within the necrobotics realm, in conjunction with specific types of animal-derived materials. Table 1.2 shows AI algorithms as control systems (Berg et al., 2022; Shen et al., 2017; Nakabo et al., 2007).

- **Reinforcement learning (RL) algorithms in control systems:** RL is a vital AI technique for crafting control systems in necrobotic applications that necessitate dynamic interaction and navigation in complex environments. Utilizing animal muscles from scorpions or spiders, known for their precise and responsive contraction mechanisms, makes them ideal candidates for RL implementation. The meticulous interplay of these muscles supports swift and agile movements, with RL algorithms enabling continuous adaptation and optimization of the necrobot's navigational strategies. However, the inherent challenges of RL, such as its data-intensive requirements and the need for durable biological materials capable of withstanding repeated use without significant degradation, must be acknowledged and mitigated.
- **Convolutional neural networks (CNNs) for visual control:** For control systems reliant on sophisticated image recognition and processing, particularly in applications like surveillance and monitoring, CNNs are fundamental. Implementing visual components from spiders, celebrated for their unique visual systems, can significantly enhance the necrobot's image processing capabilities. The integration of spider eyes, which boast acute vision capabilities, with CNNs delivers unparalleled image recognition and analysis precision. However, this integration process is intricate, necessitating careful attention to detail to ensure seamless functionality and long-term compatibility within the broader necrobotic system.
- **Recurrent neural networks (RNNs) and long short-term memory (LSTM) for sequential data processing:** When control systems require sequential data processing, particularly for sound or signal detection and analysis, RNNs and LSTMs are invaluable. Incorporating auditory components from snakes or cockroaches, known for their heightened sensitivity to environmental vibrations, facilitates the creation of systems adept at discerning and interpreting subtle acoustic signals with high fidelity. Nonetheless, the integration of these components is complex, and maintaining their functionality over time is a challenge that researchers must navigate.
- **Genetic algorithms (GAs) in optimized control:** IGAs are essential in control systems that demand optimization, especially in task allocation and scheduling.

Table 1.2 AI algorithms as control systems.

Algorithm type	Application area	Ideal dead biological materials	Key benefits	Challenges and considerations	Ethical considerations	Technical feasibility
RL	Dynamic interaction and navigation	Muscles from scorpions or spiders	Adaptive and optimized strategies	Data-intensive; needs durable materials	Animal welfare and material sourcing	Integration of biological muscles
CNN	Image recognition and processing	Visual components from spider	Acute image recognition and analysis	Delicate integration process	Handling of sensitive organs	Compatibility and functionality
RNN/LSTM	Sequential data processing	Sensory organs from snakes or cockroaches	High fidelity acoustic signal interpretation	Complex integration and sustainability	Ethical use of sensory organs	Sustained functionality
GA	Task allocation and scheduling	Materials from cockroaches	Coordinated and optimized actions	Communication and coordination protocols	Ethical material sourcing	Robust communication protocols
DRL	Learning and executing complex behaviors	Nervous systems from scorpions or spiders	Adaptable and sophisticated behaviors	Computationally intensive, careful calibration	Respect for animal life	Ensuring material resilience

No permission required.

Biological materials derived from cockroaches, known for their efficient locomotion mechanisms, are ideal for integration into systems governed by GAs. These materials support coordinated actions due to their inherent structural optimization. However, the development of robust communication protocols and the assurance of coordinated functionality among robots with integrated biological materials remain challenging.

- **Deep reinforcement learning (DRL) for complex behavior control:** DRL is crucial for necrobotic control systems that necessitate the learning and execution of sophisticated behaviors in unpredictable environments. Implementing nervous systems from scorpions or spiders, renowned for their complex neural circuitry, enhances the capability of necrobots to exhibit adaptable and sophisticated behaviors. Nonetheless, DRL algorithms are computationally demanding, requiring careful calibration, and the selected biological materials must be resilient, able to execute complex behaviors without rapid degradation.

4.1.4 Sensory feedback

Sensor integration

- **Sensor selection:** Choosing the right sensors is crucial for building a robust and sensitive necrobotic system. The selection process involves identifying sensors that can provide accurate, reliable, and instantaneous feedback about both the internal state of the robotic device and the external environment in which it operates. This multifaceted awareness is fundamental to enhancing the device's responsiveness and adaptability to varying operational contexts. Sensors to be integrated could range from basic ones like those measuring temperature, pressure, and humidity, to more advanced and specialized sensors designed to detect specific environmental conditions, changes, or stimuli. For instance, a necrobotic device meant for agricultural use might benefit from moisture and nutrient sensors, allowing it to respond dynamically to the needs of the plants. Similarly, sensors detecting changes in light, motion, or sound could be vital for devices deployed in surveillance or search and rescue missions (Shen et al., 2020). Selection should not only be based on the current needs but also considering future upgradability and compatibility with emerging sensor technologies, ensuring the necrobotic device remains relevant and adaptable to the fast-evolving tech landscape.
- **Feedback loop development:** Developing a dynamic feedback loop is essential for making sense of and utilizing the data collected by the sensors effectively. The feedback loop is a continuous process that monitors, analyzes, and interprets sensor data in real-time, providing invaluable insights that inform adjustments to control algorithms and actuation mechanisms instantaneously. This self-regulating system enhances the device's ability to adapt to changing conditions and requirements autonomously, making it more resilient and effective in its operations. For example, if a sensor detects an increase in external pressure or temperature, the feedback loop

could immediately adjust the device's movements or functions to mitigate the impact of these changes, protecting both the electronic and biological components from potential damage (Li et al., 2021). A well-designed feedback loop also facilitates a more harmonious integration between the device's biological and electronic elements, fostering a synergy that allows the necrobotic system to function as a cohesive and intelligent unit. This synergy is crucial for the longevity and efficiency of necrobotic devices, ensuring they can respond not just accurately but also intuitively to the demands placed upon them.

Adaptive control

- **Environment awareness:** Creating a control system with heightened environmental awareness is a fundamental step in enhancing the adaptability and efficiency of necrobotic devices. The control systems should be intricately designed to dynamically respond to environmental variations by adjusting the operations and responses of the device instantly and appropriately. For instance, in a scenario where the device is deployed in fluctuating temperatures, the control system must intelligently modulate its actions to maintain functionality without harming the integrated biological components (Wang et al., 2019). The cornerstone of this awareness is real-time data acquisition and analysis from integrated sensors. These sensors continuously monitor the environmental parameters, and the control system uses this data to make informed and instantaneous decisions. Environmental awareness is not only vital for the device's immediate response but is also pivotal for planning and executing future operations, allowing the system to predict and prepare for potential environmental changes, thereby ensuring smoother and more efficient functioning.
- **Self-calibration:** Self-calibration mechanisms are imperative for sustaining the necrobotic device's optimum performance, accuracy, and reliability over time. Since both the biological and electronic components of the device might experience wear, drift, or degradation over time, a self-calibrating system can autonomously adjust and fine-tune its sensors and control algorithms to compensate for these changes. For example, if a sensor begins to read data with a slight offset due to prolonged use or unexpected interference, the self-calibration mechanism would identify this discrepancy and adjust the sensor's readings or the control algorithm's interpretation of these readings accordingly. This dynamic, self-regulating approach ensures that the necrobotic system remains accurate and reliable in its operations, regardless of the inevitable changes and challenges it might face throughout its lifecycle. In addition to facilitating ongoing accuracy, the self-calibration feature significantly reduces the need for manual recalibration and maintenance, making the device more user-friendly and lowering the overall cost and effort associated with its operation and upkeep. This feature is particularly vital for devices deployed in remote, inaccessible, or hazardous locations where manual intervention is impractical or impossible (Liu et al., 2019).

Safety and reliability

- **Error handling:** A sophisticated error handling mechanism is imperative to ensure the necrobotic device operates safely and reliably in various conditions. The error handling subsystem must be designed to promptly identify, report, and manage any errors or unexpected situations that may arise during operation, minimizing risks associated with system failures or malfunctions. For instance, if there's an anomaly in the biological component's response or a sensor's reading falls outside the expected range, the system should instantly flag this as an error and initiate predefined protocols to mitigate any negative consequences. Additionally, the control system should provide detailed error logs, facilitating easier diagnosis and resolution of issues by operators or maintenance personnel (de Croon et al., 2022). This feature not only enhances the device's safety but also supports timely troubleshooting and repair, promoting longer device life and more reliable performance over time. Furthermore, integrating preventive measures, such as automatic shutdown or transition to a safe mode in the event of critical errors, helps in preventing potential damages or accidents.
- **Redundancy:** Implementing redundancy within the device's control mechanisms and sensors is a crucial strategy for bolstering reliability and fault tolerance. By incorporating redundant components, the system ensures that there's a backup available to take over if any primary component fails or malfunctions, allowing for uninterrupted operation. For example, if a critical sensor fails, the redundant sensor will immediately assume its role, ensuring that the system continues to receive the necessary data without any significant delay or disruption (Shintake et al., 2018). This redundant design also extends to the control algorithms and processing units, where duplicated systems can operate in parallel or be on standby to take over in case of a failure. Such redundancy is particularly vital in applications where the necrobotic device is responsible for critical tasks, and any interruption could lead to severe consequences. Through this approach, the system can promptly recover from component failures, providing a robust and resilient operational capability that is foundational for various applications in medical, industrial, and exploratory scenarios, among others. The redundant design contributes not only to enhanced safety but also to the overall reliability and trustworthiness of the necrobotic system in the eyes of users and operators.

5 Proposed architecture

The proposed necrobotic architecture is an intricate framework divided into four primary layers: physical, integration, control, and application. The physical layer is foundational, encompassing the biotic module, IoT devices and sensors, and network communication. The biotic module is crucial for housing and managing selected biological materials

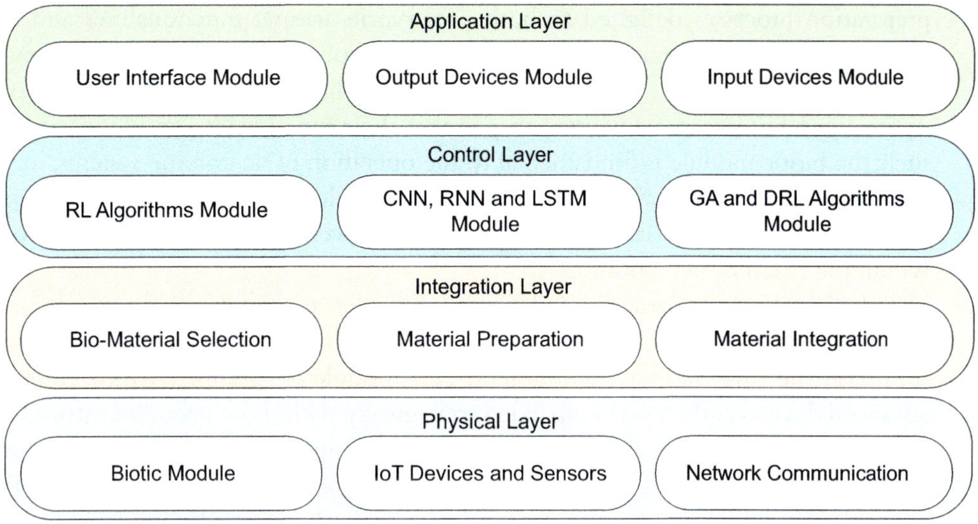

Fig. 1.3 ***Proposed necrobot architecture.*** A detailed layered necrobotic architecture shows a four-layered approach. *(No permission required.)*

from various entities, ensuring their integration and functionality within the robotic framework. IoT devices and sensors are embedded for data acquisition and environment interaction, providing real-time feedback and enabling adaptive responses. Network communication ensures connectivity for remote operation and data exchange. The integration layer is pivotal for material selection, preparation, and integration into the robotic framework, ensuring synergy between biological and artificial components. The control layer employs various algorithms modules, such as RL, CNNs, RNNs, LSTMs, and Genetic and Deep RL, each tuned to optimize necrobot performance in different environments and tasks. Lastly, the application layer features user interface, output devices, and input devices modules to facilitate user interaction and control, enact responses, and receive various input forms, making the system user-friendly and responsive. Each layer and module are meticulously designed to work seamlessly, providing a coherent and efficient system for diverse necrobotic applications shows the proposed necrobot architecture (Fig. 1.3).

5.1 Physical layer

- **Biotic module:** The biotic module serves as an integral component within the physical layer, acting as the nexus where the biological and mechanical elements within the necrobotic system converge. It is precisely structured to accommodate and manage meticulously selected biological materials, such as muscle tissues, bone structures, and sensory organs derived from various entities, including but not limited to spiders, scorpions, and snakes. Each selected material is subjected to rigorous preparation processes designed to both preserve its natural functionalities and facilitate its smooth integration within the broader robotic framework. The module provides a meticulously controlled environment to house these materials, ensuring optimal conditions are maintained to preserve their functionality and longevity. As such, the biotic module is fundamental to the operation of necrobotic systems, serving as the pivotal point where biology meets technology, and where the innate capabilities of biological materials are harnessed and leveraged for various applications within the realm of necrobotics.

- **IoT devices and sensors:** The IoT services and sensors module within the physical layer is a critical entity responsible for the seamless acquisition of data and interaction with the surrounding environment. This module is equipped with a suite of advanced devices and sensors embedded strategically within the necrobot's structure. These components work synergistically to monitor the real-time status and performance of both the biological materials and the necrobotic device in its entirety. By generating a continuous stream of vital data, this module provides invaluable feedback to the control layer, thereby facilitating the necrobot's adaptive and responsive interaction with its dynamic external environment. This continuous flow of information is crucial for enhancing the efficiency and effectiveness of the necrobot across a myriad of applications, ensuring that it can respond intelligently and promptly to various environmental stimuli and conditions.
- **Network communication:** Incorporated within the physical layer, the network communication module is essential for providing the necrobot with the capability to communicate with external systems and networks seamlessly. This module is engineered to facilitate the exchange of data between the necrobot and external entities, including centralized control units and cloud-based platforms, providing a conduit for remote operation and monitoring capabilities. Through secure, reliable, and rapid communication channels, the network communication module supports the operational needs of the necrobot in diverse and often challenging operational environments. It ensures that the necrobot remains connected, facilitating the transfer of data in real-time to and from the device, thereby supporting a range of functions including remote control, performance monitoring, and data analysis, among others.

5.2 Integration layer

- **Bio-material selection:** The bio-material selection sublayer operates as a pivotal component within the integration layer, serving as the first step in incorporating biological elements into necrobotic constructs. The significance of this phase cannot be overstated as it lays the foundation for the development of effective and ethically-compliant necrobotic devices. The procedure within this sublayer necessitates a rigorous examination of potential biological materials, evaluating their physical characteristics, mechanical strengths, and inherent biological functions. This meticulous analysis ensures that the selected materials are not only technically apt for the envisioned applications but also ethically procured and utilized. A vital aspect of the Bio-Material Selection is the compliance with ethical standards and norms, considering the sensitive nature of using biological materials derived from various organisms. Every effort must be taken to ensure that the process respects the sanctity of life, adhering to principles that prevent exploitation and misuse of biological resources.

- **Material preparation:** Following the selection, the material preparation sublayer undertakes the responsibility of readying the chosen biological materials for assimilation into the necrobotic architecture. The process within this sublayer is complex and nuanced, often requiring a series of sophisticated treatments applied to the selected biological constituents. Initially, these materials are subjected to a thorough cleaning procedure, eliminating any elements that might interfere with their integration into the robotic framework. Preservation techniques are then employed to maintain the viability and functional integrity of these biological components, extending their utility and effectiveness within the necrobotic system. Occasionally, certain modifications are imperative to align the biological materials with the technical specifications and requirements of the necrobotic device. These adjustments might include resizing, reshaping, or even molecular alterations designed to enhance compatibility and functionality when interfaced with electronic and mechanical counterparts within the system.
- **Material integration:** Material integration is the culminating sublayer in the integration layer, executing the careful incorporation of prepared biological materials into the necrobotic structure. This phase is characterized by precision and attention to detail, where the prepared biological components are strategically positioned and connected within the robotic framework. During this process, the biological materials are systematically aligned with the device's sensors, actuators, and other critical robotic elements. This alignment ensures that there is a seamless interplay and synergy between the biological and artificial components, fostering a harmonious operation within the necrobotic system. The integration process also considers the long-term functionality and sustainability of the biological materials within the device, implementing measures that support their sustained operation and contribution to the overall performance of the necrobot. The Material Integration sublayer is pivotal, setting the stage for a necrobotic device that embodies the perfect confluence of the biological and the synthetic, each enhancing the other's capabilities and functions.

5.3 Control layer

- **RL algorithms module:** The RL algorithms module plays an instrumental role in facilitating dynamic and responsive navigation and interaction for necrobots operating within intricate and unpredictable environments. This module is meticulously engineered, embedding a set of algorithms specifically designed and fine-tuned to harness and optimize the inherent capabilities of the integrated biological components. These algorithms are adept at learning and adapting in real-time from the feedback received during the necrobot's interactions with its environment, efficiently improving its performance on various tasks and operations.The RL algorithms module is crucial for scenarios requiring immediate and adaptive responses, enabling the necrobot to intelligently navigate, make decisions, and execute actions that are contingent on the dynamic contexts and stimuli it encounters, thereby enhancing the functionality and efficacy of the necrobotic system.

- **CNNs, RNNs and LSTMs module:** Serving as the cornerstone for visual perception and sequential data interpretation, the CNN, RNN, and LSTM Module is integral to the control layer of necrobotic systems.This module amalgamates various neural network architectures tailored for different sensory and processing applications. For tasks necessitating advanced image recognition and visual data processing, CNNs are employed due to their exceptional capability in handling spatial hierarchies and image classifications. Meanwhile, for applications involving the processing and analysis of sequential or time-series data, RNNs and LSTMs are utilized.These networks are adept at capturing and interpreting temporal dependencies within data sequences, making them indispensable for applications like signal processing and time-dependent decision-making in necrobots.
- **GA and deep RL algorithms module:** Incorporating GA and DRL algorithms, this module is fundamental for necrobots requiring optimization in performance and functionalities, particularly in scenarios demanding dynamic task scheduling, allocation, and the learning and execution of complex behaviors. Genetic algorithms are incorporated for their proficiency in optimizing solutions to problems through processes mimicking natural selection and genetics, thereby optimizing the task allocation and scheduling in the necrobot's operational environment. Concurrently, Deep RL algorithms are employed to facilitate the learning and execution of sophisticated, intelligent behaviors, enabling the necrobot to navigate and respond adeptly to unpredictable and complex environments.This combination of GA and Deep RL ensures that the necrobot is not only optimized for efficiency but is also capable of performing tasks with a level of sophistication and adaptability that mirrors the complexity of living organisms from which their biological components are derived.

5.4 Application layer

- **User interface module:** The user interface module is meticulously crafted, serving as an essential interactive bridge between the user and the necrobotic system. This module introduces a user-friendly, intuitive platform, enabling users to effortlessly input commands, monitor the necrobot's operational status, and receive instant, coherent feedback. Designed with the user experience at the forefront, the interface is streamlined yet comprehensive, facilitating easy navigation and control for users with varied levels of technical expertise. It incorporates visual aids, control panels, and responsive design elements that adapt to different devices and screen sizes, providing users with a seamless, accessible interaction platform regardless of their access point. Through the user interface module, users can engage with the necrobotic system efficiently, effectively directing its actions and receiving crucial information regarding its performance and status in real-time.

- **Output devices module:** Responsibly overseeing the translation of the necrobot's computational processes into tangible, physical responses, the output devices module is crucial. This module manages a suite of devices and components engineered to enact the calculated responses of the necrobot, ranging from the precise movement and manipulation of objects (leveraging integrated biological muscles or artificial limbs), to the generation of visual or auditory signals designed to communicate vital information or alerts to the user. This diverse spectrum of output capabilities ensures that the necrobot can effectively interact with its environment, execute tasks, and convey information in a manner that is immediately discernible and interpretable by users. The output devices module, therefore, plays an indispensable role in the functional loop of the necrobotic system, bridging internal computations with external actions and communications.
- **Input devices:** Designed to offer a versatile, multifaceted channel for external command and control, the input devices module is integral to the application layer. This module incorporates a variety of devices engineered to capture and interpret a wide range of user inputs, from manual controls and touch interfaces to voice commands and gesture recognition. Each input device is strategically selected and configured to offer users maximum flexibility and precision in controlling the necrobot. Through this module, users can seamlessly interface with the necrobotic system, providing specific commands and directions that are accurately received and interpreted by the system. The input devices module thus serves as the initial point of interaction between the user and the necrobot, capturing user inputs with precision and translating them into data that the system can understand and act upon, ensuring smooth, responsive control over the necrobotic operations.

6 Prospective use cases of necrobotics

This section meticulously unfolds an array of prospective use cases for necrobotics, providing a panoramic view of its versatile applications across diverse domains. Within the medical field, necrobotics is positioned as a revolutionary tool, facilitating intricate surgeries, innovative prosthetics, and advanced rehabilitation aids, thereby significantly enhancing patient care and treatment outcomes. Its application in environmental monitoring is equally promising, as necrobotic devices can be deployed for precise data collection and analysis, playing an indispensable role in conservation efforts, pollution control, and climate change mitigation.

6.1 Medical field

The integration of necrobotics in the medical field heralds a transformative potential, particularly impacting surgery, and the conception and development of prosthetics and assistive devices. Within surgery, necrobotics capitalizes on the unparalleled dexterity and sensitivity inherent in biological materials. By employing robotic surgical instruments embedded with tissues from deceased organisms, there's a realization of unprecedented precision and responsiveness, crucial for intricate procedures. This precision minimizes damage to surrounding healthy tissues, enhances patient outcomes, and replicates the nuanced movements vital during operations, offering surgeons tools that are not only precise but also intuitively aligned with the subtle complexities of human anatomy. Concurrently, the domain of prosthetics and assistive devices experiences significant advancements with necrobotics. Traditional prosthetics often fall short of mimicking the natural feel and movement of biological limbs. Necrobotics facilitates the integration of actual biological materials into prosthetics, creating devices that not only replicate the dynamics and sensations of living tissues but also provide users with a more comfortable and intuitive experience, thereby profoundly affecting their quality of life. This integration promises users enhanced mobility, ease, and confidence in daily activities while ensuring a natural interface with the user's body and shortening the adjustment period typically required with new prosthetic devices.

6.2 Industrial automation

In the realm of industrial automation, necrobotics promises to usher in groundbreaking solutions that bolster both efficiency and precision, chiefly seen in the advancements of gripping technology and the advent of biologically enhanced drones. The incorporation of specialized gripping mechanisms found in deceased organisms, like spiders, into robotic devices offers unparalleled opportunities for crafting advanced grippers. These are not only adept at delicately handling a diverse array of sensitive items, ranging from fragile food products and electronic components to slippery pharmaceutical items but also demonstrate acute precision and care. Such a technological leap proves invaluable, especially in sectors like food processing, electronics manufacturing, and pharmaceuticals, where stringent handling requirements and minimal margin for error are paramount. Simultaneously, the drone technology landscape is witnessing a transformative contribution from necrobotics. By assimilating anatomical features from organisms exemplifying optimized flying mechanisms, like birds or insects, drones equipped with improved aerodynamics and maneuverability are being developed. For example, integrating biological materials from birds facilitates the design of drones with superior gliding capabilities, thereby optimizing energy consumption during flights. Incorporating features from agile and precision-flying insects further allows the crafting of drones adept at navigating intricate environments efficiently. These biologically enhanced drones are invaluable in applications demanding sophisticated aerial maneuvering, such as surveillance and delivery, thereby offering services that are not only efficient but remarkably adaptable to a broad spectrum of operational needs.

6.3 Environmental monitoring

Environmental monitoring stands to gain substantially from the incorporation of necrobotic technology, particularly through the development of advanced bio-sensors that leverage the unique sensitivities of certain deceased organisms to environmental factors. These sensors can be crafted from biological materials derived from organisms that had heightened sensitivity to specific environmental elements while alive, providing a basis for precise and sensitive detection tools. One illustrative example is the use of biological materials from fish species known for their acute sensitivity to water pollution. Various fish species possess highly refined sensory mechanisms that allow them to detect minute concentrations of contaminants in their aquatic environments, responding to subtle changes that may be imperceptible by conventional sensors. By harnessing these biological materials and integrating them into necrobotic devices, researchers and environmental conservationists can develop sensors capable of early and accurate detection of water contaminants. Such bio-sensors could provide invaluable data on water quality, offering real-time insights into the presence and concentration of various pollutants. With this information at hand, authorities and conservation bodies can take timely and appropriate measures to address pollution, safeguard aquatic ecosystems, and ensure the provision of safe and clean water for communities. Furthermore, the potential applications of these advanced bio-sensors are vast, ranging from monitoring air quality using materials from organisms sensitive to air pollutants to soil quality assessment devices built with biological components sensitive to soil contaminants. The precise and reliable data generated by these necrobotic sensors could be instrumental in environmental conservation and public health efforts, facilitating informed decision-making and effective intervention strategies in response to environmental challenges.

6.4 Art and entertainment

Art and Entertainment is a sector that perennially seeks innovative technologies to elevate the realism and appeal of its offerings, and necrobotics may significantly contribute to this endeavor. One substantial application within this field is the development of advanced animatronics, which could greatly benefit from the integration of biological tissues into their design and function, making these creations more lifelike and engaging for audiences. Animatronics, commonly employed in theme parks, movies, and various entertainment venues, aim to replicate the movement and appearance of living creatures as closely as possible. Traditional animatronics rely extensively on synthetic materials and mechanical components, which, despite their sophistication, often fall short of mimicking the subtle nuances and complexities of living tissues. Necrobotics introduces the possibility of using preserved biological tissues in the construction of animatronics, offering a realistic texture, movement, and overall appearance unattainable with solely artificial materials. For example, the skin, muscles, or fur of deceased animals could be treated and utilized to cover the exterior of animatronic creatures, providing a visually captivating and authentic experience for viewers. The inherent flexibility and responsiveness of these biological materials could also facilitate smoother and more natural movements, enhancing the believability of these animatronic figures. Moreover, it is important to consider the ethical considerations and public perception related to the use of deceased organisms in entertainment. Engaging with ethical experts, the general public, and stakeholders in transparent discussions regarding sourcing and usage of biological materials is essential. These discussions would not only help navigate potential ethical pitfalls but also foster a public understanding and acceptance of the technology, ensuring that the deployment of necrobotics in art and entertainment is conducted responsibly and thoughtfully. With proper guidelines and considerations in place, necrobotics can revolutionize animatronics, providing audiences with unprecedented, captivating experiences while respecting ethical standards and public sentiment.

6.5 Education and research

Education and Research sectors continually strive to adopt innovative technologies that can facilitate more effective and engaging learning experiences, and necrobotics presents a unique opportunity in this regard. By integrating necrobotic devices as interactive learning tools, educational institutions can offer students invaluable hands-on experience, providing insights into both the biological and robotic aspects of these hybrid systems, thereby creating an interdisciplinary learning environment. Interactive learning tools based on necrobotics can serve as captivating visual aids that not only hold the attention of students but also encourage a more in-depth exploration of the subjects at hand. For instance, a necrobotic device that incorporates the muscle tissues and bone structures of deceased organisms can help students visually and tactilely understand the anatomy and physiology of these organisms, as well as the mechanical principles that govern their movement. This tangible interaction with biological materials, paired with robotic components, provides a concrete basis for understanding abstract concepts, thereby fostering a deeper comprehension and retention of the subject matter. Furthermore, these devices can facilitate an integrated learning approach, where students are exposed to the convergence of biology, technology, and ethics. Such an interdisciplinary perspective is crucial in preparing students for future careers in science and technology, where the boundaries between different fields are increasingly blurred. Engaging with necrobotic devices allows students to grapple with the ethical considerations and responsibilities entailed by the use of biological materials, promoting a sense of ethical stewardship and awareness that is vital in today's technologically advanced and morally complex society. Educational institutions adopting necrobotics as learning tools should also implement comprehensive educational programs to guide instructors on how to effectively integrate these devices into their curriculum. These programs might include training sessions, instructional materials, and continuous support mechanisms to ensure that educators are well-equipped to leverage the full potential of necrobotic devices in fostering an interactive and enlightening learning environment for their students. With careful planning and implementation, necrobotics can significantly enhance the educational landscape, providing students with a unique and engaging platform to explore the fascinating interplay between biology, technology, and ethics.

6.6 Military and defense

Military and Defense sectors are constantly exploring technologies that offer enhanced and stealthy surveillance capabilities. Necrobotics has the potential to contribute significantly in this area by facilitating the creation of surveillance devices that seamlessly integrate into natural environments. Devices developed with necrobotic technology can leverage the appearance and functionalities of deceased organisms, allowing these devices to closely mimic the visual characteristics and behaviors of local fauna, which makes them less conspicuous and detectable in their operational environments. The primary advantage of using necrobotic technology for surveillance devices in military and defense is the discreet observation capabilities it offers. For example, a surveillance device designed to resemble a bird or an insect can operate covertly, blending into its surroundings and avoiding detection while capturing critical data and intelligence. Such devices can be deployed in sensitive or hostile areas, providing military personnel with crucial information without exposing human lives to danger. Additionally, the physical and sensory characteristics of the biological materials integrated into these devices can offer enhanced functionality. A surveillance drone designed using the wings of birds might exploit the aerodynamics of these biological structures for more efficient and agile flight. Similarly, a device created with the eyes of animals known for their exceptional night vision could provide superior low-light imaging capabilities. These enhanced features enable the devices to operate effectively under various environmental conditions, expanding their utility on the field. However, it is imperative that the development and deployment of necrobotic devices in military and defense adhere strictly to ethical guidelines and legal frameworks. The sourcing of biological materials, even from deceased organisms, must be conducted responsibly, with respect to consent, legality, and the sensitivities surrounding the use of such materials for military purposes. Furthermore, public transparency about the use and capabilities of these devices is crucial for fostering public trust and understanding regarding the application of necrobotic technology in defense. With meticulous development and ethical considerations, necrobotic technology can provide the military and defense sectors with innovative surveillance solutions that combine the best of biological and robotic functionalities for discrete and effective operation in a myriad of contexts and environments.

6.7 Human-computer interaction

In the realm of human-computer interaction (HCI), the integration of biological tissues into interface devices through necrobotics offers the promise of creating systems that are significantly more adaptive and responsive to users. Such bio-integrated interfaces can revolutionize the manner in which individuals interact with technology, providing an experience that is more intuitive and closely aligned with natural human behavior and responses. Bio-integrated interfaces might utilize the unique properties of biological tissues, such as sensitivity to pressure, temperature, and other environmental factors, to offer responsive feedback that is not easily achieved with purely electronic devices. For example, a touchpad or touchscreen developed with biological materials may have the capability to precisely detect and respond to the nuances of human touch, providing feedback that feels remarkably lifelike and natural to users. This could result in interfaces that are more accommodating and reactive to user inputs, enhancing the effectiveness and user-friendliness of the technology. The adaptive characteristics of biological tissues can also be harnessed to craft interfaces that can learn and adjust to the specific preferences and needs of individual users. Through the incorporation of AI and machine learning algorithms, these interfaces can analyze user interaction data over time to optimize their responsiveness and functionality. This synergy between biological materials and advanced computing techniques can facilitate the development of personalized interface systems that significantly improve user engagement and satisfaction.

Furthermore, bio-integrated interfaces may offer innovative solutions for individuals with disabilities, providing them with accessible technology options that are sensitive to their unique requirements. For instance, an interface designed with biological tissues that mimic the sensitivity and flexibility of human muscles could be immensely beneficial for users with limited mobility, as it may require less force and precision to operate. However, while the potential of bio-integrated interfaces in HCI is enormous, it is also crucial to acknowledge and address the ethical concerns associated with the use of biological materials. Issues related to consent, sourcing, and the handling of these materials must be approached with the utmost responsibility and care. The development of these advanced interfaces should proceed with a commitment to ethical considerations, legal compliance, and respect for the inherent value of biological life.

6.8 Search and rescue operations

In the realm of search and rescue operations, biologically-inspired search drones developed through necrobotics offer a potent blend of biological keenness and technological precision. These drones, incorporating biological components from creatures known for acute senses and superior navigational abilities, can provide invaluable aid in locating and reaching individuals in distress across varied and challenging terrains. One key advantage of these specialized drones is their ability to leverage the natural sensors and mechanisms of the integrated biological materials. For instance, by harnessing the heightened auditory or olfactory senses found in certain animals, these drones could detect sounds or scents that are typically imperceptible to traditional technological devices. This enhancement allows the drones to more accurately pinpoint the location of people in need of rescue, even in environments with low visibility or in scenarios where victims might be concealed or trapped. Moreover, the navigational prowess of animals like birds or bats, which can adeptly maneuver through cluttered spaces, can be utilized to enhance the drone's ability to navigate through dense forests, urban rubble, or other complex environments. The fluidity and agility of their movement can be mimicked to design drones that can swiftly and efficiently traverse through challenging terrains, ensuring a quicker response time during critical rescue operations. The integration of these biological components does not only enhance sensory and navigational capabilities but also potentially allows for a more adaptive and intuitive operational control. Operators may find these drones easier to control and more responsive to commands, given their biomimetic design and function. Such intuitiveness in control mechanisms can be crucial in high-pressure rescue situations where time is of the essence, allowing for smoother and more effective missions. At the same time, it is imperative to approach the development and deployment of these necrobotic drones with careful consideration of ethical concerns. Issues related to the sourcing and use of biological materials must be meticulously addressed to ensure responsible and respectful handling of these components. This involves obtaining proper consents, adhering to legal standards, and possibly developing new regulatory frameworks that support the ethical development of this cutting-edge technology.

6.9 Agriculture and farming

In the sphere of agriculture and farming, the integration of necrobotic technology can significantly amplify the efficiency and sustainability of precision farming practices. Precision farming devices infused with deceased biological components that are inherently sensitive to crucial environmental factors, such as humidity, light intensity, and nutrient levels, are pivotal in this technological advancement. For instance, certain biological tissues or organs might possess innate sensitivities to variations in moisture and light. The tissues from organisms that have adapted to thrive in highly specific environmental conditions can be utilized as responsive sensors in precision farming devices. These biotic sensors can then continuously monitor the ambient conditions of the crop environment, providing real-time data with a level of sensitivity and accuracy that is challenging for purely synthetic sensors to achieve. This constant, accurate monitoring supports farmers in maintaining the optimal conditions necessary for maximizing crop yields and promoting sustainable farming practices. Furthermore, the inherent sensitivity of these biological components to nutrient levels in the soil can be leveraged for precise fertilization applications. Necrobotic devices could be designed to not only detect deficiencies in essential nutrients but also to respond by dispensing the exact amount of fertilizer needed to correct the imbalance. This approach to fertilization is not only efficient but also environmentally friendly as it minimizes waste and prevents the over-application of fertilizers, which could otherwise lead to runoff and contamination of nearby water sources. However, the development and application of necrobotic technology in agriculture also necessitate a robust framework for ethical considerations and practices. The sourcing of biological materials must be conducted with respect to the highest ethical standards, considering the consent, origins, and treatment of the biological tissues integrated into the devices. Furthermore, there should also be an emphasis on developing and adhering to rigorous quality control and safety standards to ensure that the necrobotic devices deployed in agricultural settings are safe, reliable, and effective in their operation. Ultimately, the infusion of necrobotic technology into precision farming devices has the potential to revolutionize agricultural practices by providing unparalleled sensitivity and responsiveness to environmental conditions crucial for crop growth. This synergy of biology and robotics not only promises enhanced yields but also fosters sustainability in the farming sector, given that the ethical considerations and quality standards are diligently observed and upheld.

6.10 Rehabilitation therapy

For patients undergoing rehabilitation therapy, the integration of necrobotic devices provides a unique and innovative approach to healing and recovery. These sophisticated devices are designed to be inherently responsive, adapting intelligently to the user's movements and exertions during therapeutic sessions. Firstly, the utilization of materials that closely resemble the characteristics of muscles or tendons is pivotal. These materials possess a natural elasticity and responsiveness that artificial materials might struggle to replicate. When integrated into therapy tools, these biotic materials can facilitate a dynamic interaction with the patient, providing a level of resistance and support that is finely tuned to the patient's needs. This nuanced interaction not only makes the rehabilitation process more comfortable but also significantly enhances its effectiveness. Furthermore, the natural feel of these necrobotic devices can substantially aid in the psychological aspects of rehabilitation. Patients may find interacting with materials that have a biological feel more comforting and less alien, potentially reducing anxiety and increasing their engagement with the rehabilitation process. This positive psychological environment, in conjunction with the physical benefits offered by the devices, creates a holistic approach to rehabilitation that addresses both the body and the mind. The precision offered by necrobotic devices ensures that each patient's unique needs are acknowledged and catered for, providing a personalized rehabilitation experience.

6.11 Space exploration

Necrobotic technology presents tantalizing possibilities in the realm of space exploration, where the demand for resilient and adaptive equipment is paramount. Bio-enhanced exploration rovers signify one of the innovative applications wherein biological materials that inherently possess the capacity to withstand or adapt to extreme environmental conditions are integrated into the design and functionality of exploration devices. Biological materials, harvested from organisms that thrive in some of Earth's most demanding environments—such as extremophiles—are prime candidates for this application. Extremophiles are organisms that can survive and thrive in environments with extreme temperatures, pressures, and radiation levels, similar to conditions that might be encountered on other planets or moons. Through careful selection and treatment, these biological materials can be preserved and incorporated into rovers, providing them with a unique set of capabilities not easily replicable with entirely synthetic materials. The incorporation of such robust biological materials can lead to the development of exploration rovers that are more resilient to the harsh and unpredictable conditions of extraterrestrial environments. For instance, these bio-enhanced rovers might demonstrate improved tolerance to temperature fluctuations, radiation levels, and physical stresses, significantly extending their operational lifespan and reliability during missions. Furthermore, these biological materials might offer novel sensory and actuation capabilities, enabling the rovers to navigate and interact with their surroundings in innovative ways. This integration not only promises enhanced durability but also opens avenues for conducting unique experiments and research in space. Scientists can leverage the distinctive properties of these biological materials to design experiments that yield deeper insights into the viability of life in outer space, the adaptability of terrestrial life to extraterrestrial conditions, and the potential for discovering new forms of life.

6.12 Consumer products

In the consumer products sector, necrobotics holds significant potential to revolutionize the creation of wearable devices through the introduction of adaptive wearable technology. The unique integration of biological materials within wearables brings forth products that not only exemplify a high degree of comfort but also showcase adaptability previously unseen in traditional devices. Biological materials, like skins, muscles, or cartilage, possess inherent flexibility and elasticity that synthetic materials often struggle to replicate accurately. When these biological materials are treated and integrated into wearable devices, they enable the creation of products that can adapt and conform intimately to the unique contours and movements of the user's body. This leads to wearables that provide an unparalleled fit that is snug, secure, yet remarkably comfortable. Furthermore, these adaptive wearables, embedded with biological tissues, can also exhibit a level of sensitivity and responsiveness to the user's physiological signals and changes. This sensitivity allows the wearable devices to monitor health metrics with precision, respond to the body's needs in real-time, and possibly even facilitate the delivery of personalized health and wellness feedback and interventions. For instance, wearables with muscle or skin tissues might detect changes in tension, temperature, or moisture levels, providing invaluable data for health monitoring and diagnostics. The adaptive wearables, while demonstrating promise, also necessitate careful design and engineering to ensure durability, functionality, and safety for everyday use. The process of integrating biological with electronic components must be meticulously carried out to foster seamless interaction between these disparate materials while maintaining the integrity and functionality of each. Attention to the aesthetic appeal and ethical considerations of using biological materials in consumer products is also vital to gain public acceptance and trust in these innovative devices. Moreover, ethical sourcing, preservation, and enhancement of the biological materials used play a pivotal role in the development of these wearables. These processes need to be refined and standardized, ensuring that the products are not only effective and reliable but also produced responsibly and sustainably, aligning with the values and expectations of consumers in the market.

6.13 Key challenges in necrobotics

This section presents key challenges and considerations in the development of necrobotics, beginning with the imperative preservation of biological materials, crucial for preventing decomposition and degradation and addressing their temperature sensitivity to maintain functionality in robotic applications (Zhalmuratova and Chung, 2020; Cianchetti et al., 2018). It discusses the importance of ensuring compatibility between biological and electronic components, requiring meticulously designed interfaces. Precise control and calibration are emphasized due to the unpredictable nature of biological materials, necessitating advanced techniques and unique control algorithms. Ethical concerns, particularly in sourcing biological materials, are outlined, highlighting the need for informed consent, legal compliance, and respect for cultural norms. The section also addresses public perception, advocating for transparent communication and public engagement to facilitate broader acceptance. Furthermore, it explores challenges in handling variability in biological components, emphasizing the need for calibration protocols and strategies to ensure long-term stability, scalability, and quality control in necrobotic systems.

6.14 Preservation of biological materials

Preservation of biological materials is a fundamental step in necrobotics, addressing critical issues like decomposition, degradation, and temperature sensitivity. Once an organism dies, its tissues start to decompose almost immediately. This rapid decomposition process not only alters the physical structure of the tissues but also impairs their functionality and utility in robotic applications. To mitigate decomposition and degradation, researchers in necrobotics employ various preservation techniques. This involves using chemical agents that can slow down or halt the decomposition process, while carefully ensuring that these agents do not adversely affect the inherent properties of the biological materials that are beneficial to the robotic system. For instance, the preservation process must maintain the elasticity of muscle tissues and the rigidity of bone structures without introducing changes that might compromise their performance in the final robotic application (Ai et al., 2021).

Temperature sensitivity is another significant challenge in preserving biological materials. These materials often necessitate specific storage environments, characterized by strictly controlled temperatures and humidity levels, to prevent degradation and loss of functionality. Responding to this challenge, necrobotics specialists are tasked with designing robotic components that can operate effectively under these constrained conditions. This might involve integrating thermal control systems within the robotic device to maintain the optimal temperature for the biological materials, or developing insulation mechanisms to shield these materials from temperature variations in the external environment. The ultimate goal is to create a stable and conducive environment within the robotic system where the preserved biological materials can function reliably and effectively over time.

6.15 Integration challenges

Compatibility between biological and electronic components in necrobotics presents a notable challenge, primarily due to the distinctive electrical, mechanical, and thermal properties of each element. These differences necessitate the careful creation of interfaces designed to accommodate and seamlessly integrate the disparate natures of biological and electronic materials, ensuring that there is no loss of functionality in either (Olsen and Kim, 2019). Developing such interfaces is no small feat, requiring innovative approaches and technologies that can bridge the gap between the living and nonliving components effectively and efficiently.

Achieving precise control and calibration of the biological materials within a robotic system adds another layer of complexity to the design and operation of necrobotic devices. The inherent variability and intricacy found in biological materials make them unpredictable and challenging to manipulate with precision. Advanced calibration techniques are essential to account for these variations and ensure that the biological components respond accurately to control inputs. Furthermore, the development of novel control algorithms may be necessary to provide reliable and precise command over the integrated system, allowing for smooth operation and functionality in various applications and environments.

6.16 Ethical and legal concerns

Navigating through the intricate web of ethical considerations, the sourcing of biological materials for necrobotics demands utmost care and sensitivity. Whether the materials are sourced from deceased humans or animals, it is imperative to obtain informed consent from the family members or appropriate entities. The procurement process must adhere strictly to the legal and cultural norms that govern the handling and utilization of dead bodies and animal remains. Beyond legal compliance, these practices should embody a deep respect for life and the deceased, acknowledging the profound ethical implications involved in utilizing biological materials postmortem for technological applications (Sarker et al., 2021).

The introduction of necrobotic technology inevitably stirs public sentiment, potentially leading to apprehension or outright rejection of these innovations due to entrenched ethical, religious, or cultural beliefs. For necrobotics to be socially accepted and integrated into various fields, it is crucial for practitioners and innovators in the field to actively engage with the public and various stakeholders. Transparent communication about the technology's purposes, potential benefits, and the measures taken to address ethical concerns is fundamental (King et al., 2009). The field needs to foster an open dialogue to understand and address public concerns and sensitivities, ensuring that the technology develops in a way that is respectful of diverse perspectives and values, thus facilitating broader social acceptance.

6.17 Reliability and consistency

Variability in biological components presents a significant hurdle in the development and deployment of necrobotic systems. Biological materials inherently exhibit a wide range of differences due to unique genetic, environmental, and developmental factors, making each sample distinct. This significant variability contrasts sharply with the standardized and predictable nature of manufactured parts typically used in robotics (Tamborini, 2023). Consequently, achieving consistent performance across various necrobotic units that incorporate biological materials necessitates the development of sophisticated calibration and adjustment protocols. These protocols must be able to accommodate and compensate for the nuances and fluctuations in the properties and behaviors of different biological samples to ensure reliable and predictable system performance.

On the other hand, the concern regarding long-term stability of biological materials within robotic systems is also pivotal. Biological tissues and structures, even when preserved and treated, might not possess the durability and longevity of synthetic materials (Cominelli et al., 2018). Their susceptibility to degradation over time, under different operating conditions and environments, can lead to a gradual loss of functionality and performance. Addressing this challenge requires innovative approaches to extend the lifespan of biological components within necrobotic systems without compromising their function. Researchers and engineers in the field must explore a combination of preservation techniques, material enhancements, and possibly the incorporation of redundant systems to provide long-term stability and reliability, minimizing the need for frequent replacements and maintenance.

6.18 Scalability and manufacturing

Scalability and mass production of necrobotic components are met with considerable challenges due to the unique nature of biological materials. Unlike traditional robotic manufacturing that relies on standardized, replicable parts, necrobotics works with biological components which vary greatly from one another. This variability complicates the process of producing large quantities of units while maintaining a level of consistency in their performance and functionality. Additionally, the procurement and preparation of biological materials involve intricate processes that are hard to standardize for mass production, which is crucial for cost-effectiveness (Shin et al., 2018). The industry would need to develop innovative manufacturing techniques and processes that not only preserve the integrity of biological materials but also allow for scalable production without inflating costs.

Quality control is a significant concern in the production of necrobotic devices, as it is essential to ensure that each unit produced is up to par with the required standards and specifications. Given the inherent variability and sensitivity of biological materials, implementing robust and reliable quality control mechanisms is inherently challenging. Each piece of biological material used must be meticulously examined and tested to ensure it meets the functional requirements without compromising the reliability of the final product (Froese and Taguchi, 2019). Establishing these quality assurance protocols requires a deep understanding of both the biological and robotic elements involved, as well as the development of new testing methodologies specifically designed for hybrid devices that integrate these diverse components. These measures are vital not only for the performance of individual units but also for building trust and reliability in necrobotic products among end-users and stakeholders.

7 Future directions of necrobotics

This section explores pivotal advancements in necrobotics, highlighting improved preservation and enhancement of biological materials essential for the system's optimized functionality and longevity. It underscores the importance of developing ethical guidelines and legal frameworks to promote responsible innovation and garner public trust (Egorov et al., 2021; Floreano and Mattiussi, 2008; Russell et al., 2015). The text also outlines diversified use cases of necrobotics in sectors like conservation, education, defense, and consumer products, underscoring its societal impact and growth potential. Additionally, it examines the synergy of AI, Machine Learning, and IoT with necrobotic devices, expanding their utility and application. The paper emphasizes the necessity of public engagement and education for fostering understanding, acceptance, and ethical advancement of necrobotic technologies.

7.1 Material advancements

Advancements in preservation techniques for biological materials are imperative for the future development of necrobotics. New and improved methods need to be explored to ensure the integrity, functionality, and longevity of biological components without relying on extensive maintenance or specific, limiting conditions. Since biological tissues begin to degrade shortly after the organism's death, finding innovative ways to slow or halt this decomposition process is crucial. Such techniques should not only preserve the tissues but also retain their essential properties and functions over time (Doncieux et al., 2015). Research and development in this area may involve experimenting with novel chemical solutions, cryopreservation technologies, or even electromagnetic fields that might influence the preservation process, providing a stable environment for the biological components within the robotic system.

Moreover, focusing on the enhancement of biological materials is pivotal. Future research should aim to augment the inherent qualities of these materials, making them more durable, resilient, and effective when utilized in necrobotic applications. This can involve manipulating the biological materials at the molecular or cellular levels, perhaps incorporating nano-technologies or advanced biochemical modifications to improve their performance (Whitesides, 2018; Rawat et al., 2018). For example, coating tissues with protective layers that not only prevent degradation but also improve mechanical strength or electrical conductivity can be a viable approach. Enhancing the biological materials would not only expand the potential applications of necrobotics but also improve the reliability and lifespan of these innovative devices, bringing the field closer to producing practical and sustainable solutions for various industries and sectors.

7.2 Ethical and legal frameworks

The crafting of comprehensive ethical guidelines for necrobotics is a cornerstone for its responsible and respectful development and implementation. As this field navigates through uncharted territories of utilizing deceased biological organisms, it is imperative to address various sensitive concerns that emerge. Ethical guidelines need to cover obtaining informed consent, especially when the biological materials sourced are human. This involves ensuring that there is a clear understanding and agreement from the donor (or next of kin) regarding the use of tissues postmortem. Furthermore, the guidelines should also encompass ethical sourcing and handling of animal tissues, considering the welfare and rights of animals. The practices should aim at minimizing suffering and ensuring that the materials are sourced sustainably and respectfully. Such guidelines should not only guide researchers and practitioners in the field but should also serve to build public trust and acceptance of necrobotics.

In addition to ethical considerations, adherence to and shaping of international legal standards and norms is vital for the advancement of necrobotics. Legal frameworks need to be developed that are able to foster innovation while ensuring that the practices within necrobotics are responsible and ethical. These policies and regulations should be crafted in collaboration with legal experts, ethicists, researchers, and the broader community to address the nuances and complexities of integrating biological materials with robotic systems. Ensuring compliance with these legal frameworks is not just about adhering to the law but also about creating an environment where the technology can be developed and used in a way that is accepted and trusted by society at large. Working within a well-defined legal structure will also facilitate international collaboration and the sharing of knowledge and technologies among researchers and institutions across the globe.

7.3 Application expansion

Exploration into diversified use cases for necrobotics is essential for the field's growth and societal impact. Researchers and developers should actively investigate a spectrum of applications where the distinct advantages of necrobotic systems can be harnessed effectively. For instance, in environmental conservation, necrobotics can be utilized to monitor delicate ecosystems without disturbing the native fauna due to their partially biological construction. In education, these technologies might serve as unparalleled tools for biology and robotics instruction, providing students with hands-on experience in both disciplines (van den Hoven van Genderen, 2018; Shuai et al., 2019). Defense applications could explore the development of biohybrid drones or reconnaissance devices, leveraging the inherent sensory and navigational capabilities found in certain organisms. Furthermore, in space exploration, necrobotic systems could offer resilient and adaptive solutions for navigating and studying extraterrestrial environments, capitalizing on the robustness and efficiency of biological materials.

Simultaneously, it is imperative to consider and develop applications of necrobotics that resonate with consumer needs and interests, thereby fostering the inception of products that integrate necrobotic technologies for everyday use. Potential consumer applications could range from biohybrid prosthetics and wearables that offer enhanced functionality and biocompatibility, to educational kits that allow enthusiasts to explore the convergence of biology and robotics in a hands-on manner. Engaging with the public through consumer products can not only drive the commercial success of necrobotics but also facilitate broader acceptance and understanding of the technology among the general population. This consumer-oriented approach will likely spur innovation, open new markets, and create opportunities for startups and established businesses alike, ultimately contributing to the economy and the advancement of technology (Ward-Cherrier et al., 2018).

7.4 Technology integration

The synergy between AI and machine learning (ML) with necrobotic devices heralds the advent of intelligent and adaptive systems that significantly broaden the utility and application of necrobotics. Through the integration of sophisticated AI and ML algorithms, necrobotic devices can learn from their environments, adapt to changing conditions, and optimize their operations in real-time, mirroring a form of "digital sentience." This amalgamation of AI and ML with biological components can lead to the creation of robotic systems that not only mimic biological functions but also possess the capability to process information, make decisions, and execute tasks autonomously. This confluence of technologies may facilitate the development of necrobotic systems for complex and nuanced applications where dynamic adaptation and continuous learning are imperative, thereby revolutionizing fields like healthcare, logistics, and disaster response among others (Mertz, 2022).

Moreover, embedding Internet of Things (IoT) connectivity into necrobotic devices is pivotal for their seamless integration into existing technological infrastructures and ecosystems. IoT connectivity ensures that necrobotic devices can communicate and collaborate with a myriad of smart devices and systems, sharing data and insights in real-time. This interconnectedness enhances the functionality and versatility of necrobotic devices, enabling them to be a part of larger networks and systems that are working synergistically toward shared objectives or tasks. For example, necrobotic devices in a smart home environment could interact with security systems, lighting, or climate control devices, providing a cohesive and intelligent living environment for users. With IoT connectivity, the potential applications of necrobotics expand exponentially, laying the foundation for a future where biological and electronic components not only coexist but collaboratively drive innovation and progress in technology.

7.5 Public engagement and education

Implementation of public awareness campaigns is pivotal for cultivating a comprehensive understanding and acceptance of necrobotics among the general populace. These campaigns should be meticulously designed to convey the myriad benefits offered by necrobotics, such as its potential in medical, environmental, and technological advancement, while simultaneously addressing and alleviating any ethical, moral, or cultural concerns associated with using deceased biological materials. Clear, accessible, and transparent information should be provided through various media channels, educational institutions, and public forums to reach a diverse audience (Wu et al., 2022). These educational initiatives would demystify necrobotics, correct misconceptions, and foster a more informed and open dialogue about its implications and opportunities, thereby building public trust and support for the field.

Furthermore, active community and stakeholder engagement is imperative for the ethical and responsible development and deployment of necrobotic technologies. A collaborative approach involving scientists, ethicists, legal experts, and representatives from various communities would ensure that the voices and concerns of different societal segments are heard and considered in the innovation process (Wang and Siau, 2019; Hang et al., 2021). This collective dialogue can inform the establishment of ethical guidelines and legal frameworks governing necrobotics, aligning the technology's development with societal values and expectations. Engaging communities also provides an opportunity for identifying new applications and markets for necrobotic technologies, fostering a sense of ownership and investment among the public and enhancing the technology's social license to operate (Laird et al., 2017; *Cyborg Futures: Cross-Disciplinary Perspectives on Artificial Intelligence and Robotics*, 2019). Through these engagements, necrobotics can be developed in a way that is not only technologically advanced but also ethically sound and socially accepted.

8 Conclusion

This research endeavor has provided a meticulous exploration and critical analysis of the innovative realm of necrobotics, wherein the extraordinary faculties of deceased biological organisms are intricately integrated with the precision and programmable capabilities of robotic systems. Through a systematic exposition, the paper elucidates the meticulous processes and criteria vital for the judicious selection and preparation of deceased biological materials, subsequently embedded into robotic devices to enhance their functionalities significantly. Necrobotics, as the discussion reveals, is more than a mere confluence of biology and robotics; it represents a forward-thinking paradigm that potentially redefines the boundaries of what is conceivable in technology, providing a unique intersection of life and artificial mechanisms. The detailed guidelines for the design and fabrication of necrobotic devices provided within the paper are not only insightful but act as a beacon that guides practical applications and implementations of theoretical concepts in the real-world, serving various domains ranging from healthcare to space exploration. We have illuminated the intricate control mechanisms pivotal to the smooth operation of necrobotic systems, offering foundational insights and guidelines that are expected to serve as a bedrock for future innovations in the field. The identified challenges and limitations within the necrobotics domain underscore the necessity for continued and focused research endeavors to navigate and mitigate these issues effectively. The roadmap of challenges and solutions delineated in this discussion is instrumental for researchers and practitioners alike, as they navigate the complex landscape of necrobotics. Moreover, the article has painted a comprehensive picture of the potential applications and far-reaching impacts of necrobotics across diverse sectors. Through this, it hopes to spark interest, discussion, and further research into this groundbreaking field that promises to blend the lines between organic and inorganic, crafting devices that are as remarkable and efficient as life itself. The future of necrobotics is not only bright but teeming with unlimited potential and possibilities, holding the promise to significantly influence and transform the technological landscape as we know it today. The paper is thus a seminal read for anyone interested in or working at the nexus of biology and robotics, providing deep insights and acting as a catalyst for future innovative projects and transformative technological solutions in the domain of necrobotics.

AI disclosure

During the preparation of this work the author(s) used ChatGPT 4 in order to The topic on necrobotics is very new, although it has basis on existing techniques. Thus, ChatGPT 4 was used to infer more futuristic aspects on how to shape necrobotics for representational and actual manifestation into the reality. However, it was used for initial drafting. Author has done necessary changes and modifications and included own content to enrich the article. After using this tool/service, the author(s) reviewed and edited the content as needed and take(s) full responsibility for the content of the publication.

References

Ackerman, E., 2022. Necrobotics: Dead Spiders Reincarnated as Robot Grippers. 2022.

Ai, Z.L., Meng, W., Liu, S.Q.X., 2021. Machine learning in robot assisted upper limb rehabilitation: a focused review. IEEE Trans. Cogn. Dev. Syst. 2021.

Auerbach, J.E., Concordel, A., Kornatowski, P.M., Floreano, D., 2019. Inquiry-based learning with RoboGen: an open-source software and hardware platform for robotics and artificial intelligence. IEEE Trans. Learn. Technol. **12** (3), 356–369. Institute of Electrical and Electronics Engineers, United States 19391382. Available from: http://ieeexplore.ieee.org/xpl/RecentIssue.jsp?punumber=4620076. https://doi.org/10.1109/TLT.2018.2833111.

Berg, R.B.S., Rusák, Z., Wu, J., 2022. OpenFish, biomimetic design of a soft robotic fish for high speed locomotion. HardwareX 12, 2022.

Bhattacharjee, N., Urrios, A., Kang, S., Folch, A., 2016. The upcoming 3D-printing revolution in microfluidics. Lab Chip 16 (10), 1720–1742. https://doi.org/10.1039/c6lc00163g. 1473-0197.

Bhaumik, A., 2018. From AI to Robotics: Mobile, Social, and Sentient Robots. CRC Press. 2018.

Blackiston, D., Kriegman, S., Bongard, J., Levin, M., Robots, B., 2023. Perspectives on an emerging interdisciplinary field. Soft Robot. 10 (4), 674–686. https://doi.org/10.1089/soro.2022.0142. 2169-5172.

Boblan, I., Bannasch, R., Schulz, A., Schwenk, H., 2007. A human-like robot torso ZAR5 with fluidic muscles: toward a common platform for embodied AI. In: Lecture Notes in Computer Science (Including Subseries Lecture Notes in Artificial Intelligence and Lecture Notes in Bioinformatics). vol. 4850. Springer Verlag Germany, pp. 347–357. https://www.springer.com/series/558. https://doi.org/10.1007/978-3-540-77296-5_31. 16113349.

Boch, A., Ryan, S., Kriebitz, A., Amugongo, L.M., Lütge, C., 2023. Beyond the metal flesh: understanding the intersection between Bio- and AI ethics for robotics in healthcare. Robotics 12 (4). Multidisciplinary Digital Publishing Institute (MDPI), Germany. 22186581. Available from: http://www.mdpi.com/journal/robotics. https://doi.org/10.3390/robotics12040110.

Cheng, Y., Chan, K.H., Wang, X.Q., Ding, T., Li, T., Lu, X., Ho, G.W., 2019. Direct-ink-write 3D printing of hydrogels into biomimetic soft robots. ACS Nano 13 (11), 13176–13184. American Chemical Society, Singapore. 1936086X. Available from: http://pubs.acs.org/journal/ancac3. https://doi.org/10.1021/acsnano.9b06144.

Cianchetti, M., Laschi, C., Menciassi, A., Dario, P., 2018. Biomedical applications of soft robotics. Nat. Rev. Mater. 3 (6), 143–153. Nature Publishing Group, Italy. 20588437. Available from: www.nature.com/natrevmats/. https://doi.org/10.1038/s41578-018-0022-y.

Cominelli, L., Mazzei, D., De Rossi, D.E., 2018. SEAI: social emotional artificial intelligence based on Damasio's theory of mind. Front. Robot. AI. 5. https://doi.org/10.3389/frobt.2018.00006. 2296-9144.

2019. Cyborg Futures: Cross-Disciplinary Perspectives on Artificial Intelligence and Robotics. Springer Nature. 2019.

de Croon, G.C.H.E., Dupeyroux, J.J.G., Fuller, S.B., Marshall, J.A.R., 2022. Insect-inspired AI for autonomous robots. Sci. Robot. 7 (67). American Association for the Advancement of Science, Netherlands. 24709476. Available from: https://www.science.org/doi/10.1126/scirobotics.abl6334. https://doi.org/10.1126/scirobotics.abl6334.

Delda, R.N.M., Basuel, R.B., Hacla, R.P., Martinez, D.W.C., Cabibihan, J.J., Dizon, J.R.C., 2021. 3D printing polymeric materials for robots with embedded systems. Technologies. 9 (4). MDPI, Philippines. 22277080. Available from: www.mdpi.com/journal/technologies. https://doi.org/10.3390/technologies9040082.

Doncieux, S., Bredeche, N., Mouret, J.-B., (Gusz) Eiben, A.E., 2015. Evolutionary robotics: what, why, and where to. Front. Robot. AI 2. https://doi.org/10.3389/frobt.2015.00004. 2296-9144.

Duncan, D.E., 2019. Talking to Robots: Tales from our Human-Robot Futures. Penguin. 2019.

Egorov, E., Pieters, C., Korach-Rechtman, H., Shklover, J., Schroeder, A., 2021. Robotics, microfluidics, nanotechnology and AI in the synthesis and evaluation of liposomes and polymeric drug delivery systems. Drug Deliv. Transl. Res. 11 (2), 345–352. Springer, Israel. 21903948. Available from: http://www.springer.com/biomed/pharmaceutical+science/journal/13346. https://doi.org/10.1007/s13346-021-00929-2.

Ensslin, A., 2017. Future modes: how 'new' new media transforms communicative meaning and negotiates relationships. In: The Routledge Handbook of Language and Media. Taylor and Francis, Canada, pp. 309–324. Available from: https://www.routledge.com/The-Routledge-Handbook-of-Language-and-Media/Cotter-Perrin/p/book/9780367735944.

Fang, J., Zhuang, Y., Liu, K., Chen, Z., Liu, Z., Kong, T., Xu, J., Qi, C., 2022. A shift from efficiency to adaptability: recent progress in biomimetic interactive soft robotics in wet environments. Adv. Sci. 9 (8). John Wiley and Sons Inc, China. 21983844. Available from: http://onlinelibrary.wiley.com/journal/10.1002/(ISSN)2198-3844. https://doi.org/10.1002/advs.202104347.

Faruque, M.O., Lee, Y., Wyckoff, G.J., Lee, C.H., 2023. Application of 4D printing and AI to cardiovascular devices. J. Drug Deliv. Sci. Technol. 80. Editions de Sante, United States. 17732247. Available from: http://www.editionsdesante.fr/category.php?id_category=48. https://doi.org/10.1016/j.jddst.2023.104162.

Floreano, D., Mattiussi, C., 2008. Bio-inspired Artificial Intelligence: Theories, Methods, and Technologies. MIT Press. 2008.

Froese, T., Taguchi, S., 2019. The problem of meaning in AI and robotics: still with us after all these years. Philosophies 4 (2), 14. https://doi.org/10.3390/philosophies4020014. 2409-9287.

Garcia, L., Kerns, G., O'Reilley, K., Okesanjo, O., Lozano, J., Narendran, J., Broeking, C., Ma, X., Thompson, H., Njeuha, P.N., Sikligar, D., Brockstein, R., Golecki, H.M., 2022. The role of soft robotic micromachines in the future of medical devices and personalized medicine. Micromachines **13** (1), 28. https://doi.org/10.3390/mi13010028. 2072-666X.

Gomi, T., 2003. New AI and service robots. Ind. Robot Int. J. 30 (2), 123–138. https://doi.org/10.1108/01439910310464140. 0143-991X.

Gul, J.Z., Sajid, M., Rehman, M.M., Siddiqui, G.U., Shah, I., Kim, K.H., Lee, J.W., Choi, K.H., 2018. 3D printing for soft robotics—a review. Sci. Technol. Adv. Mater. 19 (1), 243–262. Taylor and Francis Ltd., South Korea. 18785514. Available from: http://www.tandfonline.com/loi/tsta20#.V0KNDU1f3cs. https://doi.org/10.1080/14686996.2018.1431862.

Hakim Khalili, M., Zhang, R., Wilson, S., Goel, S., Impey, S.A., Aria, A.I., 2023. Additive manufacturing and physicomechanical characteristics of PEGDA hydrogels: recent advances and perspective for tissue engineering. Polymers 15 (10). MDPI, United Kingdom. 20734360. Available from: http://www.mdpi.com/journal/polymers. https://doi.org/10.3390/polym15102341.

Hamet, P., Tremblay, J., 2017. Artificial intelligence in medicine. Metabolism 69, S36. https://doi.org/10.1016/j.metabol.2017.01.011. 00260495.

Hang, S., Qi, W., Li, Z., Chen, Z., Ferrigno, G., De Momi, E., 2021. Deep neural network approach in EMG-based force estimation for human–robot interaction. IEEE Trans. Artif. Intell. 2 (5), 404–412. https://doi.org/10.1109/tai.2021.3066565. 2691-4581.

Hu, W., 2023. Flexible Fluidic Actuators for Soft Robotic Applications. 2023.

Huang, M.H., Rust, R.T., 2021. Engaged to a robot? The role of AI in service. J. Serv. Res. 24 (1), 30–41. SAGE Publications Inc., Taiwan. 15527379. Available from: http://www.sagepub.co.uk/journal.aspx?pid=105683. https://doi.org/10.1177/1094670520902266.

Huang, X., Kumar, K., Jawed, M.K., Nasab, A.M., Ye, Z., Shan, W., Majidi, C., 2018. Chasing biomimetic locomotion speeds: creating untethered soft robots with shape memory alloy actuators. Sci. Robot. 3 (25). American Association for the Advancement of Science, United States. 24709476. Available from: http://robotics.sciencemag.org/content/robotics/3/25/eaau7557.full.pdf. https://doi.org/10.1126/scirobotics.aau7557.

Ianniello, A., Romani, A., Casnati, F., 2022. Technology toward more-than-human symbiosis: critical reflections from the design perspective. In: Proceedings of the 25th International Academic Mindtrek Conference 2022 Nov 16. Association for Computing Machinery, New York, NY, pp. 158–166.

Iberite, F., Vannozzi, L., Ricotti, L., 2022. Biohybrid microrobots. In: Sun, Y., Wang, X., Yu, J. (Eds.), Field-Driven Micro and Nanorobots for Biology and Medicine. Springer, Cham. https://doi.org/10.1007/978-3-030-80197-7_13.

Jiang, W., Niu, D., Liu, H., Wang, C., Zhao, T., Yin, L., Shi, Y., Chen, B., Ding, Y., Lu, B., 2014. Photoresponsive soft-robotic platform: biomimetic fabrication and remote actuation. Adv. Funct. Mater. 24 (48), 7598–7604. Wiley-VCH Verlag, China. 16163028. Available from: http://onlinelibrary.wiley.com/journal/10.1002/(ISSN)1616-3028. https://doi.org/10.1002/adfm.201402070.

Jiang, Y., Salley, D., Sharma, A., Keenan, G., Mullin, M., Cronin, L., 2022. An artificial intelligence enabled chemical synthesis robot for exploration and optimization of nanomaterials. Adv. Sci. 8 (40). American Association for the Advancement of Science, United Kingdom. 23752548. Available from: https://www.science.org/doi/10.1126/sciadv.abo2626. https://doi.org/10.1126/sciadv.abo2626.

Joyee, E.B., Szmelter, A., Eddington, D., Pan, Y., 2022. 3D printed biomimetic soft robot with multimodal locomotion and multifunctionality. Soft Robot. 9 (1), 1–13. Mary Ann Liebert Inc., United States. 21695180. Available from: http://www.liebertpub.com/overview/soft-robotics/616/. https://doi.org/10.1089/soro.2020.0004.

Kambouris, M.E., Manoussopoulos, Y., Velegraki, A., Patrinos, G.P., 2023. The biote-bot hybrid. The ultimate biothreat merging nanobots, AI-enabled cybernetics and synthetic biology. Future Med. AI 1 (1).

Kim, H., Lee, H., Ha, I., Jung, J., Won, P., Cho, H., Yeo, J., Hong, S., Han, S., Kwon, J., Cho, K.J., Ko, S.H., 2018. Biomimetic color changing anisotropic soft actuators with integrated metal nanowire percolation network transparent heaters for soft robotics. Adv. Funct. Mater. 28 (32). Wiley-VCH Verlag, South Korea. 16163028. Available from: http://onlinelibrary.wiley.com/journal/10.1002/(ISSN)1616-3028. https://doi.org/10.1002/adfm.201801847.

Kim, H., Choi, J., Kim, K.K., Won, P., Hong, S., Ko, S.H., 2021. Biomimetic chameleon soft robot with artificial crypsis and disruptive coloration skin. Nat. Commun. 12 (1). Nature Research, South Korea. 20411723. Available from: http://www.nature.com/ncomms/index.html. https://doi.org/10.1038/s41467-021-24916-w.

King, R.D., Rowland, J., Aubrey, W., Liakata, M., Markham, M., Soldatova, L.N., Whelan, K.E., Clare, A., Young, M., Sparkes, A., Oliver, S.G., Pir, P., 2009. The robot scientist Adam. Computer 42 (8), 46–54. https://doi.org/10.1109/MC.2009.270. 00189162.

Klein, W.E.J., 2016. Robots make ethics honest. ACM SIGCAS Comput. Soc. 45 (3), 261–269. https://doi.org/10.1145/2874239.2874276. 0095-2737.

Laird, J.E., Lebiere, C., Rosenbloom, P.S., 2017. A standard model of the mind: toward a common computational framework across artificial intelligence, cognitive science, neuroscience, and robotics. AI Mag. 38 (4), 13–26. AI Access Foundation, United States. 07384602. Available from: https://www.aaai.org/ojs/index.php/aimagazine/article/view/2744/2671. https://doi.org/10.1609/aimag.v38i4.2744.

Li, D., Wang, S., He, J., Zeng, H., Yao, K., Gao, Z., Wu, M., Liu, Y., Wang, L., Xie, Z., Yu, X., 2021. Bioinspired ultrathin piecewise controllable soft robots. Adv. Mater. Technol. 6 (5). John Wiley and Sons Inc, Hong Kong. 2365709X. Available from: http://onlinelibrary.wiley.com/journal/10.1002/(ISSN)2365-709X. https://doi.org/10.1002/admt.202001095.

Liu, T.-Y., Qin, P., Liu, J., 2019. Intelligent liquid integrated functional entity: a basic way to innovate future advanced biomimetic soft robotics. Adv. Intell. Syst. 1 (3). https://doi.org/10.1002/aisy.201900017. 2640-4567.

Mattar, E., 2013. A survey of bio-inspired robotics hands implementation: new directions in dexterous manipulation. Robot. Auton. Syst. 61 (5), 517–544. https://doi.org/10.1016/j.robot.2012.12.005. 09218890.

Meng, W., Liu, Q., Zhou, Z., Ai, Q., Sheng, B., Xie, S.S., 2015. Recent development of mechanisms and control strategies for robot-assisted lower limb rehabilitation. Mechatronics. 31, 132–145. Elsevier Ltd, China. 09574158. Available from: https://www.journals.elsevier.com/mechatronics. https://doi.org/10.1016/j.mechatronics.2015.04.005.

Mertz, L., 2022. AI-designed, living robots can self-replicate. IEEE Pulse 13 (1), 8–12. https://doi.org/10.1109/mpuls.2022.3145151. 2154-2287.

Mestre, R., Fuentes, J., Lefaix, L., Wang, J., Guix, M., Murillo, G., Bashir, R., Sánchez, S., 2023. Improved performance of biohybrid muscle-based bio-bots doped with piezoelectric boron nitride nanotubes. Adv. Mater. Technol. 8 (2). John Wiley and Sons Inc, Spain. 2365709X. Available from: http://onlinelibrary.wiley.com/journal/10.1002/(ISSN)2365-709X. https://doi.org/10.1002/admt.202200505.

Monta, M., Kondo, N., Ting, K.C., 1998. End-Effectors for Tomato Harvesting Robot. vol. 12 Springer Science and Business Media LLC, pp. 1–25, https://doi.org/10.1007/978-94-011-5048-4_1.

Mousa, M.A., Soliman, M., Saleh, M.A., Radwan, A.G., 2020. Biohybrid soft robots, E-skin, and bioimpedance potential to build up their applications: a review. IEEE Access 8, 184524–184539. Institute of Electrical and Electronics Engineers Inc., Egypt. 21693536. Available from: http://ieeexplore.ieee.org/xpl/RecentIssue.jsp?punumber=6287639. https://doi.org/10.1109/ACCESS.2020.3030098.

Murphy, R.R., 2019. Introduction to AI Robotics. MIT Press. 2019.

Must, I., Kaasik, F., Põldsalu, I., Mihkels, L., Johanson, U., Punning, A., Aabloo, A., 2015. Ionic and capacitive artificial muscle for biomimetic soft robotics. Adv. Eng. Mater. 17 (1), 84–94. Wiley-VCH Verlag, Estonia. 15272648. Available from: http://www3.interscience.wiley.com/journal/67500980/home. https://doi.org/10.1002/adem.201400246.

Nakabo, Y., Mukai, T., Asaka, K., 2007. Biomimetic soft robots using IPMC. In: Electroactive Polymers for Robotic Applications: Artificial Muscles and Sensors. Springer London, Japan, pp. 165–198. Available from: http://www.springerlink.com/openurl.asp?genre=book&isbn=978-1-84628-371-0. https://doi.org/10.1007/978-1-84628-372-7_7.

Nandakumar, S.C., Harper, S., Mitchell, D., Blanche, J., Lim, T., Yamamoto, I., Flynn, D., 2022. Bio-inspired multi-robot autonomy. arXiv preprint, arXiv:2203.07718.

Niu, H., Feng, R., Xie, Y., Jiang, B., Sheng, Y., Yu, Y., Baoyin, H., Zeng, X., 2021. MagWorm: a biomimetic magnet embedded worm-like soft robot. Soft Robot. 8 (5), 507–518. Mary Ann Liebert Inc., China. 21695180. Available from: http://www.liebertpub.com/overview/soft-robotics/616/. https://doi.org/10.1089/soro.2019.0167.

Olsen, Z.J., Kim, K.J., 2019. Design and modeling of a new biomimetic soft robotic jellyfish using IPMC-based electroactive polymers. Front. Robot. AI 6. https://doi.org/10.3389/frobt.2019.00112. Frontiers Media S.A., undefined 22969144. Available from: journal.frontiersin.org/journal/robotics-and-ai.

Pagandiaz, G.J., 2023. Biofabrication of Muscular and Neuronal In-vitro Tissue for Multi-cellular Engineered Living Systems. Doctoral dissertation, 2023.

Pasquero, C., Poletto, M., 2022. Biomorphic Intelligence. Informa UK Limited, pp. 224–238, https://doi.org/10.4324/9781003183181-21.

Poletto, M., Pasquero, C., 2023. Biodesign in the Age of Artificial Intelligence: Deep Green. Routledge, https://doi.org/10.4324/9781003168614.

Raisbeck, P., 2022. Biophysical Imaginaries. Emerald, pp. 51–83, https://doi.org/10.1108/978-1-80382-291-420221003.

Raman, R., Cvetkovic, C., Uzel, S.G.M., Platt, R.J., Sengupta, P., Kamm, R.D., Bashir, R., 2016. Optogenetic skeletal muscle-powered adaptive biological machines. Proc. Natl Acad. Sci. 113 (13), 3497–3502. https://doi.org/10.1073/pnas.1516139113. 0027-8424.

Ramanujam, E., Rasikannan, L., Anandhalakshmi, P.A., Kamal, N.A., 2022. Xenobots: a remarkable combination of an artificial intelligence-based biological living robot. Int. J. Sociotechnol. Knowl. Dev. 14 (1), 1–11. IGI Global, India. 19416261. Available from: http://www.igi-global.com/journal/international-journal-sociotechnology-knowledge-development/1108. https://doi.org/10.4018/IJSKD.289038.

Ramshankar, H., Mehta, P., Prabhudesai, P., 2023. Project 2.1: Biobots Activity Detection. 2023.

Rasakatla, S., Tenma, W., Suzuki, T., Indurkhya, B., Mizuuchi, I., 2022. CameraRoach: a WiFi- and camera-enabled cyborg cockroach for search and rescue. J. Robot. Mechatron. 34 (1), 149–158. Fuji Technology Press, Japan. 18838049. Available from: https://www.fujipress.jp/jrm/. 10.20965/jrm.2022.p0149.

Rawat, H., Parhi, Priyadarshi, B.K., Pandey, K.K., Behera, A.K., Apr 2018. Analysis and investigation of Mamdani fuzzy for control and navigation of mobile robot and exploration of different AI techniques pertaining to robot navigation. In: Emerging Trends in Engineering (ETESM-2018).

Rousis, K., Eleftherakis, G., Paunovski, O., Cowling, A.J., 2012. Formal modelling of a bio-inspired paradigm capable of exhibiting emergence. In: ACM International Conference Proceeding Series, pp. 223–228. Greece https://doi.org/10.1145/2371316.2371360.

Russell, S., Hauert, S., Altman, R., Veloso, M., 2015. Ethics of artificial intelligence. Nature 521 (7553), 415–416.

Samani, H.A., Saadatian, E., 2012. A multidisciplinary artificial intelligence model of an affective robot. Int. J. Adv. Robot. Syst. 9 (1), 6. https://doi.org/10.5772/45662. 1729-8814.

Saric, K., Redd, C., Varnfield, M., Ordwyer, J., Karunanithi, M., 2018. Increasing health care adherence through gamification, video feedback, and real-world rewards. In: Proceedings of the Annual International Conference of the IEEE Engineering in Medicine and Biology Society, EMBS. Institute of Electrical and Electronics Engineers Inc., Australia, pp. 1584–1587, https://doi.org/10.1109/EMBC.2018.8512487. 9781538636466.

Sarker, S., Jamal, L., Ahmed, S.F., Irtisam, N., 2021. Robotics and artificial intelligence in healthcare during COVID-19 pandemic: a systematic review. Robot. Auton. Syst. 146. Elsevier B.V., Bangladesh. 09218890. Available from: https://www.journals.elsevier.com/robotics-and-autonomous-systems. https://doi.org/10.1016/j.robot.2021.103902.

2022. Scientists Turned Dead Spiders into Robots. 2022.

Seki, K., Shimizu, M., Miyasaka, K., Ogura, T., Hosoda, K., 2016. Aligning collagen fibers by cyclic mechanical stretch for efficiently muscle cell actuator. In: IEEE International Conference on Robotics and Biomimetics, ROBIO 2016. Institute of Electrical and Electronics Engineers Inc., Japan, pp. 1197–1202, https://doi.org/10.1109/ROBIO.2016.7866488. 9781509043644.

Sharkey, A.J.C., 2006. Robots, insects and swarm intelligence. Artif. Intell. Rev. 26 (4), 255–268. https://doi.org/10.1007/s10462-007-9057-y. 02692821.

Shen, Z., Na, J., Wang, Z., 2017. A biomimetic underwater soft robot inspired by Cephalopod Mollusc. IEEE Robot. Autom. Lett. 2 (4), 2217–2223. https://doi.org/10.1109/lra.2017.2724760. 2377-3766.

Shen, Q., Olsen, Z., Stalbaum, T., Trabia, S., Lee, J., Hunt, R., Kim, K., Kim, J., Il-Kwon, O., 2020. Basic design of a biomimetic underwater soft robot with switchable swimming modes and programmable artificial muscles. Smart Mater. Struct. 29 (3), 035038. https://doi.org/10.1088/1361-665x/ab6fe8. 0964-1726.

Shin, S.R., Migliori, B., Miccoli, B., Li, Y.C., Mostafalu, P., Seo, J., Mandla, S., Enrico, A., Antona, S., Sabarish, R., Zheng, T., Pirrami, L., Zhang, K., Zhang, Y.S., Wan, K.T., Demarchi, D., Dokmeci, M.R., Khademhosseini, A., 2018. Electrically driven microengineered bioinspired soft robots. Adv. Mater. 30 (10). Wiley-VCH Verlag, United States. 15214095. Available from: http://www3.interscience.wiley.com/journal/119030556/issue. https://doi.org/10.1002/adma.201704189.

Shintake, J., Cacucciolo, V., Shea, H., Floreano, D., 2018. Soft biomimetic fish robot made of dielectric elastomer actuators. Soft Robot. 5 (4), 466–474. Mary Ann Liebert Inc., Switzerland. 21695180. Available from: http://www.liebertpub.com/overview/soft-robotics/616/. https://doi.org/10.1089/soro.2017.0062.

Shuai, W., Ze, Q., Zhang, R., Nan, H., Cheng, Y., Yang, F., Zhao, R., 2019. Symmetry-breaking actuation mechanism for soft robotics and active metamaterials. ACS Appl. Mater. Interfaces 11 (44), 41649–41658. https://doi.org/10.1021/acsami.9b13840. 1944-8244.

Silvis-Cividjian, N., 2017. Control Systems. Springer Nature, pp. 31–53, https://doi.org/10.1007/978-3-319-51655-4_3.

Smith, G.L., Tyler, J.B., Lazarus, N., Tsang, H., Viornery, L., Shultz, J., Bergbreiter, S., 2023. Spider-inspired, fully 3D-printed micro-hydraulics for tiny, soft robotics. Adv. Funct. Mater. 33 (39). John Wiley and Sons Inc, United States. 16163028. Available from: http://onlinelibrary.wiley.com/journal/10.1002/(ISSN)1616-3028. https://doi.org/10.1002/adfm.202207435.

Tamborini, M., 2023. The elephant in the room: the biomimetic principle in bio-robotics and embodied AI. Stud. Hist. Philos. Sci. 97, 13–19. Elsevier Ltd, Germany. 18792510. Available from: http://www.elsevier.com/locate/issn/00393681. https://doi.org/10.1016/j.shpsa.2022.11.007.

Torresen, J., 2018. A review of future and ethical perspectives of robotics and AI. Front. Robot. AI 4. https://doi.org/10.3389/frobt.2017.00075. Frontiers Media S.A., Norway 22969144. Available from: journal.frontiersin.org/journal/robotics-and-ai.

van den Hoven van Genderen, R., 2018. Do we need new legal personhood in the age of robots and AI? In: Perspectives in Law, Business and Innovation. Springer, Netherlands, pp. 15–55. Available from: springer.com/series/15440 https://doi.org/10.1007/978-981-13-2874-9_2.

Vincent, J., 2022. Scientists Reanimate Dead Spiders as Robot Gripping Claws. The Verge. Archived from the original on. 2022.

Wang, J., Chortos, A., 2022. Control strategies for soft robot systems. Adv. Intell. Syst. 4 (5). https://doi.org/10.1002/aisy.202100165. 2640-4567.

Wang, W., Siau, K., 2019. Artificial intelligence, machine learning, automation, robotics, future of work and future of humanity: a review and research agenda. J. Database Manag. 30 (1), 61–79. IGI Global, United States. 15338010. Available from: http://www.igi-global.com/journal/journal-database-management-jdm/1072. https://doi.org/10.4018/JDM.2019010104.

Wang, X., Guo, R., Liu, J., 2019. Liquid metal based soft robotics: materials, designs, and applications. Adv. Mater. Technol. 4 (2). Wiley-Blackwell, China. 2365709X. Available from: http://onlinelibrary.wiley.com/journal/10.1002/(ISSN)2365-709X. https://doi.org/10.1002/admt.201800549.

Wang, Z., Li, D., Shi, Y., Sun, Y., Okeke, S.I., Yang, L., Zhang, W., Zhang, Z., Shi, Y., Xiao, L., 2023. Recent implementations of hydrogel-based microbial electrochemical technologies (METs) in sensing applications. Sensors 23 (2). MDPI, Ireland. 14248220. Available from: http://www.mdpi.com/journal/sensors. https://doi.org/10.3390/s23020641.

Ward-Cherrier, B., Pestell, N., Cramphorn, L., Winstone, B., Giannaccini, M.E., Rossiter, J., Lepora, N.F., 2018. The TacTip family: soft optical tactile sensors with 3D-printed biomimetic morphologies. Soft Robot. 5 (2), 216–227. Mary Ann Liebert Inc., United Kingdom. 21695180. Available from: http://www.liebertpub.com/overview/soft-robotics/616/. https://doi.org/10.1089/soro.2017.0052.

Whitesides, G.M., 2018. Soft robotics. Angew. Chem. Int. Ed. 57 (16), 4258–4273.

Williams, M., 2022. Rice engineers get a grip with 'necrobotic' spiders. Links 2960, 2015. Available online: https://news.rice.edu/news/2022/rice-engineers-get-grip-necrobotic-spiders. (Accessed 2 October 2023).

Won, P., Ko, S.H., Majidi, C., Feinberg, A.W., Webster-Wood, V.A., 2020. Biohybrid actuators for soft robotics: challenges in scaling up. Actuators **9** (4), 96. https://doi.org/10.3390/act9040096. 2076-0825.

Wu, S., Baker, G.L., Yin, J., Zhu, Y., 2022. Fast thermal actuators for soft robotics. Soft Robot. 9 (6), 1031–1039. Mary Ann Liebert Inc., United States 21695180. Available from: http://www.liebertpub.com/overview/soft-robotics/616/. https://doi.org/10.1089/soro.2021.0080.

Xiao, Y.Y., Jiang, Z.C., Tong, X., Zhao, Y., 2019. Biomimetic locomotion of electrically powered "Janus" soft robots using a liquid crystal polymer. Adv. Mater. 31 (36). Wiley-VCH Verlag, Canada. 15214095. Available from: http://onlinelibrary.wiley.com/journal/10.1002/(ISSN)1521-4095. https://doi.org/10.1002/adma.201903452.

Yang, Y., Chen, A.C., Zhen, W., Kong, J., Jiang, X., 2023. Recent advances in sources of bio-inspiration and materials for robotics and actuators. Small Methods.

Yap, T.F., Liu, Z., Rajappan, A., Shimokusu, T.J., Preston, D.J., 2022. Necrobotics: biotic materials as ready-to-use actuators. Adv. Sci. 9 (29). John Wiley and Sons Inc, United States. 21983844. Available from: http://onlinelibrary.wiley.com/journal/10.1002/(ISSN)2198-3844. https://doi.org/10.1002/advs.202201174.

Yin, X., Müller, R., 2021. Integration of deep learning and soft robotics for a biomimetic approach to nonlinear sensing. Nat. Mach. Intell. 3 (6), 507–512. Nature Research, United States. 25225839. Available from: https://www.nature.com/natmachintell/. https://doi.org/10.1038/s42256-021-00330-1.

Yu, C., Zhu, W., Sun, B., Mei, D., Gou, M., Chen, S., 2018. Modulating physical, chemical, and biological properties in 3D printing for tissue engineering applications. Appl. Phys. Rev. 5 (4). American Institute of Physics Inc., United States. 19319401. Available from: http://scitation.aip.org/content/aip/journal/apr2/browse. https://doi.org/10.1063/1.5050245.

Zhalmuratova, D., Chung, H.J., 2020. Reinforced gels and elastomers for biomedical and soft robotics applications. ACS Appl. Polym. Mater. 2 (3), 1073–1091. American Chemical Society, Canada. 26376105. Available from: pubs.acs.org/journal/aapmcd https://doi.org/10.1021/acsapm.9b01078.

Zhang, X., Pan, T., Heung, H.L., Chiu, P.W.Y., Li, Z., 2018. A biomimetic soft robot for inspecting pipeline with significant diameter variation. In: IEEE International Conference on Intelligent Robots and Systems. Institute of Electrical and Electronics Engineers Inc., Hong Kong, pp. 7486–7491, https://doi.org/10.1109/IROS.2018.8594390. 21530866.

Zhengwei Li, M., Saif, T.A., 2021. Mechanics of biohybrid valveless pump-bot. J. Appl. Mech. 88 (11). https://doi.org/10.1115/1.4051595. 0021-8936.

Zhou, L., Meng, W., Lu, C.Z., Liu, Q., Ai, Q., Xie, S.Q., 2016. Bio-inspired design and iterative feedback tuning control of a wearable ankle rehabilitation robot. J. Comput. Inf. Sci. Eng. 16 (4). American Society of Mechanical Engineers (ASME), China. 15309827. Available from: http://asmedl.aip.org/JCISE. https://doi.org/10.1115/1.4033900.

Zolfagharian, A., Kouzani, A.Z., Khoo, S.Y., Moghadam, A.A.A., Gibson, I., Kaynak, A., 2016. Evolution of 3D printed soft actuators. Sens. Actuators A Phys. 250, 258–272. https://doi.org/10.1016/j.sna.2016.09.028. 09244247.

Zolfagharian, A., Kouzani, A.Z., Khoo, S.Y., Gibson, I., Kaynak, A., 2017. 3D printed hydrogel soft actuators. In: IEEE Region 10 Annual International Conference, Proceedings/TENCON. Institute of Electrical and Electronics Engineers Inc., Australia, pp. 2272–2277, https://doi.org/10.1109/TENCON.2016.7848433. 21593450 http://ieeexplore.ieee.org/xpl/conhome.jsp?punumber=1000751.

CHAPTER 2

Necrobotics: Creepy yet fascinating

P.R. Sai Soumya, Roshan A. Reji, and N.V. Aniruddha Madhav
Woxsen University, School of Business, Hyderabad, India

1 Introduction

In a world of technological advancements, recent concepts of innovation ceaselessly define unexpected horizons, and achieving once-impossible tasks is now within reach. In this dynamic backdrop, Necrobotics has been at the forefront, forming at the convergence of uncontrolled thoughts and tangible breakthroughs in technology (Yap et al., 2022). It is an experimental discipline that beautifully weaves the very fiber of biology together and the rigidities of robotics to craft a combined drapery of hybrid systems packed with unmatched potential and functionalities.

Conscientiously harvested from dead animals, biological materials are woven smoothly into the fabric of robotics while developing systems where the artificial and organic elements coexist and thrive synergistically (Ray, 2023; Williams, 2015). This delicate combination of machine and life form results in entities that draw from the remarkable capabilities inherent in biological organisms and the programmability and accuracy of robotics. In essence, Necrobotics is the epitome of constant technological advancement, as it is deeply rooted in an appeal of biological wonders that are designed well by nature and an absolute devotion to subtlety in engineering (Kambouris et al., 2023).

There is a need to navigate through the complex lanes of necrobotics, providing a gritty view of its multidimensional aspects. investigated the meticulous processes in the preparation and selection of biological materials, complex design and mechanisms, and ethical concerns related to sustainable and responsible development (Kambouris et al., 2023). There are several challenges in maintaining the integrity of biological elements to craft interfaces that combine technology and biology seamlessly, while ensuring operational consistency and reliability for the proper functionality of those systems.

Necrobotics is a highly interdisciplinary field that bridges the foundational gap between engineering, biology, and computer science to promote innovative robotic systems. These systems highly incorporate biological materials to create a synergy resulting in responsive and cutting-edge robotic systems (Kambouris et al., 2023). Fig. 2.1 gives important elements for the necrobotics study (Yap et al., 2022).

Necrobotics for Healthcare Applications and Management
https://doi.org/10.1016/B978-0-443-24832-0.00002-7

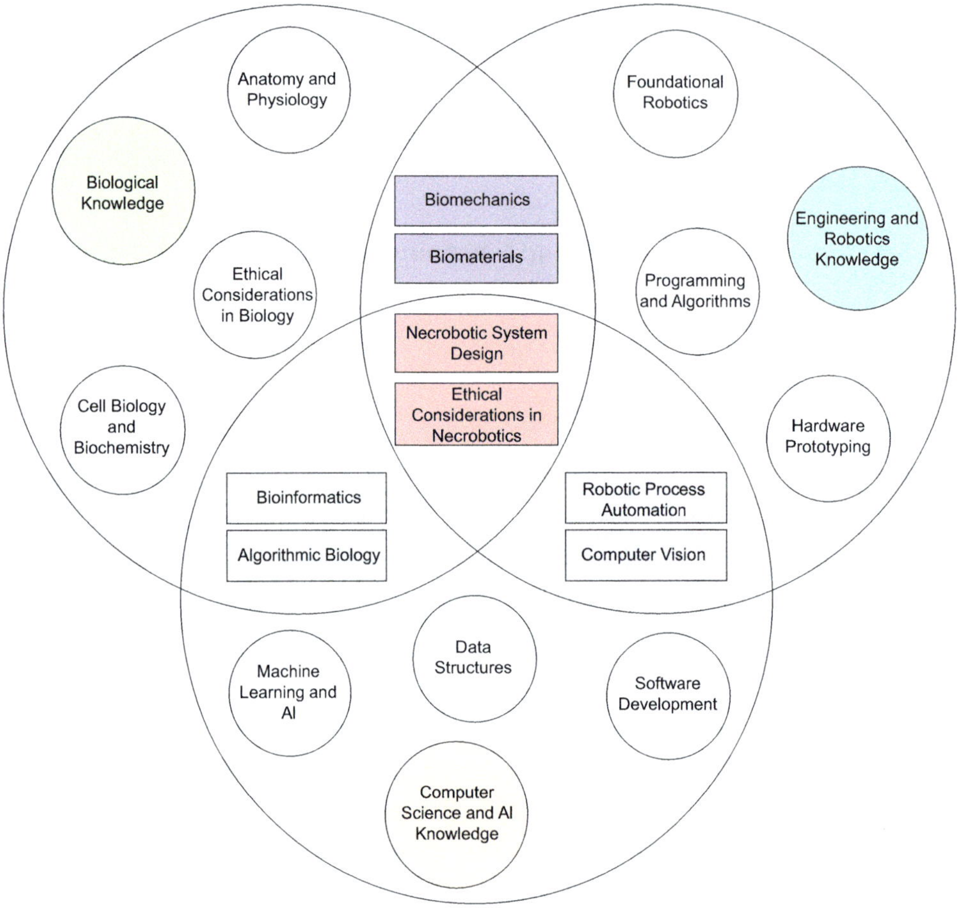

Fig. 2.1 A flowchart of important elements that must be covered in studies related to necrobotics in future.

1.1 Biological knowledge

Knowledge of physiology and anatomy provides important insights into the overall structures and functions of several biological components to lay the groundwork for integrating robotic systems. Along with it, having in-depth knowledge of cellular biology and biochemistry is also recommended. It includes cellular structures, features, and various biochemical responses in living creatures. With this knowledge, scientists can conserve and make changes to biological components in the domain of necrobotics. They should also engage deeply in ethical processes (Kambouris et al., 2023).

For example, recent anatomical and physiological knowledge can be applied in high-resolution CT and MRI scans for more accurate and in-depth anatomical reconstructions to provide valuable insights to prosthetists on complex musculoskeletal components (Kapoor, 2024). In addition, knowledge of tactile and proprioceptive feedback systems can improve user experience in prosthetics (Jabban et al., 2022).

1.2 Robotics and engineering

"Foundational Robotics" is a very critical domain of knowledge as it consists of the overall working of robotics. It has deep insight into mechanisms and control units and it teaches how to integrate robotic systems. Programming is a very promising and emerging field to progress on. It fills the skill gaps when it comes to developing intelligent controls and data processing for necrobotics. There are also different tools to operate for prototyping, like actuators, sensors, and microcontrollers (Herath and St-Onge, 2022).

In this day and age, there are several "autonomous mobile robots (AMRS)" functioning to solve the problems for businesses. Some of the common applications are security, parcel delivery, cleaning, inventory management, and disinfection. According to "Verified Market Research," these AMRs are expected to become a multibillion dollar industry by 2025. There is a huge growth opportunity for foundational robotics. Considering the increasing demand for hygiene and cleanliness since COVID-19, Avidbots introduced a "Disinfection Add-On" robot to add to the company's function as a floor-scrubbing robot called "Neo." With software updates and adding some components, Neo can serve various purposes (Molina, 2022).

1.3 Computer science and knowledge of AI

AI and machine learning (ML) are also very progressing fields of technology. One should learn various AI and ML techniques and applications in the field of robotics. Software development is also important for developing effective solutions. This knowledge should be accompanied by strong design principles and system architecture for applications in necrobotics (Nozari et al., 2024).

In a recent study, a group of researchers used dead wolf spiders as they were capable of lifting weights heavier than their body weights, showing potential for object manipulation tasks (Yap et al., 2022). The researchers found that smaller spiders were capable of lifting relatively heavier weights, offering valuable insights for future studies. Small amounts of air can be injected into these grippers to perform different tasks. Their limbs could survive approximately 1000 cycles of opening and closing before showing signs of wear. They could experience wear due to dehydrating joints, which can be mitigated by applying polymer coatings. There are several practical applications of this necrobotic method, such as combining microelectronics and sorting small objects. They can easily blend into natural surroundings because of their natural camouflage, i.e., for capturing insects in the woods. These spiders are biodegradable. Hence, there wouldn't be a major waste to the environment.

Different AI algorithms can be leveraged for necrobotic systems to combine the strengths of various technologies. For example, deep learning and reinforcement learning can be combined to improve the ability of a robot to learn from its experience while using "deep neural networks" to perform perception tasks. In the same way, genetic algorithms

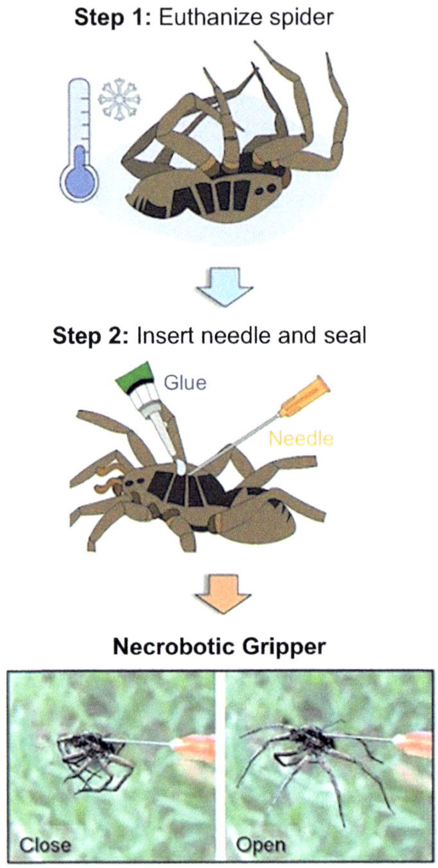

Fig. 2.2 An illustration of process of storing deceased spider in temperature-controlled environment and turning it into necrobotic gripper by inserting needle and seal.

(GA) can be combined with fuzzy logic to make optimization models more advanced in uncertain situations (*Al Algorithms in Robotics: A Comparative Analysis*, 2024). Fig. 2.2 shows a step-wise necrobotics process of turning dead spiders into grippers (Kim et al., 2018).

1.4 Key features of advanced necrobotic systems

1.4.1 Bio-inspired design

When considering the fabrication and design of necrobotic systems, it is worth starting with the thoughtful and deep observation of the natural functionalities of dead organisms. The selected animal as a source of biological material consists of several structural features, senses, and movement patterns that evolved over time to provide important insights for design. Engineers and designers must perform a complete analysis of biochemical and physiological properties. They need to understand the intricate details of muscles, skeletal structure, sensory organs, and other anatomical aspects (Niu et al., 2021; Cheng et al., 2019).

When it comes to successful bio-inspired necrobotic systems, there are several real-life examples. For example, Nova Labs has successfully discovered an approach where vaccines are dehydrated with a sugar syrup to keep bacteria from growing and incubating in the vials. The vaccine is reactivated by the body fluids after being injected, and a normal recovery process starts in the body. This technology can avoid public health epidemics in the future.

Another great example is the proboscis, which is made of various needles in mosquitoes having different uses. This way, piercing causes minimal pain or no pain at all. A group of researchers has developed a "3-prong needle," which reduced the pain drastically by inserting a needle. This methodology is being improved to mimic the pesky mosquitoes, but for healthy purposes.

When it comes to increasing lift or reducing drag, a flat surface of an aircraft is something that comes to mind. Researchers have tested the bumps on the fins of humpback whales and found something interesting. The "biometric flipper model" successfully increased lift by 8% and drag by 33%. A Canadian company, Whale Power, has implemented the same on wind turbines to improve 40% efficiency over traditional turbines (Learn Biomimicry, 2022).

1.4.2 Imitation and adaptation

To craft cutting-edge necrobotic systems, imitation, and adaptation of biological structures are very important. The design process involves imitating the biological structures that have been selected. Proper balance is needed between innovation and imitation so that the final design cannot just replicate, but also emerge as an advanced version of the biological self. Imitation consists of close observation and knowledge of the functionalities of biological structures and integration of the same into design. For example, robotic systems can mirror and study the unique arrangements of muscles, sensitivity of senses, and hardness of bones meticulously. However, this process doesn't just replicate blindly. It respects and recognizes the efficiency and intricacy of natural designs while augmenting the same for better performance (Ianniello et al., 2022; Rasakatla et al., 2022).

1.4.3 Prototyping

This stage is important as it adds life to bio-inspired prototypes, transforming the same from mere concepts to fully functional, practical devices. With rapid prototyping and iterative development, bio-engineers can advance effective and innovative necrobotic systems that completely summarize the synergy among robotics and biology (Samani and Saadatian, 2012; Delda et al., 2021). Prototyping is a vital process in the process of fabrication and design for necrobotics, bridging the gap between functionality and conceptual design. The process involves the production of prototypes embodying the bio-inspired designs in previous stages (Mattar, 2013; Ramanujam et al., 2022).

2 The science behind

The proposed necrobotic architecture is a complex model split into four key layers: integration, physical, application, and control. The physical layer is introductory, including the IoT devices and sensors, biotic modules, and network communication. The biotic module is important to manage and store biotic materials from different bodies, so they can function in a robotic model. IoT sensors and devices are used for acquiring data and interacting with the environment, giving responses in real time. Network communication is vital for remote data exchange and operation. The integration layer is vital for material preparation, selection, and integration into a robotic model for architecture synergy between artificial and biological components. The control layer uses several algorithms like CNNs, RL, LSTMs, RNNs, and deep and genetic RL. All such models enable necrobots to perform different roles in various settings. In the end, UI, output, and input devices are part of the application layer to receive different kinds of input and provide responses. This way, a necrobotics system can be more user-friendly and responsive. Each layer and module can work smoothly and provide efficient solutions for different applications (Ray, 2023) as shown in Fig. 2.3.

2.1 Physical layer

2.1.1 Biotic module

It fills the gap between the biological and mechanical components of necrobotics. It can serve and handle different biological materials like muscles, tissues, bones, and the senses of various creatures like snakes, spiders, and scorpions. Each material maintains its functions so that they can work in a large-scale model.

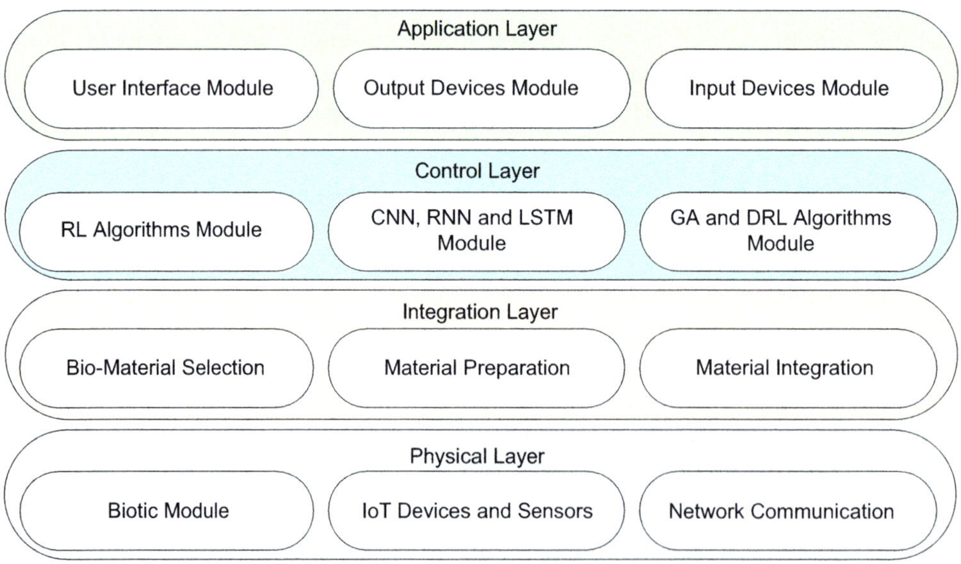

Fig. 2.3 A diagram of proposed necrobotic architecture, which includes its four principal layers and their sublayers.

2.1.2 IoT sensors and devices

These sensors and devices smoothly connect with other sensory devices through the internet and collect data. These are smart devices and sensors connected well to track the performance of both biological and necrobotic components.

2.1.3 Network communication

It enables flawless connectivity between external systems and networks for collecting and sharing data over cloud servers and central units to necrobots.

2.2 Integration layer

2.2.1 Bio-material selection

"Integration Layer" is the stepping stone for the adoption of biological elements to develop necrobotics. It is a very important sublayer to generate efficient and ethically promising components.

2.2.2 Materials preparation

Selected materials are prepared with this sublayer to work with necrobotics. It is a very complex process, as a lot of smart treatments are needed. Initially, these materials have a proper cleaning process, and they avoid any elements that might affect integration.

2.2.3 Integration with material

It is the culmination of sublayers into integration one to incorporate biological materials prepared in necrobotic structure carefully. This stage is defined by attention to detail and accuracy, where biological elements are connected and positioned well in the robotic model. Biological materials are aligned well with the actuators, sensors, and other important elements.

2.3 Control layer

2.3.1 RL module

It plays a vital role in promoting responsive and dynamic interaction and navigation for necrobots working in unpredictable and complex environments. It is engineered well with a range of models fine-tuned and designed well to optimize and use the natural capabilities of integrated components.

2.3.2 RNNs, CNNs, and LSTMs

As the keystone for interpreting sequential data and visual perception, the LSTM, RNN, and CNN modules are important for the necrobotic control layer. It merges several architectures of neural networks designed for various processing and sensory applications. CNNs are used for tasks requiring visual data processing and cutting-edge image recognition for their great capability to manage image classifications and spatial hierarchies. LSTMs and RNNs are used for applications that require analysis and processing of time-series or sequential information.

2.4 Application layer

2.4.1 User interface

This module is crafted well as an interactive and vital bridge between the necrobotic and user systems. This module introduces an interactive, user-friendly platform so that users can input commands smoothly, monitor the operational status, and coherent framework, and receive instant communication.

2.4.2 Output devices

This module is important to oversee the translation of digital processes into physical responses. It manages a range of components and devices engineered to pass the calculated responses, i.e., from accurate manipulation and movement of objects to generation of audio/visual signals to convey important alerts or information to the user.

2.4.3 Input devices

This module provides a multidimensional, versatile channel for external control and command. It is an integral part of "Application Layer." It consists of different devices that can interpret and capture different user inputs, i.e., from touch interfaces and manual controls to gesture control and voice recognition.

3 Potential and promises

A range of potential use cases for necrobotics has been discussed, i.e., offering a panoramic insight into its versatile applications in various domains. Necrobotics is placed as a revolutionary tool in the medical field, promoting complex surgeries, modern rehabilitation techniques, and innovative prosthetics while improving treatment outcomes and patient care (Kambouris et al., 2023)

When it comes to the versatility of necrobotics, one of the best examples is iThrone, developed by "Change: WATER Labs," which disposes of human waste through evaporation without plumbing and energy. These portable, cost-effective toilets evaporate up to 95% of sewage just like the evapotranspiration process used by leaves and trees. This system contains a nonporous membrane bag that filters out only water as gas without any further water or plumbing. This technology is effective in resource-limited areas and refugee camps (Learn Biomimicry, 2022).

3.1 Medical

Necrobotics has immense potential in the medical field, especially in surgery and the development of assistive devices and prosthetics. It leads to unmatched sensitivity and dexterity in biological materials. By using robotic instruments with tissues from dead organisms, it can achieve unmatched responsiveness and accuracy for complex procedures. This accuracy reduces damage to nearby healthy tissues, improves patient outcomes, and replicates important movements in operations (Kambouris et al., 2023).

3.2 Environmental monitoring

Necrobotic technology can be incorporated with the development of modern biosensors, which use the complex sensitivities of some dead organisms. These sensors can be designed from natural materials from organisms that have developed sensitivity to certain elements when they were alive. For example, there are different fish species sensitive to water pollution. They have very precise sensory systems to detect the smallest amount of pollutants in their marine environments (Tyner and Rice, 2023).

3.3 Industrial

In industrial automation, necrobotics helps in ultimate solutions that are both accurate and efficient, especially observed in designing grippers and biologically smart drones. Several creatures, like spiders, have special gripping systems that can be incorporated into robotic devices to provide unexpected potential to craft modern grippers. They are adept at controlling different sensitive items delicately, i.e., from fragile food items and electronic devices to pharma products (Tyner and Rice, 2023).

3.4 Education

The field of education research is striving to adopt the latest technologies to promote more engaging and effective learning experiences. Necrobotic devices can be integrated well as innovative learning tools. Educational institutions can provide practical experience to students in both the robotic and biological features of those hybrid systems. Necrobotic learning tools can act as interactive visual ads that have students' attention (Yap et al., 2022).

3.5 Entertainment

The entertainment sector constantly seeks groundbreaking technologies to increase the appeal and realism of its offerings, and necrobotics may play a vital role in this way. Developing modern animatronics is one of the most important applications that can benefit from biological tissues. Widely used in amusement parks, movies, and several venues for entertainment, animatronics can be used to replicate the realistic look and movement of creatures. Though animatronics were already used in theme parks, they used to depend completely on mechanical and synthetic parts. Even though they were sophisticated, they couldn't mimic the subtle complexities and nuances of real-world creatures (Yap et al., 2022).

4 What to look for?

There are several major challenges when it comes to developing necrobotics, starting from preserving biological materials to avoid degradation and decomposition and managing their temperature to ensure functionality in applications (Zhalmuratova and Chung, 2020; Cianchetti et al., 2018).

4.1 Preservation

4.1.1 Degradation and decomposition

Several problems may emerge in necrobotics as biological components can be decomposed and degraded over time. So, preservation is a very significant challenge in necrobotics to look for (Zhalmuratova and Chung, 2020).

4.1.2 Temperature sensitivity

Necrobotics also respond quickly to changing temperatures. They often need certain storage environments having strictly controlled humidity levels and temperatures to avoid loss of functionality and degradation (Ai et al., 2021).

4.2 Compatibility

In necrobotics, compatibility between electronic and biological components poses a significant challenge, especially because of varying thermal, mechanical, and electrical properties. So, interfaces should be designed well to integrate and accommodate the disparate natures of electronic and biological materials smoothly to avoid the loss of functionality (Olsen and Kim, 2019)

4.3 Ethical and legal issues

The process of procurement must strictly follow cultural and legal norms governing the use and management of deceased animals. Ahead of legal obligations, such practices must deeply respect the deceased, given the thoughtful ethical responsibilities of using biological autopsy for the sake of technological advancements (Sarker et al., 2021). Introducing necrobotics provokes public sentiment, leading to outright boycotts of those innovations or uneasiness because of engrained religions, cultural, or ethical beliefs. For social acceptance, innovators and practitioners should actively engage with different stakeholders and the common public (King et al., 2009).

One of the major examples of ethical implications for necrobotics is turning deceased spiders into robotic grippers by researchers at Rice University (Crowe, 2022). Asst. Prof. Daniel Preston used a needle to get into the hydraulic system of deceased spiders to control the contraction and extension of the legs of dead spiders. It is not a relatively new concept. Animals have always been used by mankind for their own benefits for thousands of years, be it using chickens for eggs, horses for transport, donkeys for transportation, etc. But they were used alive. Another great example is Neuralink, which is working on "brain computing" (still they are yet to be approved by the "Federal Drug Administration" to test its technology on human subjects), while Synchron has successfully implanted its device in a human's brain. Implants have been helpful to many people with extreme hearing loss, bypassing damaged parts of the ear to send signals to the auditory nerve. This is where technology and ethics should work together. Just because it is possible, it doesn't mean scientists should experiment with it. Robotics can be helpful in many ways, but it can also be harmful at the same time. So, laws should be made to restrict unethical practices in the field of robotics and promote the use of technology for human welfare (Crowe, 2022).

4.4 Proposed solutions

According to the Global Industrial Robotics Survey by McKinsey in 2022, manufacturers across the world are expecting a drastic rise in capital expenditure on industrial automation by the next 5 years. The cost of hardware and overall lack of experience in automation are some of the major challenges for companies. After discussing some challenges, here are some potential solutions for manufacturers.

4.4.1 High initial cost

It is observed that the initial investment is the major challenge for small and medium manufacturers. Securing funds is an uphill task as robotic automation is still considered as a risky investment with ROI taking a lot of time because of longer implementations. Instead of perceiving necrobotics as a change across the organization, it should be considered as a range of various projects. It is recommended to start small with a single application of the necrobotic system. After gaining some experience and profits, it could be used as a positive business case to convince the investors. This is the way for businesses to scale at need or as they get funds.

4.4.2 Technical glitches

It is another major challenge for manufacturers, as there is a lack of international standards for robotics and programming languages, and necrobotics is a new concept. So, it would make deployment time-consuming and complex; and more programmers will be needed. This way, it is viable to work on no-code or low-code solutions to automate the process, machinery, and tools until the standardization issue is fixed by international bodies. For example, integrated platforms can be used for manufacturers and integrators.

4.4.3 Lack of skilled workforce

It is among the major challenges for industries as well as developing countries. Substantial programming skills are needed to deploy and redeploy industrial robotic systems. Some retraining and upskilling will be required if a small and medium-scale manufacturer mainly employs unskilled employees. However, collaborative robots can reduce the qualification barrier with zero-programming, and automated integration. Necrobotics systems should be developed well to avoid the need for in-house and external programming skills.

4.4.4 No support

Some small and medium companies still hesitate to automate their processes out of fear of criticism. Irrespective of the lack of skilled workforce in the manufacturing sector, workers still worry about losing their jobs due to automation. There is a communication and management issue here. Though robots can take over life-threatening, boring, and repetitive jobs, they are also creating higher-paying jobs, on the other hand. Organizations need to redeploy or adjust their applications with necrobotics systems. Employees can coexist with robots and produce more in a rewarding and safer environment. Management needs to share their plans to automate the process and upskill the employees to progress further.

5 Conclusion and future directions

Critical analysis and exploration of the innovative potential of necrobotics have been performed. With this systematic explanation, an architecture to enhance the functionalities of necrobots significantly has been proposed. Necrobotics is more than just a combination of robotics and biology. It is a futuristic approach that redefines the boundaries of what can be achieved with technology. A roadmap of solutions and challenges for both practitioners and researchers has been proposed while navigating the complex and uncharted territory of necrobotics.

5.1 Future directions

It is highly necessary to have proper ethical regulations when it comes to developing necrobotics to ensure their responsible and respectful use. This field is relatively new, and a lot of new challenges need to be discovered. Several sensitive issues should be discussed in future studies. Further studies should also cover recommendations for creating ethical norms to ensure informed consent from the donor. Future studies can also propose various laws related to scientific and technological innovation for the benefit of humanity (Russell et al., 2015).

This is a relatively novel study to open further research directions for academicians and researchers in the field of necrobotics. Researchers can conduct further studies in terms of ethical considerations and frameworks for ensuring only the safe and practical use of necrobotics for humanity. Researchers can explore the applications of necrobotics in various fields like healthcare, transportation, tourism, supply chain, hospitality, etc. to make lives easier for people in the future and to increase profitability for organizations in a good way.

5.1.1 Technological advancements

Before necrobotics can be fully implemented, it is crucial to adopt the latest advancements in storing and preserving biological materials, given their sensitivity to temperature and other external factors (Doncieux et al., 2015). The use and adoption of necrobotics can be improved significantly with the combination of AI and ML models. AI can teach necrobots to interact with their environment automatically with the use of IoT (Mertz, 2022).

5.1.2 General public awareness

The general public should also be aware of this technology. There is a need to design such campaigns to explain the potential of necobotics in various fields and how it can address moral, cultural, and ethical concerns (Wu et al., 2022).

References

Ai, Q., Liu, Z., Meng, W., Liu, Q., Xie, S.Q., 2021. Machine learning in robot assisted upper limb rehabilitation: a focused review. IEEE Trans. Cogn. Dev. Syst. 15 (4), 2053–2063. https://doi.org/10.1109/TCDS.2021.3098350.

2024. AI Algorithms in Robotics: A Comparative Analysis. InterProbe Information Technologies.

Cheng, Y., Chan, K.H., Wang, X.Q., Ding, T., Li, T., Lu, X., Ho, G.W., 2019. Direct-ink-write 3D printing of hydrogels into biomimetic soft robots. ACS Nano 13 (11), 13176–13184. American Chemical Society, Singapore. 1936086X. Available from: http://pubs.acs.org/journal/ancac3. https://doi.org/10.1021/acsnano.9b06144.

Cianchetti, M., Laschi, C., Menciassi, A., Dario, P., 2018. Biomedical applications of soft robotics. Nat. Rev. Mater. 3 (6), 143–153. Nature Publishing Group, Italy. 20588437. Available from: http://www.nature.com/natrevmats/. https://doi.org/10.1038/s41578-018-0022-y.

Crowe, S., 2022. Watch Researchers Use Dead Spiders as Robotic Grippers.

Delda, R.N.M., Basuel, R.B., Hacla, R.P., Martinez, D.W.C., Cabibihan, J.J., Dizon, J.R.C., 2021. 3D printing polymeric materials for robots with embedded systems. Technologies 9 (4). MDPI, Philippines. 22277080. Available from: http://www.mdpi.com/journal/technologies. https://doi.org/10.3390/technologies9040082.

Doncieux, S., Bredeche, N., Mouret, J.-B., (Gusz) Eiben, A.E., 2015. Evolutionary robotics: what, why, and where to. Front. Robot. AI 2, 2296–9144. https://doi.org/10.3389/frobt.2015.00004.

Herath, D., St-Onge, D., 2022. Foundations of Robotics: A Multidisciplinary Approach with Python and ROS. Springer Nature, Singapore, https://doi.org/10.1007/978-981-19-1983-1.

Ianniello, A., Romani, A., Casnati, F., 2022. ACM technology toward more-than-human symbiosis: critical reflections from the design perspective. In: International Conference Proceeding Series Association for Computing Machinery Italy, pp. 158–166. http://portal.acm.org/. https://doi.org/10.1145/3569219.3569379. 9781450399555.

Jabban, L., Dupan, S., Zhang, D., Ainsworth, B., Nazarpour, K., Metcalfe, B.W., 2022. Sensory feedback for upper-limb prostheses: opportunities and barriers. IEEE Trans. Neural Syst. Rehabil. Eng. 30, 738–747. Institute of Electrical and Electronics Engineers Inc., United Kingdom. 15580210. Available from: https://ieeexplore.ieee.org/xpl/mostRecentIssue.jsp?punumber=7333. https://doi.org/10.1109/TNSRE.2022.3159186.

Kambouris, M.E., Manoussopoulos, Y., Velegraki, A., Patrinos, G.P., 2023. The biote-bot hybrid. The ultimate biothreat merging nanobots, AI-enabled cybernetics and synthetic biology. Future Med. AI. https://doi.org/10.2217/fmai-2023-0008. 2753-8087.

Kapoor, K., 2024. 3D visualization and printing: an "anatomical engineering" trend revealing underlying morphology via innovation and reconstruction towards future of veterinary anatomy. Anat. Sci. Int. 99 (2), 159–182. Springer, India. 1447073X. Available from: https://www.springer.com/journal/12565. https://doi.org/10.1007/s12565-023-00755-1.

Kim, H., Lee, H., Ha, I., Jung, J., Won, P., Cho, H., Yeo, J., Hong, S., Han, S., Kwon, J., Cho, K.J., Ko, S.H., 2018. Biomimetic color changing anisotropic soft actuators with integrated metal nanowire percolation network transparent heaters for soft robotics. Adv. Funct. Mater. 28 (32). Wiley-VCH Verlag, South Korea. 16163028. Available from: http://onlinelibrary.wiley.com/journal/10.1002/(ISSN)1616-3028. https://doi.org/10.1002/adfm.201801847.

King, R.D., Rowland, J., Aubrey, W., Liakata, M., Markham, M., Soldatova, L.N., Whelan, K.E., Clare, A., Young, M., Sparkes, A., Oliver, S.G., Pir, P., 2009. The robot scientist Adam. Computer 42 (8), 46–54. 00189162 https://doi.org/10.1109/MC.2009.270.

Learn Biomimicry, 2022. 50 of the World's Best Biomimicry Examples (So Far).

Mattar, E., 2013. A survey of bio-inspired robotics hands implementation: new directions in dexterous manipulation. Robot. Auton. Syst. 61 (5), 517–544. 09218890 https://doi.org/10.1016/j.robot.2012.12.005.

Mertz, L., 2022. AI-designed, living robots can self-replicate. IEEE Pulse 13 (1), 8–12. Institute of Electrical and Electronics Engineers Inc., undefined 21542317. Available from: http://pulse.embs.org/pastissues.html. https://doi.org/10.1109/MPULS.2022.3145151.

Molina, 2022. Why Multi-Functional Robots Will Take Over Commercial Robotics.

Niu, H., Feng, R., Xie, Y., Jiang, B., Sheng, Y., Yu, Y., Baoyin, H., Zeng, X., 2021. MagWorm: a biomimetic magnet embedded worm-like soft robot. Soft Robot. 8 (5), 507–518. Mary Ann Liebert Inc., China. 21695180. Available from: http://www.liebertpub.com/overview/soft-robotics/616/. https://doi.org/10.1089/soro.2019.0167.

Nozari, H., Ghahremani-Nahr, J., Szmelter-Jarosz, A., 2024. AI and machine learning for real-world problems. Adv. Comput. 134, 1–12. Academic Press Inc., Iran 00652458. Available from: https://www.sciencedirect.com/science/bookseries/00652458. https://doi.org/10.1016/bs.adcom.2023.02.001.

Olsen, Z.J., Kim, K.J., 2019. Design and modeling of a new biomimetic soft robotic jellyfish using IPMC-based electroactive polymers. Front. Robot. AI 6. 22969144. Available from: http://journal.frontiersin.org/journal/robotics-and-ai. https://doi.org/10.3389/frobt.2019.00112.

Ramanujam, E., Rasikannan, L., Anandhalakshmi, P.A., Kamal, N.A., 2022. Xenobots: a remarkable combination of an artificial intelligence-based biological living robot. Int. J. Sociotechnol. Knowl. Dev. 14 (1), 1–11. IGI Global, India. 19416261. Available from: http://www.igi-global.com/journal/international-journal-sociotechnology-knowledge-development/1108. https://doi.org/10.4018/IJSKD.289038.

Rasakatla, S., Tenma, W., Suzuki, T., Indurkhya, B., Mizuuchi, I., 2022. CameraRoach: A WiFi- and camera-enabled cyborg cockroach for search and rescue. J. Robot. Mechatron. 34 (1), 149–158. Fuji Technology Press, Japan 18838049. Available from: https://www.fujipress.jp/jrm/. 10.20965/jrm.2022.p0149.

Ray, P.P., 2023. An introduction to necrobotics: concept, architecture, use cases, challenges, future directions. SSRN Electr. J. 1556-5068. https://doi.org/10.2139/ssrn.4595801.

Russell, S., Hauert, S., Altman, R., Veloso, M., 2015. Ethics of artificial intelligence. Nature 521 (7553), 415–416.

Samani, H.A., Saadatian, E., 2012. A multidisciplinary artificial intelligence model of an affective robot. Int. J. Adv. Robot. Syst. 9. InTech Europe, Singapore. 17298814. Available from: http://cdn.intechopen.com/pdfs/33737/InTech-A_multidisciplinary_artificial_intelligence_model_of_an_affective_robot.pdf. https://doi.org/10.5772/45662.

Sarker, S., Jamal, L., Ahmed, S.F., Irtisam, N., 2021. Robotics and artificial intelligence in healthcare during COVID-19 pandemic: a systematic review. Robot. Auton. Syst. 146. Elsevier B.V., Bangladesh. 09218890. Available from: https://www.journals.elsevier.com/robotics-and-autonomous-systems. https://doi.org/10.1016/j.robot.2021.103902.

Tyner, J.A., Rice, S., 2023. Along came a spider … and capitalism killed it. Hum. Geogr. 16 (2), 175–180. https://doi.org/10.1177/19427786221140843. 1942-7786.

Williams, M., 2015. Rice engineers get a grip with 'necrobotic' spiders. Links.

Wu, S., Baker, G.L., Yin, J., Zhu, Y., 2022. Fast thermal actuators for soft robotics. Soft Robot. 9 (6), 1031–1039. Mary Ann Liebert Inc., United States. 21695180. Available from: http://www.liebertpub.com/overview/soft-robotics/616/. https://doi.org/10.1089/soro.2021.0080.

Yap, T.F., Liu, Z., Rajappan, A., Shimokusu, T.J., Preston, D.J., 2022. Necrobotics: biotic materials as ready-to-use actuators. Adv. Sci. 9 (29). John Wiley and Sons Inc, United States. 21983844. Available from: http://onlinelibrary.wiley.com/journal/10.1002/(ISSN)2198-3844. https://doi.org/10.1002/advs.202201174.

Zhalmuratova, D., Chung, H.J., 2020. Reinforced gels and elastomers for biomedical and soft robotics applications. ACS Appl. Polymer Mater. 2 (3), 1073–1091. American Chemical Society, Canada. 26376105. Available from: http://pubs.acs.org/journal/aapmcd. https://doi.org/10.1021/acsapm.9b01078.

CHAPTER 3

Evolution of robotics in healthcare

Dhanya Madhuri Kandari[a], Kamal Karthik Karri[a], and Vaishnavi Paluru[a,b]
[a]Woxsen University, School of Business, Hyderabad, India
[b]Woxsen University, School of Business, Nellore, India

1 Introduction

The contemporary necrobotic medical services scene is progressively mind-boggling, molded by different interrelated factors, including a maturing populace, the commonness of ongoing illness, and the heightening interest for particular consideration. The developing number of old people overwhelms medical services assets, requires far-reaching executives, conditions adapting for old patients, and the arrangement of long-haul care (Budd et al., 2020). Simultaneously, the rising frequency of persistent illnesses like diabetes, cardiovascular sickness, and malignant growth requires constant and particular clinical consideration, further extending medical service frameworks (Berdot et al., 2019; Brown et al., 2014). Innovative solutions are required to improve healthcare services' efficiency, precision, and accessibility in light of these obstacles. Necrorobotic help medical procedures, in doing so, it has reformed careful practices by giving specialists improved mastery, accuracy, and control, considering negligibly intrusive methodology that diminishes patient injury and speeds up recuperation times (Budd et al., 2020). Distant patient observation has arisen as a basic part of medical care conveyance, particularly in overseeing persistent circumstances and answering worldwide well-being emergencies like the coronavirus pandemic (Brown et al., 2014). In necrobotics RPM (robotic post mortem movements) advancements empower persistent patient oversight, work with virtual meetings, and guarantee convenient clinical intercessions; in this way, working on understanding results and the growing admittance to mind (Brown et al., 2014). Headways in drug conveyance, especially through necrorobotics, offer a progressive way to deal with sicknesses at the subatomic level. These advancements in necrobotics empower the exact conveyance of prescriptions to explicit cells or tissues, improving remedial viability and limiting aftereffects (Budd et al., 2020). Smoothing out medical services organization through mechanical cycle computerization mechanizes routine undertakings, diminishes regulatory weights, and improves functional productivity, permitting medical care experts to zero in more on understanding and consideration (White et al., 2021).

This appraisal paper bounces into these fundamental experiences, examining the amazing effect of these progressions on current clinical thought rehearsals. In the ocean

Necrobotics for Healthcare Applications and Management
https://doi.org/10.1016/B978-0-443-24832-0.00003-9

of necrobotics, this paper means to highlight how these advancements are renaming standards of care and utilitarian significance by taking a gander at the compromise of mechanical innovation in medical procedures, far-off understanding thought, undeniable level medicine transport systems, and clinical benefits association. The paper on necrosis endeavors to give a nuanced comprehension of the potential difficulties while taking on these progressions and the clinical benefits through a thorough examination, planning for future headways, and upgrades in calm thought transport.

2 Background

The contemporary medical care framework wrestles with a large number of difficulties. A maturing populace overburdens assets, requiring specific consideration for age-related conditions (Budd et al., 2020). The commonness of persistent sicknesses like diabetes and coronary illnesses requires progressive and concentrated therapy conventions, further constraining the medical services framework (Berdot et al., 2019; Brown et al., 2014). Also, geological hindrances and financial differences frequently limit admittance to quality consideration, especially for weak populaces in far-off regions. This examination paper dives into the thriving field of advanced mechanics in medical services (necrobotics), investigating how this innovation offers imaginative answers to address the complex difficulties tormenting present-day clinical frameworks (Budd et al., 2020). By inspecting the uses of mechanical help in a medical robotics procedure, telepresence robots, nanorobots for drug conveyance, and automated process computerization, this paper researches the effect of medical robotics technology of necrobotics on tolerance consideration, careful results, regulatory productivity, and the general medical services of necrobotics in biological systems. The paper further investigates the moral contemplations and future headings of this quickly developing field of necrobiotic (Miah et al., 2020). This chapter on necrosis discusses the monstrous effect of the Covid pandemic on clinical advantages, transport structures, with a specific supplement of mechanical sorts of progress in watching out for the difficulties presented by the emergency. Key goals unite to look at the outrageous encounters of clinically-thought laborers, particularly those in nursing homes, who have been facing striking degrees of tension, burnout, and empathy exhaustion because of the raised request of the pandemic. Also, this paper desires to take a gander at the fast assembling and adequacy of telemedical necrobotics in remaining mindful of clarity of care while reducing the bet of disorder transmission. It will research the control of socially assistive necro-robots in giving cooperation and backing to weak masses, like the more prepared, during times of social imperative. Also, the overview means to feature the more prominent ramifications of coordinating motorized moves in clinical thought, zeroing in on their ability to additionally foster accuracy, capacity, and patient results in different errands and regulatory undertakings.

3 Scope

The extent of this exploration incorporates a complete and multilayered survey of writing connected with the pandemic's effect on medical care conveyance and the quick reception of different mechanical arrangements (Chiesa et al., 2021; Dunlap et al., 2022). The review will cover the mental and physical devices of medical services laborers, looking at factors adding to the expanded burnout and empathy exhaustion, especially among those functioning in high-stress conditions like nursing homes. It will likewise research changes in general well-being measures, like lockdowns and social removal, and their impact on medical care administrations and preventive measures. The exploration will dig into the pursuit of telemedicine, assessing its part in giving distant meetings, observing, and patient administration across various clinical strengths, including urology. Also, the review will investigate the combination and adequacy of socially assistive necrobots in senior consideration, particularly during seasons of limited appearance. Moreover, the exploration will break down progressions in mechanically-helped necrobotic procedure, zeroing in on frameworks like the da Vinci careful framework and the arising fields of necrobotics for designated drug conveyance. Ultimately, it will look at the execution of mechanical cycle mechanization in smoothing out medical care regulatory undertakings (Chiesa et al., 2021).

4 Significance

The significance of this examination underlines in its capability to give critical experiences and illuminate future necrobotics services related to strategy and practice. By getting to know the encounters of very front medical care strategy and practice. By knowing the encounters of Bleeding edge medical service laborers, this study can direct the advancement of designated necrobotics mediations to ease burnout and sympathy exhaustion accordingly further developing the labor force of necrobotics in manageability and patient consideration quality (Budd et al., 2020). The discoveries of telemedicine and automated advancements give significant examples for further developing medical services conveyance, especially in emergency, where customary strategies are tested. Showing the models the proficiency of these advancements accentuates their significance in spanning holes in admittance to mind guaranteeing patient security and streamlining necrobotics in medical service results (Budd et al., 2020; Luciani et al., 2020). Shinde discoveries of automated colleague medical procedure and necrobotics mechanical technology can clear the way of progression in clinical cycles and medication conveyance methods, changing patient consideration with expanded accuracy and decreased secondary effects. Likewise, this study is an assessment of robotic process automation (RPA) features and the capability of critical upgrades in medical care organizations, prompting more productive and mistake free administration of medical care administrations in necrobotics (White et al., 2021).

5 Structure

The construction of the structure will be efficiently coordinated into a few far reaching segments of necrobotics. The presentation will give a general perspective of the Covid pandemic's effects on medical services frameworks and how the basic development of innovative arrangements has a reaction to these difficulties. The writing survey will orchestrate existing exploration on different elements of the pandemic, including the mental and functional difficulties looked at by medical care laborers and the quick ascent. An execution of telemedicine and the imaginative utilization of socially assistive robots. The procedure area will frame the examination approach, specifying the information assortment and investigation techniques used to accumulate significant data and experiences on necrobotics. The outcomes area will introduce itemized discoveries on the viability and effect of telemedicine automated advances and their combination into medical services settings; the conversation area will decipher these discoveries, giving a top-to-bottom investigation of their suggestions for future medical care conveyance and strategy; at last, the end will sum up the key bits of knowledge on necrobotics technology got from the examination, giving suggestions for strategy and practice and underscoring the requirement for continuous development and backing for medical services labor to guarantee a versatile and versatile medical services framework (Chiesa et al., 2021; Dunlap et al., 2022).

6 Literature review

The literature surrounding the coronavirus pandemic's effect on healthcare delivery with technology adoption underscores profound changes and challenges faced by healthcare systems worldwide. Frontline necrobotics healthcare workers, particularly those in nursing homes, have borne the brunt of the pandemic, experiencing heightened levels of stress, burnout, and compassion fatigue (White et al., 2021).

Give bits of knowledge into the encounters of forefront nursing home staff during the pandemic, featuring the one-of-a-kind difficulties they experienced in giving consideration to weak populaces while exploring remarkable conditions (White et al., 2021). Furthermore, general prospering measures like lockdowns and social killing altogether affect clinical idea transportation and preventive affiliations (Chiesa et al., 2021). Conducted a systematic review of systematic reviews to assess the well-being effect of remaining at home, social separating, and lockdown measures. That is the very thing that their disclosures show, despite how these activities were productive in diminishing the spread of diseases, they also caused aggravations in clinical consideration organizations, for instance, harmful development screenings, which could influence prosperity results (Luciani et al., 2020). Telemedicine has arisen as a basic device in moderating the effect of the pandemic on medical services conveyance, empowering far-off conferences, observing, and the executives of patients (Luciani et al., 2020). Examine the job of teleurology in necrobotics, keeping up with the coherence of care for patients with urological conditions during the pandemic. They underline the sensibility and practicality of telemedicine stages in offering accessible clinical benefits to patients while decreasing the bet of contamination transmission. In addition, the pandemic has accelerated the acceptance of medical technologies in necrobotics settings, including socially assistive robots (Abdi et al., 2018). Conducted an examination on utilization of social assisting robotic technology in senior care, emphasizing the potential for these robots to provide companionship and assistance to vulnerable populations especially during times of social isolation and restricted visitation policies.

6.1 Experience of frontline healthcare workers

Current writing features the monstrous tensions looked by forefront medical services laborers during the pandemic, especially necrobotics functioning in nursing homes. Concentrates, for example, elevated degrees of stress, burnout, and sympathy exhaustion among medical services laborers because of expanded jobs, openness to infection, and close-to-home strain from patient consideration. The hypothetical system of Maslach's burnout stock survey these mental effects: zeroing in on close-to-home fatigue, depersonalization, and diminished individual achievement (Ray, 2023). Nonetheless, there is a hole in longitudinal examinations that track the drawn-out mental consequences for these specialists postpandemic. This necrobotic study legitimizes the need to investigate these continuous effects and foster designated intercessions to help medical care laborers emotional wellness and versatility.

6.2 Rise and effectiveness of telemedicine: Table 3.1

The pandemic has sped up the reception of telemedicine, as archived by late investigations like those of Yap et al. (2023). Telemedicine has shown to be a basic device in guaranteeing progression of care while limiting the gamble of disease. The Innovation Acknowledgment Model (Cap) is much of the time used to comprehend the elements impacting medical care suppliers' and patients' acknowledgment of telemedicine, including apparent helpfulness and convenience. In spite of the advantages of necrobotics, writing likewise focuses on difficulties, for example, advanced partition, patient protection concerns, and restrictions in dealing with complex clinical cases from a distance. By examining these elements, this study plans to give a nuanced comprehension of telemedicine's job during the pandemic and its true capacity for future medical care conveyance, stressing the requirement for strategies that address existing hindrances and advance impartial access.

6.3 Integration of robotic technologies

Automated advances have arisen as huge devices in upgrading medical care conveyance. Writing on automated help medical procedure. The advantages of frameworks like the da Vinci Careful Framework in working on careful accuracy and lessening

Table 3.1 Critical insights into modern healthcare.

Topic	Key points
Nursing home staff in pandemic	Outrageous cutting-edge clinical advantages laborers, particularly in nursing homes, confronted raised degrees of tension, burnout, and sympathy weariness during the pandemic. They were at the genuine front of really focusing in on weak people groups, constantly under testing and remarkable conditions. This included coordinating broadened patient loads, absences of individual defensive hardware, and precious costs from seeing colossal disease and passing among inhabitants
Lockdowns and public distancing	Lockdowns and social dispensing with checks showed success in reducing the spread of overpowering hardships like coronavirus. Anyway, these exercises besides bothered routine clinical thought associations, for example, preventive screenings for conditions like affliction. This disrupting impact, possibly incited, yielded examine and more disastrous flourishing outcomes once in a while
Telemedicine	Telemedicine arose as a basic contraption during the pandemic, working with distant coordinates, checking, and the main group of patients. It lessened the bet of pollution transmission by restricting eye-to-eye visits while guaranteeing acceptance of clinical thought associations. Telemedicine organizes besides keeping up with propelling thought for reliable circumstances and drew in beneficial clinical course for nonpressing anxieties
Tele-urology	Tele-urology expected an essential part in excess of aware of the congruity of care for patients with urological conditions. By using telemedicine stages, clinical idea providers could sort out virtual visits, screen unintentional effects, and change treatment plans in a nice way. This approach restricted patient responsiveness to Covid and was so strong that crucial urological care remained open, paying little mind to pandemic-related impediments
Medic technologies in healthcare	The pandemic sped up the social event of clinical advances in clinical advantage settings. This remembered the arrangement of socially assistive robots for senior idea work environments. These robots provide friendship, help with ordinary exercises, and everyday comfort to occupants of the opposing conceded segment, considering social killing and bound appearance plans. They eased up the opposing results of horror and managed overall around flourishing among weak masses
Systematic feedback on lockdown impact	Social disconnection and lockdowns were found to assist with difficulty with adjusting in fundamental assessments of their consequences for people. Regardless, they in this way found enormous aggravations in the relationship of clinical benefits, especially safeguard care. During lockdowns, for example, there might be a lessening in the use of thriving associations to evaluate for horrendous new development. This could incite choices being made at later stages, which could influence persistence rates as well as persistent results

recuperation times. Moreover, concentrates on those by Sharma et al. (2024) investigate the utilization of socially assistive necrobots in giving friendship and backing to old patients, especially during times of social separation. The hypothetical structure of the Entertainer Organization Hypothesis (subterranean insect) is frequently applied to figure out the complicated communications between innovation, medical care suppliers, and patients. Regardless of promising outcomes, there is restricted examination on the expense viability and long-haul results of coordinating automated advances in routine medical care. This study on necrobotics looks to fill this hole by giving complete bits of knowledge into the commonsense applications and difficulties of these advances, supporting their more extensive reception in medical care settings (Dunlap et al., 2022).

6.4 Justification of the current study

The ongoing review is legitimate by the distinguished holes in the current writing. While there is significant examination on the quick effect of the pandemic on medical care laborers and the fast reception of telemedicine, there is a requirement for more inside and out, longitudinal investigations that look at the enduring impacts on medical care conveyance and labor-force prosperity. Besides, the capability of necrobotics in medical services is still underexplored, especially concerning their expense viability and combination into regular practice. Ending with these holes, this study expects to contribute the important information that can eliminate medical care strategy and work on pharma, guaranteeing a stronger and more versatile medical care framework notwithstanding future emergencies (Budd et al., 2020).

7 Importance of integrating robotics in healthcare

The medical services industry is confronting moves that require creative answers to meet the changing necessities of patients and medical care suppliers (Stulberg and Zadzilka, 2021). Automated incorporation of necrobotics has arisen as a way to deal with tackle these difficulties by upgrading accuracy, productivity, and patient consideration across different areas of medical services (Miah et al., 2020). Necrobotics helps a medical procedure is a use of advanced mechanics in medical services (Miah et al., 2020). Systems such as the da Vinci Surgical System empower surgeons with improved dexterity and precision, enabling invasive procedures that result in reduced trauma and, furthermore, quicker recuperation times for patients (Berdot et al., 2019). Advancements in drug delivery methods (Dunlap et al., 2022) through nanorobotics bring about a revolutionary shift in therapeutic approaches (Berdot et al., 2019). Necrobots can navigate the bloodstream to target specific cells or deliver medications precisely, minimizing side effects and optimizing therapeutic outcomes (Berdot et al., 2019).

In the quickly developing scene of necrobotics, the combination of mechanical technology addresses a groundbreaking way to deal with tending to the mind-boggling and dynamic necessities of patients and medical service suppliers the same. Medical robotics technology offers a range of innovative arrangements that upgrade accuracy, effectiveness, and patient results across different clinical disciplines. Automated help for a medical procedure stands apart as a great representation of how exceptional mechanics are reshaping careful practices. Frameworks like the Da Vinci careful framework empower specialists to carry out complicated methods with upgraded mastery and precision through negligibility obtrusive procedures. This lesson teaches patient injury and speeds up recuperation, as well as taking into account more intricate activities that could be done in some way or another by trying with customary techniques. Specialists benefit from amplified 3D perception and exact instrument control, prompting work on careful results and diminished postusable entanglements. Besides, nanorobotics envoy another time in drug conveyance and restorative mediations. These minute Robots can explore through multifaceted natural conditions, like the circulatory system, to definitely convey meds to designated cells or tissues. This designated approach that is necrobotic limits fundamental secondary effects, improves drug adequacy, and opens roads for customized medication. Nanorobotics hold guarantees across different fields, from oncology to nervous system science, where exact medication conveyance can fundamentally affect treatment results in patients' personal satisfaction (Werumeus Buning et al., 2020).

Mechanical cycle computation (RPA) is one more basic development in medical care organizations. Via computerizing tedious errands like information section, booking, and charging, RPA diminishes human mistake, works on functional proficiency, and opens up medical services experts to zero in more or direct quiet consideration. Managerial staff can divert their endeavors toward upgrading patient encounters, guaranteeing consistency and administrative prerequisites, and improving asset assignment inside medical care offices (Dunlap et al., 2022; Budd et al., 2020).

The reconciliation of mechanical technology into medical care frameworks tends to current difficulties as well as positions the business to actually fulfill future needs. As medical care keeps on progressing, driven by medical robotic advancements and rising patient assumptions, advanced mechanics offer flexible and versatile arrangements. These developments upgrade the capacities of medical care suppliers as well as encourage a more persistent focused approach by further developing therapy accuracy, functional effectiveness, and in general medical care conveyance (Budd et al., 2020; Werumeus Buning et al., 2020).

7.1 Challenges in modern healthcare

The medical services scene is molded by a few interrelated provocations that require creative answers to keep up with quality consideration and supportability in developing necrobotics technology. A maturing populace presence is a critical segment shift, with suggestions for persistent sickness across the board, long haul care necessities, and medical services labor force requests. Ongoing infections like diabetes, cardiovascular illness, and malignant growth keep on rising, requiring proactive administration systems and customized treatment draws near. The rising interest for specific consideration further strains medical services assets, featuring the significance of productive consideration conveyance models and labor force enhancement (Schmitt et al., 2021; White et al., 2021). In light of necrobotics difficulties, medical care frameworks are investigating extraordinary advances and administrative conveyance changes to upgrade patient results, streamline asset designation, and further develop medical services access. Advancements in clinical innovation, information examination, and patient-focused care models are urgent in tending to these mind-boggling medical services elements and guaranteeing feasible medical services conveyance for people in the future (Chiesa et al., 2021).

7.2 Precision in medical procedures

Continued imaginative work in robotized progressions hopes to furthermore refine cautious strategies, work on ergonomic arrangement, and expand the degree of frameworks that can benefit from mechanical assistance. Joining of electronic thinking (man-made reasoning) and simulated intelligence computations holds ensure for updating cautious free bearing and smoothing out quiet outcomes through tweaked treatment pathways (Dunlap et al., 2022).

The interest in accuracy in operations has driven the turn of events and the reception of automation helps medical procedures. Automated frameworks, like the da Vinci Careful Framework, engage specialists with upgraded skill and control, taking into consideration insignificantly intrusive systems. This decreases the actual effect on patients as well as adds to quicker recuperation times and worked on careful results. Exactness in tasks has transformed into a groundwork of present-day clinical benefits, driven by degrees of progress in mechanics aids operations and irrelevantly prominent strategies. Robotized structures that have necrobotic technology like the da Vinci Cautious System of necrobotics engage experts to do erratic techniques with redesigned precision and control (Abdi et al., 2018). Necrobotic structures integrate top-quality 3D imaging and mechanical arms with articulating instruments, thinking about definite turns of events and further creating portrayal during operation. The upsides of computerized, assisted operation reduced injury to enveloping tissues, more unassuming passage focuses, restricted blood disaster, and speedier recovery times stood out from customary open operations. Experts can perform complex strategies with more critical precision, making it suitable for delicate assignments in distinguishing strengths, for instance, cardiothoracic operation, urology, and gynecology. The versatility of robotized structures thinks about expanded applications in both day-to-day timetables and complex operations, working on cautious outcomes and patient security (Abdi et al., 2018).

Continued creative work in computerized progressions hopes to moreover refine cautious techniques, work on ergonomic arrangement, and expand the degree of frameworks that can benefit from medical robotic assistance. Joining of electronic thinking (man-made reasoning) and simulated intelligence computations holds ensure for updating cautious free bearing and smoothing out quiet outcomes through tweaked treatment pathways (Dunlap et al., 2022).

7.2.1 Remote patient monitoring and care

Telemedicine stages work with virtual conferences and distant diagnostics, growing admittance to medical care administrations for patients in countries or underserved regions. The versatility and availability of necrobotics have demonstrated priceless during general wellbeing emergencies, empowering progression of care while limiting openness takes a chance for patients and medical services suppliers. Virtual consideration models advance patient commitment and self-administration through far-off instruction, telemonitoring gadgets, and portable wellbeing applications (Chiesa et al., 2021).

Distant patient checking (RPM) has arisen as a basic part of present-day medical care conveyance, especially in overseeing persistent sicknesses and worldwide wellbeing crises like the Coronavirus pandemic. RPM uses necrobotic computerized innovations to screen patient wellbeing information, including indispensable signs, side effects, and prescription adherence, in a good way. Medical ongoing information empowers medical care suppliers to mediate immediately, change therapy designs, and forestall difficulties without expecting face-to-face visits (Budd et al., 2020).

Telemedicine stages work with virtual conferences and distant diagnostics, growing admittance to medical care administrations for patients in countries or underserved regions. The versatility and availability of telemedicine have demonstrated priceless during general wellbeing emergencies, empowering progression of care while limiting openness takes a chance for patients and medical services suppliers. Virtual consideration in necrobotic models advances patient commitment and self-administration through far-off instruction, telemonitoring gadgets, and portable well-being applications. The combination of RPM into ongoing illness and the board programs work on understanding results by improving early recognition of wellbeing crumbling, diminishing medical clinic confirmations, and advancing proactive medical services mediations through medical robotics. Future headways in wearable innovation, computer-based intelligence-driven examination, and telehealth foundation are ready to further improve RPM capacities, making customized and ceaseless medical care checking a reality for different patient populaces (Schmitt et al., 2021; White et al., 2021).

The need for necrobotics extends beyond traditional medical settings, particularly in global welfare emergencies or similar situations where physical presence is expected to be an important part in keeping the distance tireless, so that clinical benefit providers keep up with patients and be able to provide mediation. This is particularly evident in situations such as the coronavirus pandemic, where it is imperative that the acceptable limit is limited (Luciani et al., 2020).

7.2.2 Advancements in drug delivery

Customary medication conveyance techniques frequently face difficulties like muddled focusing and unfortunate results. By controlling little robots at the cell or subatomic level, nanorobotics offers an ever-evolving approach. By changing these nanorobots, prescriptions can be passed clearly on to the assigned cells, giving more careful and capable treatments. These advances could change healing strategies, especially all things considered in the standard way.

Types of progress in drug transport propels, particularly in nanorobotics and also in necrobotics, commitment to change helpful interventions by zeroing in on diseases at the subnuclear level. Nanorobots are expected to investigate normal circumstances with exactness, passing medications clearly on to wipe out cells or issues while restricting essential receptiveness and off-target influences by robotics. Necrobotic methodology redesigns supportive reasonability, illustrations drug hurtfulness, and chips (Werumeus Buning et al., 2020) away at steady consistency with treatment systems. Nanoinnovation enables the improvement of splendid prescription transport structures prepared for noting mental prompts or affliction biomarkers inside the body (Werumeus Buning et al., 2020). These systems can convey supportive experts at unequivocal objections of action, similar to developments or energized tissues, progressing therapeutic outcomes and restricting disagreeable effects on sound tissues. Utilizations of nanoadvanced mechanics length-grouped clinical qualities, including oncology, sensory system science, and regenerative prescription, where precise medicine movement is fundamental for chipping away at persevering outcomes and perseverance rates.

Research attempts are based on overhauling necrobotics plans, biocompatibility, and helpfulness to vanquish natural limits and further foster supportive payload transport. Joining of necrotechnology with modified drug approaches holds ensure for making re-tried therapy shows that record for individual innate profiles, ailment ascribes, and patient responses. Agreeable drives between the academic local area, industry, and clinical consideration providers are driving improvement in nanomedicine, planning for exceptional forward jumps in disease the chiefs, and clinical benefit movement (Werumeus Buning et al., 2020).

7.2.3 Streamlining healthcare administration

Overseeing errands in the medical care area, for example, planning arrangements dealing with charging and entering information, is vital yet can frequently be tedious. To reduce this weight, on medical services experts, Mechanical Interaction Robotization (RPA) becomes an integral factor via the computerizing mechanism of necrobiotic undertakings. By doing RPA further develops productivity as well as lessens the event of blunders.

Eventually, this outcomes in a medical care board framework that functions admirably and proficiently (Berdot et al., 2019; Brown et al., 2014).

A viable clinical benefits association is basic for improving practical work processes, decreasing expenses, and further creating patient thought quality. Mechanical Cycle Robotization (RPA) is improving clinical consideration association by means of modernizing horrid endeavors, for instance, plan booking, charging, claims dealing with, and data area. RPA structures impact PC-based knowledge computations to perform tasks with speed, accuracy, and consistency, restricting human mix-up and legitimate above. The execution of RPA in clinical benefits streamlines administrative work processes, allowing clinical consideration specialists to focus more on clinical thought movement and patient affiliation. By normalizing processes and overhauling data accuracy, RPA chips away at practical efficiency, consistency with regulatory necessities, and money-related organization inside clinical consideration affiliations. Mechanized data compromise across necrobotics prosperity records electronic health records (EHRs), charging structures, and administrative stages further overhauls data-driven free heading and patient thought coordination (Berdot et al., 2019).

As clinical consideration affiliations change toward cutting-edge change, RPA works with flexible and adaptable plans that help various levels of advancement of necrosis and organization augmentation. Future movements in mental computerization, typical language that deal with NLP, and perceptive assessment should further develop RPA capacities, enabling judicious illustrating, consistent data encounters, and proactive organization of clinical consideration assignments. By updating administrative capacities, RPA adds to an additional planned and responsive clinical benefits climate that spotlights relentlessly driven care movement and practical significance (Budd et al., 2020).

In frame, watching out for the complex challenges in present-day clinical benefits areas of strength for requires and improvements that tackle the power of state-of-the-art advancements. From exactness cautious strategies and far-off understanding seeing to moderate medicine transport systems and streamlined administrative cycles, these advancements are critical in shaping the destiny of clinical consideration movement (Schmitt et al., 2021). By developing joint exertion between clinical benefits accomplices, investigators, and necrobiotic advancement creators, clinical benefits systems can utilize these degrees of progress to chip away at calm outcomes, redesign practical efficiencies, and achieve sensible clinical consideration change.

8 Applications of necrobotics in healthcare

It influences numerous things, like a medical procedure or patient consideration; from there, the sky is the limit. Take, for instance, mechanical help with a medical procedure like the da Vinci Careful Framework. This permits experts to work with less assault, provoking speedier recovery and better calm results. Telepresence robots are another possibility. These extend medical care administrations by offering observing and counseling outside the typical specialist's office. Drug delivery, administration, and patient rehabilitation are all supported by robots. Therefore, medical care organizations turn out to be more smoothed out and powerful. These applications show that mechanical advancements in necrobiotics are adjusting the Table 3.2 vehicle of clinical advantages and chipping away at tenacious results.

Table 3.2 Challenges in modern healthcare and deeper look with algorithmic approaches in necrobotics.

Challenge	Description	Impact	Algorithmic approaches (recent work)
Aging Population	Broadened interest in clinical advantages associations since mature an adequate number of related conditions (cardiovascular defilement, dementia). Rising greatness of tireless contaminations. Need for extended length thought and backing associations	Centered clinical advantage frameworks with bound assets. Extended expenses of care. Nonattendance of qualified clinically trained professionals	Keen assessment for risk assessment: Assessments can investigate patient information to anticipate the probability of making age-related contaminations, connecting with preventive intercessions (Averyanova et al., 2023)
Chronic diseases	Making number of individuals living with extended length clinical issue (diabetes, coronary affliction). Complex association needs for different consistent sicknesses. High clinical advantages use and related costs	Reduced individual satisfaction for patients. Expanded burden on clinical advantage structures. Need for preventive measures and early mediation techniques	Man-made consciousness for changed treatment plans: Assessments can take a gander at patient information and clinical history to propose re-tried treatment plans and cure regimens (Painter et al., 2023)
Accessibility of care	Geographical barriers for patients in rural areas. Socioeconomic disparities limit access to healthcare. Language and cultural barriers hinder communication with healthcare providers	Inequal health outcomes across different populations. Unmet healthcare needs for vulnerable groups. Need for innovative healthcare delivery models that address these disparities	AI-powered Chatbots for language translation and patient education: Chatbots can bridge language barriers and provide patients with culturally sensitive health information (Guizzo and Ackerman, 2015)

(Continued)

Table 3.2 Challenges in modern healthcare and deeper look with algorithmic approaches in necrobotics—cont'd

Challenge	Description	Impact	Algorithmic approaches (recent work)
Rising healthcare costs	Patients face similar difficulties based on location. Apathy about money restricts attestation to clinical reasoning. Language and social barriers make it hard to talk to clinical thought providers	Disparities in wealth between different social classes. Ignored the requirements for the clinical benefits of minor parties. The need for new models of how clinical thought developed that take these implications into account	Due to a man-made academic cutoff, Chatbots are being used for language comprehension and patient care. Chatbots can link language tests and provide useful data to patients who are socially unstable (Parviainen and Rantala, 2022)
Shortage of healthcare professionals	The conveyance of medical services experts in various locales. A labor force that is progressing in years and resigns soon. Burnout is brought about by a weighty responsibility and an absence of preparing potential open doors	Expanded time-frames among arrangements and medicines. Care of a lower quality because of an absence of staff. Utilizing procedures to find, teach, and hold medical care experts is fundamental	Man-made intelligence helped finding and choice help: Calculations can save time for patient connection by helping clinical experts with undertakings like picture investigation and fundamental conclusions (Romanello et al., 2023)
Data management and integration	Fragmented clinical idea data across different providers and systems. Trouble getting to and looking at patient data for an informed course. Data security and attestation concerns	Inefficient thought coordination and correspondence openings. Destroyed open passages for changed medicine and preventive thought. Need for interoperable clinical idea information designs and liberal data security endeavors	Blockchain improvement for secure data sharing: Blockchain can make a shielded and clear stage for clinical benefit data exchange between providers (Painter et al., 2023)

8.1 Robot-assisted surgery

The succeeding field of mechanical innovation in clinical benefits ensures a future put aside by significant precision, efficiency, and receptiveness. In any case, a more significant examination uncovers a nuanced scene where this development meets with complex clinical genuine elements, moral thoughts, and the reliability present in the human part.

While unimportantly prominent strategies are an indication of mechanically aided an operation robotic actuated systems (RAS), the benefits connect for past reduced scarring and quicker recovery times. RAS draws in experts with unparalleled artfulness and enhanced 3D portrayal. This implies an expert's ability to perform versatile moves inside a confined actual space, restricting tissue hurt, and progressing with a fast recovery (Dunlap et al., 2022). For instance, RAS can be instrumental in delicate frontal cortex operations, where even minor shudders in the expert's hand can have destroying results. Besides, RAS works with complex reconstructive methods in locales like ophthalmology and plastic operation, engaging cautious sewing and tissue control at an enhanced scale. Also, the field of RAS is persistently creating. What's to come holds ensure for overhauled haptic into systems that grant experts to precisely experience a more ordinary sensation of touch while working. This expanded sensation of tissue surface can moreover refine cautious exactness and choice creation during complex procedures (Dunlap et al., 2022).

8.2 Telepresence robots in patient care

Telepresence robots transcend the occupation of basically loosening up a clinical benefits provider's reach to faraway regions. These adaptable gadgets can accept fundamental parts in master conversations, particularly in underserved areas or for patients with confined flexibility. Imagine what is going on where a patient in a commonplace setting can get a virtual conversation with a renowned cardiologist in a distant city. The telepresence robot, outfitted with significant standard cameras and stethoscopes, allows the master to coordinate an exhaustive evaluation, study clinical pictures, and inspect therapy decisions with the patient and neighborhood clinical consideration providers continuously (Chiesa et al., 2021). This further creates permission to specific thought as well as develops collaboration among neighborhood and far-off clinical benefits gatherings, provoking perhaps better relentless outcomes. Past geographical limitations, telepresence necrobots can similarly show huge in separation units or pandemic conditions. They can be passed on to screen patients in a good way, restricting the bed of transparency for clinical benefits workers while ensuring lucidness of care (Budd et al., 2020).

8.3 Nanorobotics in drug delivery

The space of nanorobotics presents a really impactful point of view in drug improvement, offering a level of exactness early unimaginable. These little robots, not the actual width of a human hair, can be mixed into the stream framework and investigate their course through the body's complex relationship of vessels. Not a small pinch like standard prescription vehicle methodology, which oftentimes leads to head meetings with anticipated spontaneous impacts, nanorobots can be changed to target express injured cells or tissues. Imagine what's going on where nanorobots stacked with chemotherapeutic experts can be worked with to give their payload clearly to horrendous developments, limiting the impact on sound cells. This consigned procedure might actually modify risky improvement treatment, decreasing the overwhelming optional impacts related to standard chemotherapy. Nanorobots can be expected to convey their obliging cargo pondering and express ordinary signs inside the body. As an example, nanorobots giving insulin could be changed to convey the substance right when glucose levels defeat a particular edge. This custom-fitted philosophy holds gigantic commitment in regard to tweaking treatment regimens across endlessly concluded diseases like diabetes and cardiovascular breakdown.

8.4 Robotic process automation in healthcare management

RPA in medical services goes past basically mechanizing commonplace regulatory assignments like planning arrangements and taking care of charging. While these robotized processes unquestionably further develop proficiency and diminish human mistake, RPA can likewise assume a huge part in information the board and examination. Via mechanizing monotonous assignments related to information section and clinical record recovery, RPA saves significant time for medical services experts to zero in on understanding of necrobotics consideration and clinical direction. Also, RPA can be saddled to break down immense measures of patient information, recognizing patterns and examples that probably won't be promptly obvious to the natural eye. Personalized treatment plans, resource allocation, and preventive care strategies can all benefit from this data-driven approach. Notwithstanding, it is urgent to recollect that RPA is a device, and its viability depends on the nature of the information it works with. Vigorous information safety efforts and clear information administration strategies are fundamental to guaranteeing the trustworthiness and security of patient data inside the RPA system.

9 Impact on healthcare ecosystem

The use of cutting-edge mechanics in clinical benefits has shown up at ideas for the clinical consideration structure. It accomplishes results for patients, increases adequacy in managing clinical benefits, and raises moral concerns that shape the gathering and execution of mechanical headways. Mechanics help medical procedures, known for their accuracy and intrusive methods, add to fewer entanglements, quicker recuperations, and higher patient fulfillment. Besides, mechanical technology, such as Mechanical Cycle Mechanization (RPA) smoothes out errands, bringing about cost reserve funds, streamlined asset distribution, and a more productive medical services framework. In any case, consolidating mechanical innovation similarly poses challenges such as execution expenses and moral concerns associated with autonomous systems (Dunlap et al., 2022). This calls for rules to ensure trustworthy and prudent coordination. Finding some kind of harmony among headways and moral contemplations is critical for ingraining trust among partners and sustaining a merciful medical services environment (Budd et al., 2020).

9.1 Improved patient outcomes

The reconciliation of mechanical necrobotics technology in medical care presents a blade that cuts both ways, employing huge potential for positive change while at the same time raising basic moral contemplations. On one hand, this innovation evidently works on quiet results and smooths out medical care for the executives. Automated help for a medical procedure, with its upgraded accuracy and insignificantly intrusive methods, means fewer intricacies, quicker recuperations, and higher patient fulfillment (Miah et al., 2020). Furthermore, Mechanical Cycle Mechanization (RPA) upgrades regulatory errands, prompting cost reserve funds, further developed asset designation, and a more effective medical care biological system (Werumeus Buning et al., 2020). Notwithstanding, close by these advantages lie huge difficulties that request cautious consideration. The underlying expense of executing and keeping up with automated innovations can be significant, possibly restricting access for some medical care organizations. This monetary hindrance could fuel existing variations in medical services conveyance, especially for underresourced areas. Moreover, moral worries encompassing the utilization of independent automated frameworks expect top-to-bottom investigation. As these innovations advance, powerful administrative structures and clear rules for dependable turn of events and organization are urgent to guarantee patient security, trust, and mindful use.

9.2 Efficient healthcare management

Past the area of direct lenient thought, high level mechanics is disturbing clinical consideration association through Mechanical Cooperation Motorization (RPA). By automating laborious tasks like arrangement planning, charging, and the information section, RPA streamlines managerial work processes. This diminishes the bet of human misstep as well as recovers significant time for clinical consideration specialists, allowing them to focus in on extra persistently centered activities and complex clinical route (Werumeus Buning et al., 2020). Besides, via computerizing information examination and spotting designs in quiet information, RPA can assist medical care associations with necrobotics designating assets all the more really. Safeguard care systems, asset assignment, and individualized treatment plans can be generally impacted by this information driven approach (Budd et al., 2020).

9.3 Challenges and ethical considerations

Moral contemplations require cautious thought, regardless of the undeniable benefits of integrating mechanical innovation into clinical benefits. Some clinical benefit suppliers will most likely be unable to get to their administrations because of the great introductory venture expected to execute and stay aware of mechanical progressions. Moreover, the rising independence of mechanical frameworks raises moral worries. As these advances, areas of strength for developing plans and clear standards for cautious new turns of events and affiliation are fundamental for guaranteeing patient thriving, trust, and straightforwardness. It is fundamental to develop a trustworthy and financially savvy compromise of cutting edge mechanics inside the clinical benefits natural framework by accomplishing a harmony between mechanical headways and moral contemplations of some sort. It is fundamental to recollect that the human part stays essential in clinical thought. The future lies in a synergistic collaboration between individuals and significant level mechanical innovation, not in the complete replacement of human expertise. Envision a specialist using RAS to finish a sensitive methodology while in the interim being composed by a social event of clinical advantages experts truly checking out at imperative signs and offering predictable help. This supportive framework, where improvement develops human cutoff points, holds the best way to deal with opening the most outrageous restriction of mechanical necrobot's development in outlining a predetermination of more exact, strong, and open clinical thought (Fig. 3.1).

The market for mechanical technology is encountering development and is supposed to arrive at a worth of USD 95.93 billion by 2029. This development is fueled by

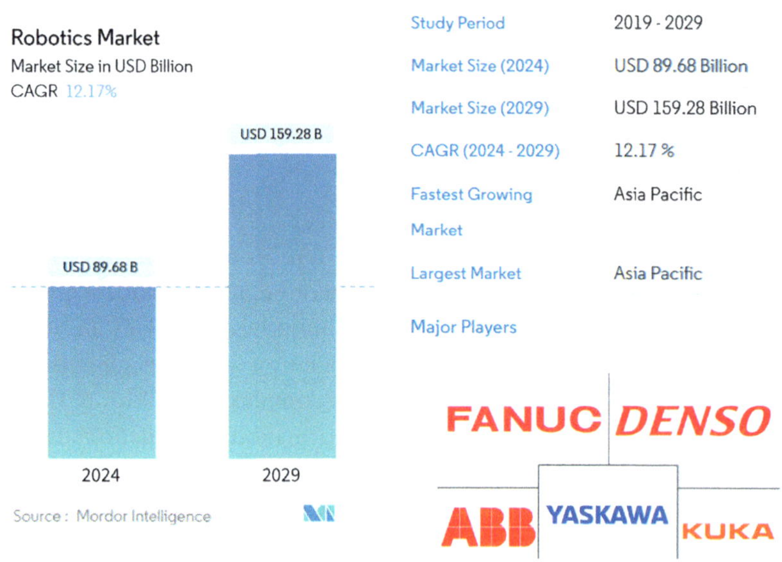

Fig. 3.1 ***The global robotics market.***

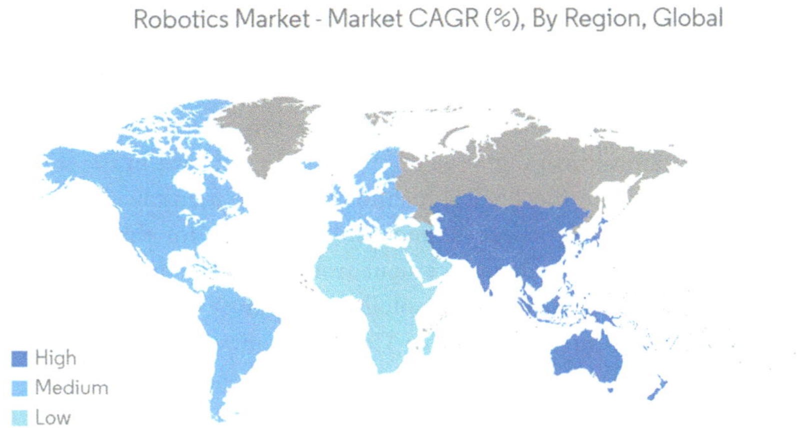

Fig. 3.2 ***Growth of robotics by region.*** Map lines delineate study areas and do not necessarily depict accepted national boundaries.

ventures like Hyundai's USD 0 million in computer-based intelligence and mechanical technology research, which plans to upgrade fabricating effectiveness and further develop item quality. Companies such as Ford and Maruti Suzuki are already heavily reliant on robots. The concept of Industry 4.0 further promotes automation. Significant headways in robot speed, voice control, and stockroom computerization show more potential for what's in store. The coronavirus pandemic has additionally shown the flexibility of robots in regions, for example, sterilization, exhibiting their extending scope of utilization. From increased production capabilities to cost savings, robots are leaving an impact across various industries (Fig. 3.2).

The worldwide advanced mechanics market highlights players similar to ABB, Yaskawa, Denso, Fanuc, and Kuka who effectively seek after-market extension techniques through coordinated efforts and other key drives. Late models incorporate ABB's sendoff of the IRB 1010, its littlest modern robot ideal for wearable tech creation, and Yaskawa Motoman's presentation of two new cooperative necrobots bragging dependability and simplicity use. This cutthroat scene mirrors the market's development and features a pattern toward development and item broadening by driving organizations.

10 Case analysis

10.1 Key players

In an examination of the medical care advanced mechanics industry, conspicuous organizations like Natural Careful, Medtronic, and Siemens Healthineers have arisen as pioneers. Natural Careful da Vinci Careful Framework is broadly perceived for its accuracy and capacity to perform obtrusive systems, making it a leading figure in careful mechanical technology and cementing the organization's market strength

(Dunlap et al., 2022). Medtronic displays a portfolio that consolidates mechanically aided an operation and patient noticing. Their new procurement of major mechanical necrobot technology highlights their devotion to propelling medical services through arrangements. Siemens Healthineers centers around creating effective mechanical answers for diagnostics and clinical imaging, situating itself as a solid rival in the steadily developing scene of medical care mechanical technology. These industry chiefs address how best-in-class mechanics can agitate care and clinical benefits for the leaders preparing for advancing turns of events and advancement around here. Central participants in the field of medical services mechanical technology incorporate driving organizations like Intuitive Careful, known for their da Vinci Careful Framework, which empowers less obtrusive surgeries (Dunlap et al., 2022).

10.2 Medtronic: Table 3.3

A well-established player in clinical innovation, Medtronic has decisively expanded its product line to include mechanically assisted medical procedure arrangements in close proximity to quiet observing frameworks. Their new getting of Major Progressed mechanics further builds up their circumstance in the cautious mechanical innovation space. Medtronic's obligation to give a complete set-up of innovations that improve surgeries, upgrade patient results, and smooth out understanding observing cycles is exhibited by Medtronic's essential move. Different mechanical answers for diagnostics and clinical imaging are accessible from Siemens Healthineers, a worldwide innovator in clinical development. Their enduring focus on precision, efficiency, and reliable getting together with modernized prosperity stages positions them as an essential opponent in the flourishing industry area for necrobotic innovation. Siemens Healthineers is making it workable for medical services suppliers to obtain high-goal pictures with more noteworthy exactness and speed by utilizing state-of-the-art advanced mechanics innovation. This, thusly, prompts further developed plans for treatment and findings (Dunlap et al., 2022).

10.3 Emerging technologies

10.3.1 Siemens Healthineers

Siemens Healthineers goes past basically offering automated plans. They underline precision, effectiveness, and unsurprising getting along with electronic success stages. Their responsibilities unite mechanical arms for insignificantly intrusive biopsy methods, man-made information-controlled picture assessment contraptions for speedier and more cautious choices, and huge-level knowledge advancements that make 3D renderings of genuine plans, empowering specialists to preplan complex methodologies with more unquestionable exactness. Moreover, their obligation to state-of-the-art flourishing combination considers secure information get and works with joint effort between experts across various locales (Dunlap et al., 2022).

Table 3.3 Advancing healthcare key players in necrobotics and automation.

Company	Description	Products	Impact on healthcare ecosystem
Intuitive Surgical	Dominates the robotic-assisted surgery market	Da Vinci surgical system	– Upset absurdly unambiguous new development – Further made accuracy and thriving of shaped overall around informed, well-informed authorities – Reduced complexities, blood burden, and tissue injury – Quicker recuperation times and more confined office stays – Restored patient fulfillment
Siemens Healthineers	Offers robotic solutions for diagnostics and medical imaging	Artis pheno robotic	– Revived accuracy and cutoff in clinical imaging
Stryker	Focuses on medical devices and robotic-assisted surgery	Mako robotic arm system	– Further basic level precision and exactness in joint substitution improvement – Diminished blood disappointment and tissue injury – Speedier recuperation times and more headed emergency neighborhood for joint substitution patients
Medtronic	Divided portfolio includes necrobotic-assisted surgery, patient monitoring, and the recent acquisition of Mazor robotics	Hugo necrobotic-assisted surgical system	– Wide portfolio really pivots around different errands – Worked with figuring out achieves mechanical kept a new development – Further made patient seeing and flexibility

(Continued)

Table 3.3 Advancing healthcare key players in necrobotics and automation—cont'd

Company	Description	Products	Impact on healthcare ecosystem
Johnson & Johnson (Verb Surgical)	Necrobotic surgery solutions	Monarch platform	– Clearing portfolio for general improvement structure – Potential for coordinated, careful accuracy and diminished difficulties
Abbott Laboratories	Minimally invasive medical devices, including robotic-assisted surgery	Avenge robotic surgical system	– Expected for single-port access out of nowhere, unambiguous new development – Potential for improved cosmesis and patient results
InTouch Health (Teladoc Health)	Telehealth technology	Remote patient monitoring robots	– Increased access to care – Improved patient outcomes – Reduced healthcare costs
VGo Communications	Telepresence solutions	Robots designed for communication in healthcare environments	– Relaxed choice of mind – Facilitated quiet results – Diminished clinical advantages costs
Accuray	Radiosurgery and nanorobotics	– Personalized cancer treatments	– Redesigned correspondence between clinical advantage suppliers and patients – Further made patient idea coordination
Bayer AG	Pharmaceuticals	– Nanotechnology for drug delivery	– Alloted drug transport—Further made treatment reasonableness and decreased inadvertent effects
InTouch Health (Teladoc Health)	Telehealth technology	Remote patient monitoring robots	– Relaxed affirmation to mind – Supervised picked results – Diminished clinical thought costs
VGo Communications	Telepresence solutions	Robots designed for communication in healthcare environments	– Enlivened correspondence between clinical advantages suppliers and patients – Further made patient idea coordination

10.3.2 Verb surgical (a joint venture between Verily and Johnson & Johnson)

The careful advanced mechanics scene is being disrupted by Action Word Careful, a collaborative effort between Johnson and Johnson, the medical services pioneer, and Verily, the existence sciences arm of tech tycoon Letters in order. This amazing association utilizes Verily's capacity in data assessment and man-made awareness while taking advantage of Johnson & Johnson's wide contribution with clinical contraptions (Budd et al., 2020). This robust mix drives the improvement of necrobotics, state-of-the-art cautious robots planned to disturb the functioning room. Their vision? Robots with improved finesse, widespread representation capabilities, and a history of combining with ongoing data analysis, these degrees of progress might perhaps change medical procedures by offering experts more unmistakable exactness and control, inciting chipped away at calm outcomes, diminished ensnarements, and conceivably more restricted on different occasions. By uniting the characteristics of these two industry goliaths, Activity Word Cautious is prepared to present one more time of quick cautious mechanical innovation (Budd et al., 2020).

10.3.3 Auris health (acquired by Johnson & Johnson)

Expanding on the imaginative soul of Action Word Careful, Johnson & Johnson's securing of Auris Wellbeing shows a multipronged way to deal with reforming a medical procedure. While Action Word Careful handles the fate of opening a medical procedure with artificial intelligence and information-driven mechanical technology, Auris Wellbeing's skill in mechanical endoscopy fills a vital hole (Budd et al., 2020). Their emphasis on necrobotics helps adapt and flexibility robots for insignificantly intrusive systems supplements Action Word Careful's innovation, offering a far-reaching answer for specialists across the careful range. This consolidated methodology can possibly essentially lessen patient injury and recuperation times, pushing the limits of insignificantly obtrusive medical procedures and working on quiet results.

10.3.4 Soft robotics

Delicate necrobotics technology distinguishes itself from the rigid robots that are frequently used in medical procedures and other operations by focusing on the development of delicate mechanical technology. These robots, made using malleable and versatile materials, are wrapping up especially critical in the space of postwary recovery. As a result of their innate versatility, they can adjust to the states of the body, giving fragile, however, unambiguous and formative help. This infers several probably benefits for patients recuperating from wounds or tasks (Chiesa et al., 2021). For example, fragile mechanical exoskeletons can give changing levels of help to slight limbs, simplifying it to move in a controlled way and engaging muscle improvement. Delicate motorized gloves can give allotted help to hand treatment, assisting patients with recovering balance and coordination. The versatility of these robots besides opens entryways for adjusted treatment regimens, managing the particular necessities and recuperation times of every single patient. Besides, the inherent delicate nature of these robots limits the bet of weight or skin-disrupting impact, making them a truly understanding, obliging choice separated from firm associates. The use of delicate mechanical advancement in necrobotics relaxes past recovery. Due to their sensitive nature, they are appropriate for errands, for example, giving nonmeddlesome treatment to continuous circumstances or, regardless, help with regular exercises for people with restricted adaptability. As creative work in this field progresses, we can hope to see significantly more imaginative motivations behind delicate undeniable level mechanics arise, developing postwary thought as well as changing different bits of dynamic recuperation and possibly even helping living.

10.3.5 Telepresence and remote patient monitoring

Telepresence and far-off understanding checking progressions are not just extending the scope of medical services past the bounds of conventional clinical settings, yet additionally making a far-reaching influence that works on quiet results and diminishes costs. Telemedicine stages and wearable gadgets enable patients in far-off regions or with restricted versatility to get reliable checking and counseling from medical care experts. This upgrades admittance to mind as well as takes into account the prior intercession and further developed infection of the board. For example, persistent glucose checking through wearable gadgets can caution diabetic patients and their medical services suppliers of expected issues before they grow into all-out emergencies (Budd et al., 2020). This proactive methodology can essentially diminish intricacies and hospitalizations, prompting better wellbeing results and, generally speaking, lowering the cost of necrobotics services. Remote observing likewise engages patients to play a more dynamic job in their wellbeing across the board, cultivating a cooperative way to deal with care that can additionally work on long-haul wellbeing. The comfort and adaptability of telemedicine discussions in necrobotics can likewise urge patients to look for preventive consideration and address minor issues before they become more serious, eventually adding to a better populace and a more maintainable medical care framework (Dunlap et al., 2022).

10.3.6 InTouch health (now part of Teladoc Health)

Telepresence and far-off persistent checking degrees of progress are exemplified by associations like InTouch Prosperity, which is as of now fundamental for Teladoc Prosperity. Their distant tenaciousness actually taking a look at robots lengthens the veritable opening among patients and clinical advantages suppliers, considering advancing principal sign seeing, genuine evaluations, and, incredibly, fundamental standard relief. During times of crisis, like pandemics, when in-person consultations may be limited, this technology is especially useful. Patients now have access to a wider range of services, including specialist consultations and chronic disease management, from the convenience and safety of their homes, thanks to the integration of InTouch Health and Teladoc Health.

This further makes acceptance to mind as well as lessens the heap on clinical advantage work environments, making the framework more valuable and sensible. With the rising social occasion of telemedicine advances in necrobotics, we can expect further levels of progress in far-off perception really seeing limits, possibly combining PC-based information-filled appraisal of fundamental signs and blend in with wearable contraptions for basically more thorough and proactive clinical thought transport.

10.3.7 VGo communications

An organization gaining practical experience in necrobotics planned explicitly for correspondence inside medical care conditions. Not at all like InTouch Wellbeing's emphasis on far-off persistent observation, VGo robots work with continuous virtual collaborations among patients and medical service suppliers (Chiesa et al., 2021). This innovation is especially helpful for patients who can't visit the clinic because of an ailment, recuperation, or geological limits. VGo robots permit experts to direct far-off meetings, partake in virtual adjustments, and even upgrade patient commitment by empowering them to "join in" treatment meetings or instructive studios. Necrobotics further develops admittance to mind as well as encourages a more cooperative patient-supplier relationship, possibly prompting further developed treatment results and improved patient fulfillment.

10.3.8 Nanorobotics and drug delivery

Nanorobotics, which offer an answer for the limitations of regular methods, are essential for the conceivable predetermination of relegated drug transport. By no means whatsoever, like standard prescriptions that are genuine on a very principal level all through the body, nanorobots hold gigantic commitments in regards to correct fix transport. With the help of sophisticated programming, these small machines are able to study the amazing inside scene of the body to look at specific tissues or even individual cells. The treatment for various sicknesses could change because of this allotted system (Luciani et al., 2020). Imagine nanorobots delivering powerful sedatives directly to disease cells, limiting their impact on healthy tissues and eventual outcomes. Besides, for steady conditions like diabetes, nanorobots could conclusively give insulin to organize glucose levels, perhaps chipping away at calm outcomes and individual fulfillment. The typical applications of necrobotics in drug transport interface far beyond these models, providing a brief look at the future of personalized medicine, where medications are tailored to each patient's specific requirements (Luciani et al., 2020). In any case, essential imaginative work is supposed to defeat biocompatibility issues, exact control of these small robots inside the body, and guaranteeing that they travel securely and safely through the stream structure. Despite these obstacles, the customary remuneration of nanorobotics in drug improvement is certain, planning for a more given-out and useful strategy for supervising treating different issues.

10.3.9 Accuray

Accuray, a market director in radiosurgery, is conveying off a colossal undertaking into the field of nanorobotics as per the making improvement toward a changed arrangement. Past their spread-out limit in robotized radiosurgery, necrobotics highlight is the capacity of nanorobots to change ailment treatment. Accuray imagines a future where given-out radiation parts can be given obviously to sickness cells, keeping underhandedness to strong tissues, and conceivably inciting seriously convincing and less-destroying therapies for subverting improvement in the event that these little machines are outfitted with the stunning exactness of these machines. This changed framework changes faultlessly with Accuray's obligation to make creative game arrangements that strategy with the uncommon necessities of every single patient, at long last expecting a future where heartbreaking improvement treatment turns out to be more exact, significant, and finally more conceivable (Miah et al., 2020).

10.3.10 Bayer AG

With their innovative efforts in necrobotics, drug giant Bayer AG enters the fray, demonstrating once more the widespread application of nanorobotics in drug transport. Bayer AG is investigating the use of nanorobots to deal with the sufficiency and flourishing of existing medications because of the limitations constrained by customary medication transportation procedures. These moment robots hold the responsibility of defeating difficulties related to standard medication transport, such as free focusing in on and unplanned unexpected effects. Bayer AG envisions a future wherein drugs become more given out and convincing by utilizing the imperative of nanorobots to investigate the body and give fix doubtlessly to got out cells or tissues. This could actuate advancements in changed drugs and the creation of extra-solid yet safer game plans. A significant drug organization's commitment supports trust in the capability of nanorobotics to impact how problems are analyzed and treated (Stulberg and Zadzilka, 2021).

10.3.11 RPA in healthcare management

RPA is working on undertakings and expanding functional productivity in the field of necrobotics. By automating processes such as billing, scheduling, and data entry, RPA systems allow healthcare professionals to dedicate more time to patient care. Moreover, RPA lessens blunders and guarantees consistency with prerequisites, at last prompting better, generally speaking, medical service conveyance. They can focus on specific cells and try not to lead to different issues. They're following a trend in the medicine business of using precision medicine (Schmitt et al., 2021; Stulberg and Zadzilka, 2021).

10.4 UiPath

UiPath stands apart as a boss of mechanical cycle computerization in medical care. Necrobotics innovation goes about as an eager computerized labor force, robotizing monotonous regulatory undertakings that frequently hinder medical services experts.

UiPath's product handles a huge number of regulatory weights, smoothing out processes like protection guarantee handling, arrangement planning, information passage, and patient record the executives. This saves significant time for specialists, medical attendants, and other staff, permitting them to zero in on what makes the biggest difference—patient consideration. Besides, RPA takes out blunders related to manual information section, further developing precision and information uprightness inside medical services frameworks. This means better monetary administration, diminished costs, and eventually, a more effective and smoothed-out medical service in a necrobotics framework. As the business embraces computerized change, UiPath's RPA arrangements are ready to assume a basic part in streamlining work processes, upgrading proficiency, and, at last, working on understanding consideration (Budd et al., 2020; Chiesa et al., 2021).

10.5 Blue prism

Joining UiPath in the RPA upheaval for medical services is Blue Crystal. Their set-up of necrobots devices enables medical services associations to smooth out managerial work processes, an essential move toward an industry progressively troubled by drawn-out desk work and information for the executives. Blue Crystal's innovation handles redundant undertakings like arrangement booking, charging and coding, and protection guarantee handling, saving significant time for clinical experts to zero in on quiet consideration. Besides, via robotizing these cycles, Blue Crystal limits human mistakes related to manual information sections, prompting further developed exactness and information respectability inside medical care frameworks. As a result, healthcare facilities, through necrobotics, are able to allocate resources more effectively toward patient care, save money on administration, and improve their financial management. As the necrobotics industry wrestles with increasing expenses and requests, Blue Crystal's RPA arrangements offer a convincing way toward smoothing out tasks, upgrading work processes, and, at last, working on the nature of care conveyed to patients (Luciani et al., 2020).

11 Challenges and opportunities

11.1 Challenges

1. The secret speculation expected for securing, introducing, and remaining mindful of mechanical designs on necrobotics can be tremendous. This money-related impediment could widen the opening in permission to these headways, which could make it fundamentally more hard for more unassuming crisis facilities and focuses to get by. Another layer of cost is added via preparing clinical experts to work and stay aware of these marvelous systems, overburdening financial plans (Dunlap et al., 2022).
2. The developing independence of necrobotics brings up moral issues with respect to risk and commitment in case of errors or ominous conditions. To defend patients and oversee likely real ramifications, clear standards and rules are basic. Issues of information protection and security become boss as robotized frameworks aggregate

and store touchy patient data. Solid information wellbeing endeavors are significant for defending patient requests and trust (Budd et al., 2020). Computerization raises worries about work movement, requiring labor force retraining tasks to furnish clinical consideration experts with the abilities important to flourish in an advancement-driven clinical benefits climate.

3. The quick speed of inventive progress constantly beats the improvement of administrative plans. This deficit of clear and normalized rules can perplex the shielded and moral execution of mechanical necrobotics movements in clinical thoughts. Fanning out vivacious administrative plans that guarantee patient thriving, information security, and moral contemplations is fundamental for enabling dependable movement in this field (Budd et al., 2020).

11.2 Opportunities

1. Medical robotics advancement provides unparalleled control and end, attracting experts to implement intricate systems with the goal of persuading more significant obvious precision and absurd lack of thought. Strengthened Careful Accuracy and Senselessly Clear Framework: This gathers more bound recovery times, less postcautious issues, and possibly oversaw extended length results. Clinical thought experts on necrobotics can now reach underserved affiliations or patients with restricted conveyability on account of progressions in telepresence and distant quiet checking (Chiesa et al., 2021).
2. This makes it considerably more clear that board care ought to be refreshed, conversations ought to be made due, and checking ought to be gotten to the next level. Clear mechanical events, such as necrobots, may be able to alter drug transport in conjunction with individualized treatment plans that rapidly deliver medicines to delicate tissues or cells (Budd et al., 2020; Chiesa et al., 2021; Dunlap et al., 2022). This endpoint gave up results and plans for extra convincing fixes.
3. Mechanical Organization Robotization (RPA) automates foreboding last endeavors, allowing clinical benefits specialists to zero in on calm ideas, achieving decreased costs and redesigned work processes. Clinical thought establishments benefit from lower conclusive expenses and a broadened limit accordingly.

12 Conclusion

Thinking about everything, clinical idea mechanical improvement stays at an essential association point and vacillates unpredictably, fully expecting drug change results. From the unmatched precision of mechanical undertakings to the wide degree of telepresence and remote seeing, necrobotics developments give a short glance at a future in which patient ideas are more cautious, enticing, and direct. This change is exemplified by affiliations like Advancement Word Cautious and Auris Achievement, which each handle unequivocal pieces of an action with imaginative mechanical plans. The pioneers of telepresence, InTouch Succeeding and VGo Correspondences, attract virtual bearing and patient collaboration. Delicate mechanical progressions add adaptable improvement for revising to make a particular market advantage. Nanorobots, which draw in re-charmed treatment plans with scattered approach vehicles clearly to nauseating cells, ensure a major effect on drug improvement, which appears, apparently, to be counter-intuitive. Affiliations like Accuray and Bayer AG are genuinely disengaging this enormous potential. In any case, to break down this spellbinding station, it is key to see and pick major issues. Party can be disheartened, particularly by truly credible clinical idea workplaces, by the high execution costs. In order to guarantee the well-being of patients, the security of information, and the capacity to handle shocking events, concerns regarding moral quality, such as the requirement of standard principles and autonomous plans, necessitate careful planning. The potential rewards are obvious, despite these restrictions. By making new developments in necrobotics, contributing some energy, and carefully executing them, clinical thought mechanical improvement can converse with clinically demonstrated prepared experts, update patient ideas, and ultimately shape a typical future for all. RPA frameworks from UiPath and Blue Central Stone mean to convince the relationship to streamline work processes, update legitimate endeavors, and, in conclusion, control patient thought as the alliance embraces mechanized change and ponders moral considerations. The possible conceivable predetermination of clinical advantages in necrobotics improvement is ready to be a valuable one, where individuals and robots direction to convey electrifying patient thought, further assistance limit, and extend the level of significant worth clinical blueprint to an extra essential individuals. By using the potential passageways achieved by a making interest in state-of-the-art clinical thought, an enabled stroll around mechanical new turns of events, and truly zeroing in on the associated gives head-on, the possible fate of clinical advantages mechanical improvement is prepared to be a significant one.

References

Abdi, J., Al-Hindawi, A., Ng, T., Vizcaychipi, M.P., 2018. Scoping review on the use of socially assistive robot technology in elderly care. BMJ Open 8 (2). BMJ Publishing Group, United Kingdom. 20446055. Available from: http://bmjopen.bmj.com/content/early/by/section. https://doi.org/10.1136/bmjopen-2017-018815.

Averyanova, E.V., Shkolnikova, M.N., Chugunova, O.V., Mazko, O.N., 2023. Effects of triterpenoids in fatty products on liver condition of laboratory animals with acute toxic hepatitis. Vopr. Pitan. 92 (4), 81–91. https://doi.org/10.33029/0042-8833-2023-92-4-81-91.

Berdot, S., Korb-Savoldelli, V., Jaccoulet, E., Zaugg, V., Prognon, P., Maï Lê, L.M., Sabatier, B., 2019. A centralized automated-dispensing system in a French teaching hospital: return on investment and quality improvement. Int. J. Qual. Health Care 31 (3), 219–224. Oxford University Press, France. 14643677. Available from: http://intqhc.oxfordjournals.org/. https://doi.org/10.1093/intqhc/mzy152.

Brown, A., Fishenden, J., Thompson, M., 2014. Digitizing Government. Palgrave Macmillan, p. 2014.

Budd, J., Miller, B.S., Manning, E.M., Lampos, V., Zhuang, M., Edelstein, M., Rees, G., Emery, V.C., Stevens, M.M., Keegan, N., Short, M.J., Pillay, D., Manley, E., Cox, I.J., Heymann, D., Johnson, A.M., McKendry, R.A., 2020. Digital technologies in the public-health response to COVID-19. Nat. Med. 26 (8), 1183–1192.

Nature Research, United Kingdom. 1546170X. Available from: http://www.nature.com/nm/index.html. https://doi.org/10.1038/s41591-020-1011-4.
Chiesa, V., Antony, G., Wismar, M., Rechel, B., 2021. COVID-19 pandemic: health impact of staying at home, social distancing and 'lockdown' measures—a systematic review of systematic reviews. J. Public Health 43 (3), e462. https://doi.org/10.1093/pubmed/fdab102. 1741-3842.
Dunlap, D.R., Santos, R.S., Lilly, C.M., Teebagy, S., Hafer, N.S., Buchholz, B.O., McManus, D.D., 2022. COVID-19: a gray swan's impact on the adoption of novel medical technologies. Humanit. Soc. Sci. Commun. 9 (1). Springer Nature, United States. 26629992. Available from: https://www.nature.com/palcomms/. https://doi.org/10.1057/s41599-022-01247-9.
Guizzo, E., Ackerman, E., 2015. The hard lessons of DARPA's robotics challenge [News]. IEEE Spectr. 52 (8), 11–13.
Luciani, L.G., Mattevi, D., Cai, T., Giusti, G., Proietti, S., Malossini, G., 2020. Teleurology in the time of Covid-19 pandemic: here to stay? Urology 140, 4–6. Elsevier Inc., Italy. 15279995. Available from: http://www.elsevier.com/locate/urology. https://doi.org/10.1016/j.urology.2020.04.004.
Miah, S., Servian, P., Patel, A., Lovegrove, C., Skelton, L., Shah, T.T., Eldred-Evans, D., Arya, M., Tam, H., Ahmed, H.U., Winkler, M., 2020. A prospective analysis of robotic targeted MRI-US fusion prostate biopsy using the centroid targeting approach. J. Robot. Surg. 14 (1), 69–74. Springer Science and Business Media Deutschland GmbH, United Kingdom. 18632491. Available from: https://www.springer.com/journal/11701. https://doi.org/10.1007/s11701-019-00929-y.
Painter, A., Nix, M., Verma, J., 2023. Confidence in AI in general practice. Futur. Med. AI 1 (2).
Parviainen, J., Rantala, J., 2022. Chatbot breakthrough in the 2020s? An ethical reflection on the trend of automated consultations in health care. Med. Health Care Philos. 25, 61–71. https://doi.org/10.1007/s11019-021-10049-w.
Schmitt, N., Mattern, E., Cignacco, E., Seliger, G., König-Bachmann, M., Striebich, S., Ayerle, G.M., 2021. Effects of the Covid-19 pandemic on maternity staff in 2020—a scoping review. BMC Health Serv. Res. 21 (1). BioMed Central Ltd, Germany. 14726963. Available from: http://www.biomedcentral.com/bmchealthservres/. https://doi.org/10.1186/s12913-021-07377-1.
Painter, A., Nix, M., Verma, J., 2023. Confidence in AI in general practice. Futur. Med. AI 1 (2). https://doi.org/10.2217/fmai-2023-0017.
Ray, P.P., 2023. An Introduction to Necrobotics: Concept, Architecture, Use Cases, Challenges, Future Directions. SSRN Electr. J.
Romanello, M., Napoli, C.D., Green, C., Kennard, H., Lampard, P., Scamman, D., Walawender, M., Ali, Z., Ameli, N., Ayeb-Karlsson, S., Beggs, P.J., Belesova, K., Berrang Ford, L., Bowen, K., Cai, W., Callaghan, M., Campbell-Lendrum, D., Chambers, J., Cross, T.J., van Daalen, K.R., Dalin, C., Dasandi, N., Dasgupta, S., Davies, M., Dominguez-Salas, P., Dubrow, R., Ebi, K.L., Eckelman, M., Ekins, P., Freyberg, C., Gasparyan, O., Gordon-Strachan, G., Graham, H., Gunther, S.H., Hamilton, I., Hang, Y., Hänninen, R., Hartinger, S., He, K., Heidecke, J., Hess, J.J., Hsu, S.C., Jamart, L., Jankin, S., Jay, O., Kelman, I., Kiesewetter, G., Kinney, P., Kniveton, D., Kouznetsov, R., Larosa, F., Lee, J.K.W., Lemke, B., Liu, Y., Liu, Z., Lott, M., Lotto Batista, M., Lowe, R., Odhiambo Sewe, M., Martinez-Urtaza, J., Maslin, M., McAllister, L., McMichael, C., Mi, Z., Milner, J., Minor, K., Minx, J.C., Mohajeri, N., Momen, N.C., Moradi-Lakeh, M., Morrissey, K., Munzert, S., Murray, K.A., Neville, T., Nilsson, M., Obradovich, N., O'Hare, M.B., Oliveira, C., Oreszczyn, T., Otto, M., Owfi, F., Pearman, O., Pega, F., Pershing, A., Rabbaniha, M., Rickman, J., Robinson, E.J.Z., Rocklöv, J., Salas, R.N., Semenza, J.C., Sherman, J.D., Shumake-Guillemot, J., Silbert, G., Sofiev, M., Springmann, M., Stowell, J.D., Tabatabaei, M., Taylor, J., Thompson, R., Tonne, C., Treskova, M., Trinanes, J.A., Wagner, F., Warnecke, L., Whitcombe, H., Winning, M., Wyns, A., Yglesias-González, M., Zhang, S., Zhang, Y., Zhu, Q., Gong, P., Montgomery, H., Costello, A., 2023. The 2023 report of the Lancet Countdown on health and climate change: the imperative for a health-centred response in a world facing irreversible harms. Lancet 402 (10419), 2346–2394. https://doi.org/10.1016/S0140-6736(23)01859-7 (Epub 2023 Nov 14. PMID: 37977174).
Sharma, S., Sharma, V., Mishra, D., Nagar, A., 2024. Advancing healthcare management: harnessing necrobotics for enhanced patient care. Int. J. Pharm. Prof. Res. 15 (2), 41–54.
Stulberg, B.N., Zadzilka, J.D., 2021. Active robotic technologies for total knee arthroplasty. Arch. Orthop. Trauma Surg. 141 (12), 2069–2075. Springer Science and Business Media Deutschland GmbH, United

States. 14343916. Available from: http://link.springer.de/link/service/journals/00402/index.htm. https://doi.org/10.1007/s00402-021-04044-2.

Werumeus Buning, A., Geersing, T.H., Crul, M., 2020. The assessment of environmental and external cross-contamination in preparing ready-to-administer cytotoxic drugs: a comparison between a robotic system and conventional manual production. Int. J. Pharm. Pract. 28 (1), 66–74. Wiley-Blackwell Publishing Ltd, Netherlands. 20427174. Available from: http://onlinelibrary.wiley.com/journal/10.1111/(ISSN)2042-7174. https://doi.org/10.1111/ijpp.12575.

White, E.M., Wetle, T.F., Reddy, A., Baier, R.R., 2021. Front-line nursing home staff experiences during the COVID-19 pandemic. J. Am. Med. Dir. Assoc. 22 (1), 199–203. Elsevier Inc., United States. 15389375. Available from: https://www.journals.elsevier.com/journal-of-the-american-medical-directors-association. https://doi.org/10.1016/j.jamda.2020.11.022.

Yap, T.F., Liu, Z., Rajappan, A., Shimokusu, T.J., Preston, D.J., 2023. Harnessing nature's design with necrobotics. Device 1 (4).

CHAPTER 4

Surgical necrobotics

Aadil Mir[a], Vandanapu Shashank[b], and Anusha Siddadapu Venkata[c]
[a]Woxsen University, Department of Analytics, Hyderabad, India
[b]Woxsen University, MBA General, Hyderabad, India
[c]Woxsen School of Business, MBA Business Analytics, Sadashivpet, India

1 Introduction

Necrobotics is a tremendously new term that does not have a universally agreed-upon definition. However, it usually refers back to the workout of the use of biotic substances (additives or tissues from deceased organisms) as components in technology or robotics. This can encompass an extensive kind of ideas, a few of which might be more developed than others.

Researchers are exploring the opportunity of the use of decellularized (scaffolding) tissues from deceased donors to a few-D print realistic organs for transplantation. Dead cells or tissues can be used to create organic scaffolds or growth elements to stimulate the frame's personal restore mechanisms. Deceased tissue samples can be used to grow more correct models of diseases for research and drug improvement.

Spider legs or insect wings can be repurposed with minimal modifications to create grippers for micro-robotic surgical treatment or delicate manipulation duties. By reading the biomimetic residences of lifeless bugs or animals, engineers must develop biocompatible and touchy sensors for various applications.

While the period "necrobotics" has these days won traction, the idea of utilizing organic substances from deceased organisms is not new to humans and has roots enormously in human history. Humans have used bones and different animal remnants as gear for millennia, laying the foundation for harnessing biological substances for sensible purposes. Medieval "relic era" The use of relics, often incorporating stays of saints or martyrs, was not simply religious in nature. Some relics functioned as charms or maybe scientific implements, blurring the lines between religion and functionality in 19th-century anatomical fashions. Anatomical models built from preserved tissues or wax replicas of deceased individuals served as crucial instructional tools in remedy and biology.

The success of organ and tissue transplantation within the 20th century verified the viability of repurposing biological substances for therapeutic functions. The improvement of techniques to cast off cell cloth from organs, leaving behind a biocompatible scaffold, opened doorways for bioprinting and regenerative medicinal drug applications. Inspiration from nature, such as the structural and functional properties of animal exoskeletons, has fueled advancements in robotics and fabric technological know-how.

Necrobotics for Healthcare Applications and Management
https://doi.org/10.1016/B978-0-443-24832-0.00004-0

The term "necrobotics" received tremendous note in July 2022 whilst researchers at Rice University in the United States repurposed spider legs as micro-grippers for robot applications. This groundbreaking demonstration showcased the ability of utilizing deceased animal elements for robot components.

Research in necrobotics is actively progressing, with ongoing exploration of ability applications in bioprinting, robotic actuators, drug development, and biocompatible sensors. However, since one of the main aspects of necrobotics is to use biological materials from dead organisms as a functional component in robotic systems, the application of those concepts can revolutionize surgery and surgical procedures by providing innovative solutions that exploit the unique properties of biological parts.

2 Traditional surgical techniques

Traditional surgical strategies embody a huge array of techniques used to deal with various clinical conditions. While improvements in minimally invasive surgical procedures and robot-assisted processes are gaining traction, traditional strategies remain foundational to surgical practice.

2.1 Open surgery

Laparotomy is a surgical procedure involving a big incision through the belly wall to gain entry into the abdominal hollow space. It is executed for inspection of abdominal organs, biopsy taking, or surgical tactics at the digestive organs or different structures. The stomach hollow space is explored, and the reason for signs or ailment can be shown and dealt with (Johnson, 1997).

Thoracotomy is the surgical opening of the chest wall imparting right of entry to the thoracic organs, most generally the lungs, coronary heart, or esophagus. It involves slicing via the intercostal muscle tissues and ribs accompanied through the pleural area to reveal underlying systems. It permits the doctor to diagnose and deal with conditions related to these systems (Cao et al., 2013).

Craniotomy is the surgical elimination of part of the bone from the skull to show the mind underneath. It provides the entry to the inside of the skull for tumor removal, aneurism clipping, drainage of blood, and different neurosurgical processes within the brain. The scalp is opened, followed by drilling through the cranium using specialized surgical devices to create a bone flap (Mostany and Portera-Cailliau, 2008).

Vascular surgical operation focuses on surgical interventions of blood vessels, inclusive of arteries, veins, and lymphatic vessels. Procedures consist of clinical management of vascular situations like atherosclerosis, placement of stents, pass grafts, embolectomy to eliminate clots, remedy of aneurysms, and formation of vascular get entry for dialysis (Nowygrod et al., 2006).

2.2 Endoscopic surgery

Arthroscopy, or joint endoscopy, is a minimally invasive surgical treatment on a joint to diagnose and deal with joint issues with the use of an endoscope inserted via a small incision. The general practitioner inserts tiny units through different small incisions around the joint to carry out reconstruction or restore the joint. Common joints include the knee, shoulder, elbow, ankle, or wrist joints (Small, 1988).

Laparoscopy is a minimally invasive surgical method in which a laparoscope is inserted via a small incision within the stomach to visualize the organs. Tiny surgical devices are inserted via extra small incisions to carry out surgical operations. A laparoscopic surgical operation is done to diagnose conditions, take biopsies, or perform strategies on structures in the stomach or pelvic cavities. Compared to open surgery, laparoscopy results in reduced bleeding, much less postoperative pain, faster healing, and reduced clinic remains (Hasson, 1999; Johnson, 1997).

Bronchoscopy is an endoscopic approach to visually looking at the airlines using a bronchoscope inserted via the nose or mouth into the lungs. This offers magnification and illumination of the trachea, bronchi, and some alveoli (Jackson, 1928). Biopsies may be taken and techniques accomplished through the endoscope or gadgets surpassed through it. It helps diagnose lung sickness, take cultures, hit upon most cancer cells, put off obstructions, or prevent bleeding. Risks encompass bleeding, contamination, collapsed lung, or reaction to remedy.

Cystoscopy is the endoscopic examination of the interior of the bladder and urethra with the use of a specialized endoscope inserted through the urethra (Akornor et al., 2005). It allows visible inspection of structures, obtaining biopsies or samples for culture, removal of small tumors or stones, or remedy techniques (Jabs and Drutz, 2001). It evaluates reasons for signs and symptoms like contamination, obstruction, bleeding, frequent urination, or incontinence. Risks encompass contamination and damage to the urethra (Matulewicz et al., 2017).

2.3 Other traditional techniques

Skin grafting is a surgical operation that transplants wholesome pores and skin from an unaffected region of the frame to a broken or lost skin area (Sun et al., 2014). The pores and skin graft replaces lacking or broken skin with functioning pores and skin that regrows blood vessels and reattaches at the transplant website online. Skin grafts are accomplished to update pores and skin misplaced in burns, trauma, infections, or surgical removal of skin cancers. Anesthesia is used for graft harvest and switch (Sun et al., 2014; Billingham and Medawar, 1951).

Bone-grafting is the transplant of bone tissue. Bone grafts restore lacking bone fragments or provide structural balance (Finkemeier, 2002). There are exceptional bone-graft options: autograft makes use of bone from some other site of the affected person, allograft is from a bone financial institution, and synthetic versions. Grafts deal with fractures, joint irritation, tumors, infections, or boom disorders. The bone fuses because the graft fabric contains it into the body over time. Risks include pain, bleeding, and contamination (Finkemeier, 2002).

Nerve blocks are injections of local anesthetics, steroids, or different medications close to nerves to numb or block an ache from a specific location of the frame (Jeng et al., 2010). They can prevent ache indicators from achieving the mind throughout or after surgical treatment, allowing procedures to be performed underneath sedation without general anesthesia and its aspect effects. Nerve blocks are also used diagnostically to evaluate the source of pain or to deal with continual pain conditions whilst other strategies have failed. Risks include excessive numbness or weak spot, bleeding/bruising, nerve harm, or infection at the injection website online (Ilfeld, 2011).

Tracheostomy is the surgical advent of an opening through the front of the neck into the trachea (windpipe) to vicinity a tube to facilitate respiratory and airway protection. It may be carried out while mechanical ventilation is needed, however impaired in conditions like respiration failure, disturbing higher airway swelling, facial/mouth injuries, or prolonged anesthesia use. The tracheostomy tube permits removal of secretions through suctioning as well as unobstructed management of oxygen and inhaled medicinal drugs at once into the lungs. Tubes can be brief-time periods or permanent. Risks encompass contamination, bleeding, and displacement (Cao et al., 2013).

The preference of surgical method depends on different factors, just like the unique circumstance being handled, the patient's health, and the doctor's know-how. Each conventional method has its benefits and barriers, and surgeons carefully weigh those elements to decide the most appropriate technique for each affected person.

While conventional techniques continue to be enormous, improvements in minimally invasive techniques and robot-assisted surgical operations are offering much less invasive and potentially quicker healing alternatives for positive procedures. However, conventional surgical operation nevertheless plays a crucial function in healthcare and continues to evolve with progressed instruments, strategies, and anesthesia regimens.

3 Robotic surgery

Robotic surgery has revolutionized present-day surgical tactics, providing a collection of benefits over conventional open and laparoscopic methods. Imagine surgeon consoles resembling video game controllers, controlling excessive-precision robot fingers geared up with miniature devices. These palms maneuver within the body via tiny incisions, presenting magnified three-D views of the surgical subject. The healthcare provider's hand movements are translated, scaled, and filtered via the robot machine, making sure tremor-unfastened and extremely-particular manipulation of units.

3.1 The benefits of robotic surgery

Minimally invasive surgical treatment pursuits to restrict incision number and length, tissue dissection, and trauma with the aid of making use of small incisions, specialized contraptions like laparoscopes or endoscopes collectively with video technology, and other advances in place of one big incision and open exposure. Benefits as opposed to conventional open surgical operation doubtlessly consist of drastically reduced blood loss, postoperative ache, danger of contamination, health center stays, healing instances, and effect on immunity. Limitations center around technical issues, loss of direct publicity/ tactile remarks, longer running instances, and elevated charges. As techniques develop, more complex approaches are becoming feasible using minimally invasive procedures.

Enhanced precision describes the general capability to perform surgical strategies with greater accuracy, control, consistency, and repeatability, leading to advanced patient results. This is enabled by means of advances such as robotics, PC steerage, high-resolution imaging, microsurgery contraptions, augmented truth structures, superior simulator training, and greater. Such technologies permit increasingly delicate, complex, and minute strategies far surpassing the natural boundaries of unassisted human arms, eyesight, and stamina. They additionally reduce the possibility of human errors while standardizing procedural ability degrees.

Improved ergonomics means optimizing devices and the general surgical environment to lessen bodily and mental fatigue, pain, repetitive pressure, long-term damage, and errors to improve healthcare professionals overall performance and affected person protection. Ergonomic advances consist of adjustable equipment, chairs assisting proper posture, simplified controls, energy saving tools, minimum twisting motions, reduced repetitive responsibilities, suspended imaging screens, speech reputation dictation of notes, and workflow changes to comprise micro-breaks. These changes' goal is to reduce stresses and workloads that can otherwise compromise complex, prolonged tactics.

New, relatively advanced techniques have enabled extra complex surgical processes no longer previously feasible or distinctly excessive-danger, such as multiorgan transplantation, revolutionary organ-maintaining cancer operations, rather complex neuro procedures, the usage of augmented reality steerage, novel professional implants, and prosthetics, and pioneering bench-to-bedside regenerative treatments, amongst others, on the leading edge of medical advancement. While widening therapeutic alternatives, such complicated techniques also carry a greater risk of bleeding, contamination, pain, tool failure or rejection, prolonged hospitalization, severe disability, or loss of life, requiring careful affected person evaluation and fully informed consent regarding benefits versus substantially extended risks (Ghezzi and Corleta, 2016).

3.2 Applications of robotic surgery

General surgical operation entails operations and care of patients across stomach organs, endocrine devices, skin, smooth tissues, breast, and blood vessels—essentially surgical treatment regarding all frame areas outside the cardiovascular, neurological, orthopedic, and obstetric specialties (Akornor et al., 2005). General surgeons are widely trained to assess and stabilize patients throughout disorder types, perceive an extensive range of acute and persistent situations, interpret diagnostic results, and decide the most appropriate surgical versus nonsurgical options. They perform lifesaving interventions for trauma, injuries, cancer, obstruction, infection, bleeding problems, plus elective approaches for hernias, thyroid, gallbladder, appendix, and so forth (Giulianotti et al., 2003).

Urologists concentrate on clinical and surgical care of the male and female urinary tracts plus the male reproductive machine. This encompasses areas that include kidney feature, kidney stones, bladder management/incontinence, pelvic issues, urinary tract infections, erectile disorder, infertility, sexual fitness, adrenal disorders, urethral troubles, and cancers of the bladder, kidney, or prostate (Akornor et al., 2005). They contain clinical control, office-based totally minimally invasive methods, open and robotic-assisted laparoscopic surgical procedures, and use specialized strategies like ureteroscopy, cystoscopy, and LASER era to diagnose and deal with urological conditions (Hemal and Menon, 2004).

Gynecologists' recognition of women's reproductive health with expertise associated with lady anatomy, hormones, and associated issues. They offer habitual girls' health screenings, diagnose situations ranging from hormonal imbalances to cysts, cancers, and infertility, and provide each clinical and surgical treatment alternative as appropriate. Routine approaches encompass minimally invasive laparoscopic hysterectomy, cyst drainage, or tubal ligation at the same time as complicated regions consist of gynecologic oncology with radical debulking strategies and rising alternatives in robotics (Ahmed et al., 2012).

Cardiothoracic surgery covers operations on organs within the chest, especially the heart and lungs. This includes strategies to treat blockages in heart vessels, restore/update coronary heart valves, deal with arrhythmias, cast off fluid/air around the coronary heart, accurate congenital defects, biopsy or get rid of hundreds in the lung, treat emphysema or esophageal problems, amongst others (Diodato and Damiano, 2003). Traditional strategies require open coronary heart surgical operation and the usage of a heart-lung skip device. Newer methods emphasize minimally invasive/robotic alternatives consisting of VATS (video-assisted thoracoscopic surgical operation through micro-incisions among ribs) and specialized microsurgeries (Mohr et al., 2001).

4 The future of robotic surgery

Continued advancements in the robotic era are anticipated, together with more autonomy and AI integration for superior surgical choice-making. Broader accessibility via price reduction and expanded schooling packages is vital for wider adoption of robot surgery. Necrobotics can be the next inline revolution that changes the landscape and level of surgery. The possibilities are enormous, and the potential benefits most likely shall surpass any costs and legal complications.

4.1 Necrobotics in surgical operation

Necrobotics, while nonetheless in its nascent tiers, offers charming possibilities for the destiny of surgery. The trade of using robotics and bioengineering as one for exploiting the use of biological organisms (or parts of them) as a robot or robotic component feels creepy; however, the applications and possibilities are limitless.

Decellularized tissues ought to serve as scaffolds for bio-printed organs, alleviating the important organ donor scarcity and imparting personalized transplants. Microscopic grippers and sensors crafted from animal exoskeletons ought to allow surgeons to navigate sensitive tissues with unrivaled precision, minimizing risk and recuperation time. Implants and scientific devices created from treated bones or cartilage could seamlessly integrate with the body and biodegrade over the years, casting off the want for destiny surgeries. Deceased tissue samples may be used to expand more correct models of diseases, accelerating research and drug development.

Insights from animal exoskeletons may want to lead to more efficient, biocompatible, and dexterous robots with applications in numerous fields, from healthcare to environmental monitoring. Sensors constructed from insect antennae should offer surprisingly touchy remarks in confined areas, whilst actuators powered through repurposed muscular tissues could offer unique benefits in robotics and prosthetics. Biodegradable implants and gear derived from necrobotics could provide a greater sustainable alternative to standard materials, decreasing environmental effects.

Advancements in organ alternative, tissue engineering, and minimally invasive processes may want to substantially improve the quality of existence for patients suffering from various conditions. The potential for biocompatible implants and personalized medicine ought to cause price and financial savings inside the healthcare system because of fewer complications and shorter sanatorium stays. Necrobotics should sell an extra respectful and sustainable approach to death and encourage conversations about organ donation and autopsy usage of organic materials.

Although the steps and procedure shall remain somewhat same for the majority of the surgeries, using necrobotics in surgical robotics shall provide different actuators. That in itself shall bring multifacet advantages and benefits like:

4.1.1 Precision and dexterity

The use of natural movements possible only with biological tissues can allow more precision and accuracy. That can be harnessed to perform delicate surgeries, as the flexibility and responsiveness of necrobots are much better than the mechanical robotic parts. An additional advantage being the availability of soft biological parts to create soft robots for the cases of surgeries on soft tissues that need to be handled gently rather than by a metal or plastic component, e.g., in the case of brain and nerve surgeries or cardiovascular surgeries where handling and manipulation of delicate tissues like brain tissue, veins, and nerves is required.

4.1.2 Biocompatibility and biodegradability

The possibility of infection and inflammation and the rejection ratio of synthetic material are high. That can reduce significantly in the case of using biological materials, as they are expected to be more compatible with human tissues. The risk of autoimmune reactions that are related to synthetic materials also gets reduced by using a biocompatible material.

In surgeries where parts need to be placed temporarily until healing is complete, biodegradable necrobiotic components can be used that shall get dissolved by themselves after the required time. This shall reduce postoperation and re-surgery risks and costs.

4.1.3 Minimally invasive procedures

Necrobotics tools promise better navigation between small incisions with greater flexibility, which results in reduced trauma as well as promotes better recovery. Repurposed spider legs or insect wings might be converted into dexterous grippers for robot surgical treatment, permitting surgeons to navigate sensitive tissues in the mind, blood vessels, or different difficult-to-reach areas with remarkable precision and minimal invasiveness. Microscopic sensors built from insect exoskeletons ought to offer real-time remarks on tissue residences, temperature, and other important parameters during surgical procedures, leading to greater particularity and controlled techniques.

4.1.4 Personalized organ repair and replacement

Decellularized tissues from deceased donors ought to serve as scaffolds for bioprinting customized organs, addressing the critical scarcity of organ donors and supplying sufferers' custom-made replacements for broken organs. Decellularized matrices might be used to engineer precise tissues or blood vessels needed for reconstruction or transplantation, imparting patients with biocompatible and purposeful replacements. Adaptive devices and replacements like dynamic stents and sutures that expand or contract based on different physiological parameters provide better support and healing. Also, necrobotic can provide scaffolds that support the growth, regeneration, or integration of new tissues into a damaged, weak organ.

4.1.5 Enhanced surgical tools and implants

Exploiting multiple biological parameters of a patient, customized surgical tools can be created that suit different anatomies of individual patients. Spider silk, recognized for its notable energy and elasticity, could be repurposed as sutures for sensitive surgeries, minimizing tissue damage and promoting quicker recuperation. Implants crafted from mainly dealt-with bones or cartilage could seamlessly integrate with the frame and biodegrade through the years, putting off the want for future elimination surgical procedures and ability headaches.

4.1.6 Advanced surgical assistances

Necrobiotic materials could be used to develop biocompatible robot assistants for surgeons, imparting extra help and manipulation skills all through complex methods. Sensors constructed from insect antennae or different biological structures may be included in surgical tools, providing surgeons with superior tactile comments and a higher know-how of tissue properties at some point of operations. Enhanced precision using biological actuators and sensors can also aid in the precision of graft placements, transplantation procedures, and general surgeries. Moreover, necrobotic models can be used in surgical training and simulation that shall help surgeons to advance their skills, providing a dynamic and realistic practice environment.

5 Challenges in surgical necrobotics

Many advancements have been made in robot-based and robot-assisted surgeries; however, ethical issues concerning necrobotics, the function of robots in surgery, and the human-system interface stay essential subjects for dialogue and development (Ghezzi and Corleta, 2016). Obtaining human consent for postmortem examination of their use of natural substances leads to situations disturbing and raises questions regarding affiliation and ownership postclosing. There are concerns about the potential for disposal or commodification of human remains, and sensitive social and religious issues regarding the treatment of the dead have to be addressed. It is important to ensure objective affirmation of advances in necro-biology and to prevent abuse of the capabilities of powerless populations. Clear guidelines and regulations are needed to address the sourcing and usage of biological materials in necrobotics.

Also, safety and efficacy regulations for surgeons, the compliance of necrobotic devices or material to medical device regulations, and extensive clinical trials are necessary. The control system of the necrobots needs to be highly precise and responsive and should have multiple sensors and feedback mechanisms.

Proper disinfection and adaptation of expired human organics is the first step in avoiding the transmission of pathogen-functional diseases or irreversible diseases. A mismatch between the transplanted or combined necrobotics and the recipient's safety system must cause migraines in good condition. Long-term protection and degradation of reused natural substances requires extensive research to ensure the stability and capacity of inserts and different packages.

Necrorobotics may also face public resistance to thinking about the truth about ethical issues and society, which will most likely sabotage one's significant advancement and appropriation. The mental impact of using expired material in inserts or surgical treatments on patients and their families requires careful consideration and recommendation support.

The use of plant-based materials from the dead exposes philosophical and existential questions about the boundary between way of life and passage. Ensuring the supply of low-value natural substances is important to avoid natural damage or asset consumption. In cases where exoskeletal or organism-specific materials are used, ethical issues related to their processing and sourcing must be addressed.

6 Conclusion

A transformative intersection of biology and robotics offers groundbreaking potential to revolutionize surgical techniques and improve patient outcomes. The biological tissues from deceased organisms used as functional components in robotic systems provide a different landscape for surgical efficiency. Necrobotics harnesses the unique properties of such tissues to create advanced surgical tools and methods that are more precise, biocompatible, and adaptable. Natural biological movements and flexibility allow delicate tissue manipulation, critical in surgeries involving soft tissues like the brain, nerves, and cardiovascular system. The use of biodegradable materials also offers significant benefits, such as reducing the risk of postoperative complications and the need for additional surgeries. Necrobotics shall pave the way to minimally invasive procedures, provide biocompatible implants, and adaptive surgical devices, all of which can pave the way to navigate through small incisions with greater ease and flexibility and fewer rejections and infections, thereby reducing patient trauma and promoting faster recovery.

Despite the immense potential, surgical necrobotics faces significant challenges, including ethical concerns and regulatory hurdles. Technically, a lot of issues related to the preservation and functionality of biological tissues and the selection and processing of the same pose a big challenge. Addressing these issues requires rigorous research, clear guidelines, and comprehensive regulatory frameworks that shall ensure a safe and ethical use of necrobotic technologies.

Prospects for necrobotics in surgery are promising. The ongoing research is likely to yield innovative solutions that shall further enhance the precision, safety, and efficacy of surgical procedures. As the field evolves, necrobotics has the potential to become an integral part of modern surgery. This can lead to new possibilities and procedures for patient care and medical advancements. Continued research and development, coupled with ethical considerations and regulatory compliance, shall be important in realizing the full potential of necrobotics in transforming surgical practices.

References

Ahmed, K., Ibrahim, A., Wang, T.T., Khan, N., Challacombe, B., Khan, M.S., Dasgupta, P., 2012. Assessing the cost effectiveness of robotics in urological surgery—A systematic review. BJU Int. 110 (10), 1544–1556. https://doi.org/10.1111/j.1464-410X.2012.11015.x. 1464410X.

Akornor, J.W., Segura, J.W., Nehra, A., 2005. General and cystoscopic procedures. Urol. Clin. N. Am. 32 (3), 319–326. https://doi.org/10.1016/j.ucl.2005.04.002.

Billingham, R.E., Medawar, P.B., 1951. The technique of free skin grafting in mammals. J. Exp. Biol. 28 (3), 385–402. https://doi.org/10.1242/jeb.28.3.385. 0022-0949.

Cao, C., Manganas, C., Ang, S.C., Peeceeyen, S., Yan, T.D., 2013. Video-assisted thoracic surgery versus open thoracotomy for non-small cell lung cancer: a meta-analysis of propensity score-matched patients. Interact. Cardiovasc. Thorac. Surg. 16 (3), 244–249. https://doi.org/10.1093/icvts/ivs472. 15699285.

Diodato, M.D., Damiano, R.J., 2003. Robotic cardiac surgery: overview. Surg. Clin. N. Am. 83 (6), 1351–1367. W.B. Saunders, United States. 00396109. Available from: http://www.elsevier.com/inca/publications/store/6/2/3/1/9/6/index.htt. https://doi.org/10.1016/S0039-6109(03)00166-X.

Finkemeier, C.G., 2002. Bone-grafting and bone-graft substitutes. J. Bone Joint Surg. 84 (3), 454–464. Journal of Bone and Joint Surgery Inc., United States. 00219355. Available from: https://journals.lww.com/jbjsjournal/pages/default.aspx. https://doi.org/10.2106/00004623-200203000-00020.

Ghezzi, T.L., Corleta, O.C., 2016. 30 years of robotic surgery. World J. Surg. 40 (10), 2550–2557. Springer New York LLC, Brazil. 14322323. Available from: http://link.springer.de/link/service/journals/00268/index.htm. https://doi.org/10.1007/s00268-016-3543-9.

Giulianotti, P.C., Coratti, A., Angelini, M., Sbrana, F., Cecconi, S., Balestracci, T., Caravaglios, G., 2003. Robotics in general surgery: personal experience in a large community hospital. Arch. Surg. 138 (7), 777–784. https://doi.org/10.1001/archsurg.138.7.777.

Hasson, H.M., 1999. Open laparoscopy as a method of access in laparoscopic surgery. Gynaecol. Endosc. 8 (6), 353–362. https://doi.org/10.1046/j.1365-2508.1999.00316.x.

Hemal, A.K., Menon, M., 2004. Robotics in urology. Curr. Opin. Urol. 14 (2), 89–93. https://doi.org/10.1097/00042307-200403000-00007.

Ilfeld, B.M., 2011. Continuous peripheral nerve blocks: a review of the published evidence. Anesth. Analg. 113 (4), 904–925. Lippincott Williams and Wilkins, United States. 00032999. Available from: http://journals.lww.com/anesthesia-analgesia/toc/publishahead. https://doi.org/10.1213/ANE.0b013e3182285e01.

Jabs, C.F.I., Drutz, H.P., 2001. The role of intraoperative cystoscopy in prolapse and incontinence surgery. Am. J. Obstet. Gynecol. 185 (6), 1368–1373. https://doi.org/10.1067/mob.2001.119072. 00029378.

Jackson, C., 1928. Bronchoscopy; past, present and future. N. Engl. J. Med. 199 (16), 759–763. https://doi.org/10.1056/NEJM192810181991603. 0028-4793.

Jeng, C.L., Torrillo, T.M., Rosenblatt, M.A., 2010. Complications of peripheral nerve blocks. Br. J. Anaesth. 105. i97, Elsevier Ltd, United States. 14716771. Available from: https://www.journals.elsevier.com/british-journal-of-anaesthesia. https://doi.org/10.1093/bja/aeq273.

Johnson, A., 1997. Laparoscopic surgery. Lancet 349 (9052), 631–635. https://doi.org/10.1016/s0140-6736(96)10032-5. 01406736.

Matulewicz, R.S., DeLancey, J.O., Meeks, J.J., 2017. Cystoscopy. JAMA 317 (11), 1187.

Mohr, F.W., Falk, V., Diegeler, A., Walther, T., Gummert, J.F., Bucerius, J., Jacobs, S., Autschbach, R., 2001. Computer-enhanced "robotic" cardiac surgery: experience in 148 patients. J. Thorac. Cardiovasc. Surg. **121** (5), 842–853. Mosby Inc., Germany. 00225223. Available from: http://www.elsevier.com/inca/publications/store/6/2/3/1/5/1/index.htt. https://doi.org/10.1067/mtc.2001.112625.

Mostany, R., Portera-Cailliau, C., 2008. A craniotomy surgery procedure for chronic brain imaging. J. Vis. Exp. (12). Journal of Visualized Experiments, United States. 1940087X. Available from: http://www.jove.com/details.php?id=680. https://doi.org/10.3791/680.

Nowygrod, R., Egorova, N., Greco, G., Anderson, P., Gelijns, A., Moskowitz, A., McKinsey, J., Morrissey, N., Kent, K.C., 2006. Trends, complications, and mortality in peripheral vascular surgery. J. Vasc. Surg. 43 (2), 205–216. https://doi.org/10.1016/j.jvs.2005.11.002.

Small, N.C., 1988. Complications in arthroscopic surgery performed by experienced arthroscopists. Arthroscopy 4 (3), 215–221. https://doi.org/10.1016/s0749-8063(88)80030-6. 07498063.

Sun, B.K., Siprashvili, Z., Khavari, P.A., 2014. Advances in skin grafting and treatment of cutaneous wounds. Science **346** (6212), 941–945. American Association for the Advancement of Science, United States. 10959203. Available from: http://www.sciencemag.org/content/346/6212/941.full.pdf. https://doi.org/10.1126/science.1253836.

CHAPTER 5

Human-robot interaction in healthcare

Radhakrishanan Niranchana[a], P. DeepanRamkumar[b], and Antony Kovilpillai Judeson[c]
[a]Alliance University, Department of Computer Science & Engineering, Bengaluru, India
[b]Vellore Institute of Technology, Amaravathi, India
[c]Alliance University, CSE, Bengaluru, India

1 Introduction

1.1 Brief overview of bio-robotics in healthcare

The field of healthcare is constantly evolving, with advancements in technology playing a crucial role in improving patient care and outcomes. One such emerging area of research is bio-robotics, which combines the principles of biology and robotics to create innovative solutions for healthcare applications. Bio-robotics involves the use of organic materials, derived from living organisms, as ready-to-use actuators in robotic units. By repurposing inanimate bodies of organisms and leveraging their unique biological mechanisms, bio-robotics aims to develop robotic components that can mimic and enhance natural movements and functions (Yap et al., 2022).

Interaction between humans and robots, known as human-robot interaction, is an expanding field within healthcare, offering considerable promise. This field involves the deployment of robots to aid both patients and healthcare providers across diverse healthcare settings. The effectiveness of these interactions often hinges on the personality traits exhibited by both the patient and the healthcare provider. Research findings indicate that specific personality traits, such as agreeableness, conscientiousness, and extroversion, tend to have a positive influence on patient outcomes and their overall experiences. Conversely, personality traits like neuroticism and openness to experience may lead to less favorable impacts. Similarly, personality traits exhibited by healthcare providers, such as agreeableness and extroversion, tend to have a positive effect on their performance and interactions with patients.

The use of organic materials in bio-robotics offers several advantages, including increased flexibility, adaptability, and efficiency in performing specific tasks. For example, the unique walking mechanism of spiders, which relies on hydraulic pressure for leg extension, can be harnessed to create necrobiotic grippers that naturally reside in a closed state and can be opened by applying pressure. These grippers have the ability

Necrobotics for Healthcare Applications and Management
https://doi.org/10.1016/B978-0-443-24832-0.00005-2

to grasp objects with irregular geometries and can be used as handheld devices in various healthcare settings. Bio-robotics has the potential to revolutionize healthcare by providing innovative solutions for tasks such as object manipulation, surgical procedures, prosthetics, and more. By incorporating organic materials with similar hydraulic mechanisms from other organisms, bio-robotics can further expand its applications and create more advanced robotic components for healthcare purposes. However, it is important to address challenges such as large data requirements, data interpretation, and ethical concerns associated with the use of AI and organic materials in healthcare (Ujager et al., 2023).

1.2 Importance of human and robot interaction in this context

Human-robot interaction (HRI) is crucial in the field of bio-robotics to ensure effective collaboration and communication between humans and robots. This enables robots to understand human intentions and adapt their behavior accordingly, leading to improved performance and safety in necrobotic applications and facilitates intuitive and natural interaction between humans and robots, allowing for seamless cooperation in tasks such as surgical procedures and rehabilitation. Effective HRI bio-robotics can enhance the overall user experience and acceptance of robotic units, leading to increased adoption and integration of robots in various healthcare settings. HRI research in bio-robotics focuses on developing advanced sensing, perception, and control algorithms to enable robots to interpret human gestures, emotions, and intentions accurately. The design of user-friendly interfaces and intuitive control mechanisms is essential in HRI to ensure that humans can easily interact with and control necrobotic units (Yap et al., 2022).

1.2.1 Objectives of the chapter

Elucidate the concept and development of bio-robotics in healthcare

This chapter aims to provide a comprehensive overview of bio-robotics, detailing its evolution, underlying principles, and the integration of organic materials into robotic units. Specifically, the objective is to explore how bio-robotics utilizes biological mechanisms from organisms to develop innovative solutions for various healthcare applications, such as surgical aids, rehabilitation devices, and patient care systems.

Analyze the role and impact of human-robot interaction in bio-robotics

The chapter will investigate the significance of HRI within the bio-robotics domain. It will focus on how effective interaction between humans and robots can improve healthcare outcomes, enhance user experiences, and promote the adoption of robotic units in different healthcare settings. The analysis will include specific examples and case studies to illustrate these interactions in real-world scenarios.

Review historical and contemporary developments in HRI

A detailed examination of existing literature will be conducted to map the historical context and current advancements in HRI in healthcare. This will involve exploring early developments, significant technological innovations, and key milestones that have shaped the field. The review will also highlight the evolution of specific technologies and methodologies that have contributed to the current state of HRI.

Discuss technological foundations enabling HRI in healthcare

The chapter will delve into the key technologies that facilitate effective HRI in healthcare, such as artificial intelligence (AI), the Internet of Things (IoT), and wearable sensor technology. It will explain how these technologies contribute to the development of intuitive and responsive robotic units, providing concrete examples of their applications in healthcare, such as AI-driven diagnostic tools, IoT-enabled patient monitoring systems, and wearable sensors for real-time health data analysis.

Evaluate applications of HRI in healthcare

By presenting various case studies and practical implementations, the chapter aims to demonstrate the diverse applications of HRI in healthcare. These examples will cover areas like surgery, where robots assist in precision tasks; rehabilitation, where robots provide personalized exercise routines; and patient care, where robots support daily activities and monitor health conditions.

Investigate ethical, social, and design considerations in HRI

The chapter will address critical ethical, social, and design considerations in developing and deploying HRI systems in healthcare. This will include discussions on user experience, interaction design, ergonomics, and accessibility, with an emphasis on ensuring that robotic systems are safe, effective, and acceptable to users. Examples of ethical dilemmas and solutions in real-world applications will be provided.

Identify challenges and future directions in bio-robotics and HRI

The objective is to pinpoint current challenges in implementing HRI in healthcare, such as technical limitations, data privacy concerns, and user acceptance issues. The chapter will also explore future directions and emerging trends in bio-robotics, highlighting potential innovations and their implications for healthcare services.

Synthesize findings and reflect on the impact on healthcare

Finally, the chapter aims to synthesize the key findings from the literature review and case studies to provide a comprehensive reflection on the impact of HRI and bio-robotics on healthcare. This section will focus on how these technologies enhance patient care, improve efficiency, and transform the overall healthcare experience. Specific metrics or case study outcomes will be used to illustrate these impacts.

1.2.2 Background and literature review

Historical context: Early development of HRI in healthcare

The early development of HRI in healthcare began with basic mechanized tools and has since evolved into sophisticated systems due to significant advancements in AI, machine learning, and material science (Halder Roy et al., 2023). Initially, HRI in healthcare focused on improving surgical precision. Over time, the scope expanded to include rehabilitation and patient care, significantly enhancing healthcare efficiency. A comprehensive survey of healthcare robots categorizes them into various types and explores their applications, challenges, and potential for improving healthcare services (Kyrarini et al., 2021).

Adapting robot designs for specific needs

Research highlights the importance of adapting robot designs to meet the specific needs and preferences of different populations, such as the elderly (Broadbent, 2017; Broadbent et al., 2023). For instance, socially assistive robots (SARs) are crucial for assisting individuals with special needs. These robots must adapt their behavior to the user's specific conditions to be effective (Ahmad et al., 2017). The evolution of social robots, their application areas, and the importance of user-centric design in healthcare robotics are also emphasized (Lambert et al., 2020).

Integration and adaptability in healthcare settings

Further studies discuss the integration of robots in healthcare settings, focusing on adaptability and usercentered design (Ruhland et al., 2015). Advanced strategies in physical HRI are essential for intuitive interactions, especially in rehabilitation therapy (Losey et al., 2018). The significance of communication behaviors, both verbal and nonverbal, in healthcare HRI is underlined, suggesting areas for future research (Johanson et al., 2021). Finally, the role of robot personalities in enhancing interactions, particularly in healthcare, is discussed, emphasizing the alignment of robotic personalities with user expectations (Mou et al., 2020).

These comprehensive reviews collectively underscore the ongoing development and potential of HRI in healthcare, highlighting the positive attitudes and willingness to interact with social robots in healthcare contexts (Hari and Kujala, 2009; Naneva et al., 2020).

1.3 Bio-robotics evolution: Literature on the evolution of bio-robotics in healthcare settings

Researchers have introduced a groundbreaking concept in the field of robotics, particularly in healthcare, by leveraging organic materials, specifically deceased spiders, as components for robotic actuators. This pioneering approach in bio-robotics utilizes the inherent hydraulic leg movement mechanisms of spiders to create effective and efficient gripping devices. It represents a significant step in biohybrid robotics, showcasing the potential of utilizing natural mechanisms and materials in designing robotic units, especially for delicate tasks in healthcare and other fields. This innovation could lead to more sustainable and effective solutions in robotic design and application (Yap et al., 2022).

Furthermore, the application of biological principles, particularly inspired by jellyfish, to ocean exploration technology has led to significant advancements. Researchers have explored how the characteristics of jellyfish, such as their energy efficiency and adaptability to various ocean environments, have been incorporated into the design of underwater exploration technologies. This includes the development of biohybrid systems combining live jellyfish with microelectronics to enhance their swimming capabilities. These innovations represent a substantial step in oceanographic research, providing new, efficient, and sustainable ways to explore and understand ocean environments (Xu and Dabiri, 2022).

The concept of bio-robotics, which involves using organic materials or deceased organisms as robotic components, has emerged as an innovative approach in robotics. Researchers have demonstrated the feasibility of repurposing dead spiders as robotic grippers, leveraging their natural hydraulic systems. This approach offers advantages such as natural morphology, compliance, adaptability, and environmental sustainability, making bio-robotics a versatile and sustainable approach in robotics (Ray, 2023).

In addition to bio-robotics, biomimetic soft robotics has gained attention. This field focuses on creating soft robots inspired by biological entities, exploring various fabrication methods, actuation mechanisms, and application fields. The aim is to mimic natural organisms in form and function, emphasizing adaptability, efficiency, and versatility in robotic units. These biomimetic robots find applications in healthcare and other industries, showcasing their potential impact (Dong et al., 2022).

Overall, the literature highlights the transformative potential of bio-robotics, biomimetic soft robotics, and bio-inspired technologies in healthcare and various fields, emphasizing sustainability, efficiency, and adaptability as key factors in these innovations (Naraharishetty, 2022; Rodriguez-Guerra et al., 2021; Stahl, 2022; Vincent, 2022; Williams et al., 2022).

Recent research (Table 5.1) continues to expand on these concepts, exploring new materials and methods to further integrate biological principles into robotic design.

1.3.1 Recent research works in bio-robotics

To provide a clearer picture of the recent advancements in bio-robotics, here are some notable research studies.

These studies highlight the innovative use of organic materials and biological principles in developing robotic systems. From necrobotic grippers for precision tasks to biohybrid jellyfish for ocean exploration, the research underscores the potential for these technologies to revolutionize healthcare and other industries through sustainability, efficiency, and adaptability.

Table 5.1 Recent research works in bio-robotics.

Research study	Authors	Publication year	Key findings	Applications
Necrobotic grippers	Yap et al. (2022)	2022	Utilized hydraulic leg mechanisms of deceased spiders to create efficient grippers	Precision surgery, handling small objects
Biohybrid jellyfish for ocean exploration	Xu and Dabiri (2022)	2022	Combined live jellyfish with microelectronics to enhance swimming capabilities	Oceanographic research, underwater exploration
Soft robotics inspired by octopus	Dong et al. (2022)	2022	Developed soft robotic arms with enhanced flexibility and control inspired by octopus tentacles	Minimally invasive surgery, delicate manipulation
Bio-robotics using organic materials	Yang et al. (2023)	2023	Emphasized the use of natural morphology and hydraulic systems of deceased organisms	Versatile robotic components, environmental sustainability
Biomimetic soft robotics	Naraharishetty (2022)	2022	Explored various fabrication methods and actuation mechanisms inspired by biological entities	Healthcare applications, industrial use

No permission is required.

2 Technological foundations

2.1 Technology overview: Literature covering the key technologies enabling HRI in healthcare

Key technologies enabling human and robot interaction in healthcare include AI, the IoT, and wearable sensor technology. AI has the potential to improve the quality of healthcare services by enhancing preliminary detection and diagnosis, as well as assisting in drug discovery and disease prediction (Kitsios et al., 2023). The IoT plays a crucial role in the development of smart healthcare systems, enabling personalized and proactive care using intelligent devices that monitor human physical parameters. Additionally, wearable sensor technology and emotion detection software contribute to the development of emotion intelligence systems, allowing for more natural interactions between humans and robots in healthcare settings. These technologies offer opportunities for improved decision-making, cost reduction, and round-the-clock services in healthcare (Ujager et al., 2023).

AI technologies can identify emotions using human expressions and interactions. Wearable tech and IoT advancements aid emotion detection in healthcare. Emotional intelligence software development involves balancing cost and functionality. Extensive datasets and reliable models are crucial for developing EI software. The study analyzes EI systems' impact on human-computer interaction in healthcare (Vanmathi et al., 2023). Smart healthcare utilizes AI technologies for intelligent healthcare systems. Natural language processing (NLP) is essential for analyzing human language in smart healthcare. Zhou et al. (2024) focus on natural language processing (NLP) for smart healthcare. It reviews existing studies on NLP techniques and applications in smart healthcare. The authors review existing studies that concern NLP for smart healthcare from the perspectives of technique and application, discuss the limitations of current works, and identify the directions for future works.

The impact of AI in healthcare is extensive and multifaceted, encompassing AI-driven data acquisition, image reconstruction, and risk prediction for various diseases (Kitsios et al., 2023). A key focus is on elderly care, where integrated intelligent systems offer personalized services like nutrition recommendations, health monitoring, and education, leveraging wearable IoT for real-time data analysis (C. et al., 2022). These systems employ hierarchical data fusion architecture in smart healthcare, enhancing data accuracy and timeliness. Moreover, AI's role in healthcare extends to monitoring physiological data, medication management, and daily activity tracking, with a focus on responsible AI principles like autonomy and explainability. Additionally, AI technologies facilitate the detection of abnormalities in medical images, risk stratification, and operational decision-making in healthcare. NLP emerges as a crucial tool in this landscape, applied across various healthcare domains including clinical practice and public health, particularly relevant in addressing challenges like the COVID-19 pandemic and mental health

(Zhou et al., 2024). In the evolving landscape of human and robot interaction in healthcare, a multitude of technologies play a pivotal role. Dahl and Boulos highlight the integration of robots into healthcare settings, emphasizing the need for advanced sensing technologies like cameras, depth sensors, LiDAR, and biosensors (Dahl and Boulos, 2014). These technologies enable robots to perceive their environment accurately and gather vital health data, contributing to personalized care delivery. In parallel, Simões et al. (2022) discuss the emergence of robotic aides, telepresence robots, and rehabilitation robots in healthcare, revolutionizing patient care and clinical workflows through advanced sensing and AI capabilities.

Children's perspectives on HRI are explored by Vallès-Peris et al. (2018), emphasizing the design of social robots tailored for pediatric patients. These robots aim to reduce pain and anxiety, underscoring the importance of adaptive behavior and responsiveness to emotional cues. The development of Lio, a personal robot assistant detailed by Miseikis et al. (2020), further exemplifies the advancements in HRI technology. Lio's design includes a variety of sensors and safety mechanisms, highlighting its adaptability to healthcare needs, including during the COVID-19 pandemic.

The role of personality in H-HRI is critically examined by Esterwood and Robert (Naraharishetty, 2022), identifying the need for more comprehensive studies on the interaction between human and robot personalities. This aspect is crucial for enhancing interaction quality and healthcare outcomes. Focusing on elderly care. The importance of robot appearance, interaction with wearable technologies, and protocols for emergencies, enhancing robots' acceptance and utility in elderly healthcare (Park et al., 2021).

Shamsuddin et al. (2018) contribute to this field by developing a simple mobile robot for patient assistance in healthcare environments. This robot features a user-friendly interface and wireless communication capabilities, demonstrating its potential to enhance patient care. Lastly, De Santis et al. (2008) provide an in-depth analysis of physical human-robot interaction, emphasizing safety, dependability, and control strategies essential for effective and safe interactions in healthcare settings.

Collectively, these studies (Shamsuddin et al., 2018) illustrate the multifaceted development of HRI in healthcare, ranging from technical advancements and personalized care approaches to considerations of safety, ethics, and user experience. This comprehensive approach is vital for the successful integration of robotics into healthcare systems, ultimately enhancing patient care and healthcare efficiency.

2.2 Bio-robotics technology: Specific literature on necrobotic technologies for HRI

In the field of bio-robotics and human and robot interaction, recent advancements focus on developing advanced sensing, perception, and control algorithms to enhance robot performance in healthcare applications like surgery and rehabilitation. Central to this development is the design of user-friendly interfaces and intuitive controls, aiming to improve user experience and safety in these critical settings (Yap et al., 2022).

Significant strides in bio-robotics involve using organic materials, particularly dead organisms, as components in robotic units. This approach leverages the naturally evolved features of organic materials, such as dead spiders, for tasks like locomotion and manipulation. Researchers have effectively controlled spider cadavers using pressurized air for object manipulation, offering advantages in sustainability and robust design due to the inherent properties of the organic materials (Ackerman, 2022).

At the Preston Innovation Lab at Rice University, research has advanced this concept by transforming dead spiders into functional robotic grippers. This innovative use of pneumatic actuation in the spider's legs allows for gripping actions and weight lifting, presenting a more sustainable alternative to traditional actuators and grippers (Stahl, 2022; Vincent, 2022; Williams et al., 2022).

The concept of "bio-robotics," which employs organic materials as ready-to-use actuators, is gaining traction. This innovative approach uses a dead spider's body as an actuator, repurposing its natural hydraulic leg mechanism for robotic applications, showcasing a unique fusion of biohybrid systems and robotics (Yap et al., 2022).

Additionally, the exploration of bioinspired technologies in ocean exploration emphasizes the efficiency and sustainability of designs based on biological principles (Xu and Dabiri, 2022). In contrast, the in-depth exploration of biomimetic soft robotics covers various fabrication techniques and actuation strategies inspired by biological organisms (Dong et al., 2022).

Furthermore, the use of dead biological organisms in robotics, with a focus on a dead spider as a robotic gripper, provides insights into the fabrication, potential applications, and ethical considerations of bio-robotics. This research highlights the efficiency, ease of fabrication, and potential uses in precision tasks, addressing durability and ethical issues in the development of biohybrid robotic units.

Collectively, these studies (Yap et al., 2022; Farjana, 2023; Dong et al., 2022; *The Rise of Necrobotics and the Evolution of Biohybrid Gripping Systems*, 2023; Stahl, 2022; Williams et al., 2022; Vincent, 2022; Dahl and Boulos, 2014; Simões et al., 2022; Vallès-Peris et al., 2018; Miseikis et al., 2020) showcase diverse and groundbreaking approaches in bio-robotics and HRI. They range from advanced algorithms for improved human-robot interaction to the unique application of organic materials, opening new avenues for sustainable and efficient robotic units.

3 Case studies literature review: Analysis of documented cases where HRI has been implemented in healthcare

Human and robot interaction in healthcare has witnessed significant advancements and applications in recent years. This literature survey presents key insights from various studies and case analyses in the field of HRI in healthcare. The following paragraphs highlight noteworthy research contributions and their implications for the evolving landscape of healthcare robotics.

HRI in healthcare extends to diverse applications, including interactive services provided by healthcare service robots. In one case study, a healthcare service robot named "Cafero" was designed and implemented with a multidisciplinary approach. This approach emphasized user-friendly and responsive interactions, involving iterative design processes and the integration of interactive and social cues (Kuo et al., 2011). Personality plays a crucial role in HRI in healthcare. A systematic review of human and robot personalities in healthcare settings revealed that personality traits impact interaction quality and overall user experience significantly. The review emphasized the importance of aligning human and robot personalities for improved healthcare outcomes (Esterwood and Robert, 2021). Understanding the perspectives of various user groups is essential in designing effective healthcare robots. In the context of child-robot interactions in healthcare, a study explored children's imaginaries of robots. The findings underscored the significance of considering children's views on robot design to create healthcare robots that resonate with their imaginaries and needs (Vallès-Peris et al., 2018).

Compatibility between human and robot personalities is vital for successful HRI in healthcare. Another literature review highlighted the importance of personality in healthcare HRI, emphasizing the need for personality matching between humans and robots to optimize interaction quality (Kyrarini et al., 2021). HRI in healthcare extends beyond hospital settings to private homes. A case study investigated the implementation of a robotic health assistant, "Max," in private homes for elder care. The study provided valuable insights into the practical challenges and effectiveness of HRI in domestic healthcare settings (Esterwood and Robert, 2020). Physical human-robot interaction (pHRI) is a critical aspect of healthcare robotics. An analysis of pHRI in healthcare case studies demonstrated the mechanics, control, safety, and dependability of these interactions. This research highlighted technological advancements and challenges in implementing pHRI systems in healthcare (De Santis et al., 2008).

The preferences and expectations of older adults in healthcare HRI were explored in a qualitative research approach. The study provided insights into designing social robots that better meet the needs and preferences of elderly users in healthcare contexts. Long-term HRI dynamics are crucial in healthcare, particularly in home-based settings. A study focused on understanding the effects of ongoing interactions with robots in domestic environments. The research shed light on changes in human behavior and attitudes toward robots, contributing to a deeper understanding of HRI in home-based healthcare and assistance (Kidd and Breazeal, 2008).

The integration of robots into healthcare settings continues to evolve, with numerous documented cases demonstrating their effectiveness and potential. A comprehensive review of documented cases showcased the multifaceted nature of HRI in healthcare.

Robots have been applied in various healthcare scenarios, from assisting in surgeries to providing emotional support to patients. This research highlights the transformative potential of HRI in healthcare, enhancing patient care, medical procedures, and communication between healthcare providers and patients (Benos et al., 2020). In summary, the field of human and robot interaction in healthcare is witnessing significant developments, driven by multidisciplinary approaches, personality considerations, and practical implementations in diverse healthcare settings. These studies underscore the importance of designing user-friendly and responsive interactions, aligning human and robot personalities, and considering the unique perspectives of various user groups. HRI in healthcare extends from hospital environments to private homes, with a focus on safety, dependability, and the evolving dynamics of long-term interactions. The documented cases reviewed in this literature survey illustrate the diverse applications and potential transformative impact of HRI in healthcare.

4 Challenges and considerations

4.1 Technical challenges: Literature on the technical difficulties in implementing HRI

In the literature survey on technical challenges in implementing human and robot interaction, various aspects are highlighted across different papers. De Santis et al. (2008) discuss multifaceted challenges in HRI, including safety, mechanics, control, and dependability, emphasizing the need for safe physical interactions and dependable design. Papanastasiou et al. (2019) focus on challenges in industrial HRI, such as safety in shared workspaces and the integration of multimodal interaction. Porfirio et al. (2019) address the complexities in bodystorming HRI, including rapid prototyping and program synthesis. Buerger and Hogan (2007) propose novel control strategies for improved interaction performance and stability. Weiss and Spiel (2022) delve into integrating robots into human environments, highlighting participatory design for responsive robot behavior.

Murphy (2004) discusses HRI in rescue robotics, emphasizing navigation and operation in unpredictable environments. Yoon and Lee (2019) explore AI and robotics in healthcare, focusing on user-friendly interfaces and technology-based service encounters. Lastly, Sim and Loo (2015) review methodologies for assessing social robots, emphasizing user acceptance and behavioral adaptation. Each study contributes unique insights into the evolving field of HRI, underscoring the importance of addressing these technical challenges for successful human-robot collaborations. Alaiad and Zhou (2014) identify factors influencing the adoption of home healthcare robots, such as performance expectations and legal issues, to ensure successful integration into healthcare services.

4.2 Ethical and social considerations: Review of ethical and social implications discussed in the literature

In the context of health and social care, the ethical and social implications of robot integration have been a focal point of recent research. Papadopoulos et al. (2020) scrutinize the use of socially assistive humanoid robots in healthcare, highlighting privacy, safety, and emotional impacts on patients and healthcare workers. They also discuss the integration challenges of these robots into healthcare systems, stressing the need for culturally sensitive and personalized care.

Stahl and Coeckelbergh (2016) explore ethical and social concerns in healthcare robotics, such as the replacement of human caregivers with robots and the dehumanization risk. They advocate for responsible research and innovation, emphasizing stakeholder engagement and ethical reflexivity in the development of healthcare robotics.

Broadbent's work (2017) examines human-robot interactions, revealing insights into human behaviors and societal norms. The paper discusses the potential for emotional attachment to robots, ethical considerations in care settings, and the social impact of robots, highlighting the psychological aspects of these interactions.

Vanmathi et al. (2023) focus on the ethical implications of using care robots in elderly care, discussing autonomy, dignity, deception, and social isolation. They emphasize the need for a balanced and ethical integration of robotics in aged care, considering various ethical frameworks.

Lastly, Dahl and Boulos (2014) investigate the implications of using robots in healthcare settings. They examine the potential benefits and risks, such as patient privacy, user safety, and the dehumanization of care, underscoring the importance of integrating robots in a way that respects user dignity and autonomy (Dahl and Boulos, 2014).

4.3 User experience and interaction design

4.3.1 Design principles review: Literature on user-centered design in HRI

In the field of HRI, various studies have explored the integration of user-centered design (UCD) principles. One such study emphasizes the evolution of human-robot collaboration in industrial settings, highlighting the need for natural, adaptive, and intuitive interfaces, with safety and operator-centric programming as key factors. Another research introduces a structured method to include user experience (UX) in HRI, focusing on communication exchange between humans and robots, involving a multidisciplinary team for interface development (Prati et al., 2021).

Further, a study discusses the design of dynamic-autonomy remote interaction for service robots assisting elderly people, advocating for a user-centric and iterative methodology, including remote human operators for tasks beyond autonomous capabilities (Mast et al., 2012). Additionally, a paper on engaging teens in the codesign of a social robot presents a participatory design methodology, stressing the importance of real-world contextual validity and leveraging teens' social nature (Björling and Rose, 2019).

In healthcare service robots, a multidisciplinary approach is advocated, using UML for detailed robot behavior modeling and an iterative design process for identifying HRI requirements (Kuo et al., 2011). For designing HRC workspaces in industrial settings, a systematic review categorizes guidelines based on complexity levels, focusing on the cognitive and social processes of human operators (Simões et al., 2022).

Moreover, a detailed exploration of UCD principles in manufacturing HRI emphasizes a UX-oriented methodology for analyzing human-robot dialogue (Prati et al., 2021). Lastly, a study on supervisory control of multiple robots discusses managing human performance issues, including operator multitasking and situation awareness, emphasizing user-interface design for effective collaboration (Chen et al., 2011).

4.3.2 Ergonomics and accessibility: Studies and articles focusing on these aspects in HRI design

The integration of ergonomics and accessibility in HRI design is a crucial aspect, especially in industrial and agricultural settings. Studies such as Simões et al. (2022) and Prati et al. (2021) emphasize the importance of designing HRI systems that are not only efficient but also prioritize the safety, comfort, and physical and cognitive well-being of human workers. These papers introduce methodologies and considerations for ergonomic design, focusing on factors like minimizing physical strain, ease of use, and creating intuitive interfaces for effective collaboration. In the agricultural context, Benos et al. (2020) and their subsequent study (Khan and Germak, 2018) highlight the significance of designing HRI systems that address ergonomic risk factors such as exposure to vibration, awkward postures, and repetitive movements. These factors are essential in ensuring the health and safety of workers interacting with robots in dynamic and sometimes unpredictable agricultural environments.

Furthermore, Michalos et al. (2015) discuss the critical elements of ergonomics and safety in collaborative industrial HRI settings. Their work focuses on adaptable and flexible design strategies to ensure the safety and comfort of human workers while maintaining productivity. Similarly, Lunghi et al. (2019) concentrate on designing interfaces for HRI in hazardous environments, emphasizing user-friendly, intuitive, and adaptable interfaces suitable for various complex tasks and prioritizing operator safety and comfort. Additionally, the social dimensions of HRI in industrial settings, explored by Moniz and Krings (2016), bring attention to the impact of robotics on work organization, necessitating a reevaluation of job roles, productivity, and the nature of work. The study by Onnasch and Roesler (2021) presents a comprehensive taxonomy for evaluating and categorizing various aspects of HRI, contributing to a more structured understanding of these interactions. Overall, these studies underscore the need for a balanced approach in HRI design, considering both human ergonomic needs and the efficiency of robotic units, thereby enhancing the overall effectiveness and safety in various environments.

5 Future directions

5.1 Emerging trends: Literature on emerging trends and predictions for the future of HRI in bio-robotics

The future of human and robot interaction in bio-robotics promises transformative changes, particularly in healthcare. This evolution is propelled by several key trends:

Utilization of Biotic Materials: Bio-robotics is advancing through the use of organic materials like dead spiders, fostering the development of more sustainable, efficient, and adaptable robotic units. This approach leverages natural properties for enhanced functionality in tasks like locomotion and manipulation (Ray, 2023; Stahl, 2022; Vincent, 2022; Williams et al., 2022; Yap et al., 2022).

Enhanced sensing and perception: Improvements in sensing, perception, and control algorithms are vital for more intuitive and effective HRI, crucial in healthcare applications such as surgery and rehabilitation (Esterwood and Robert, 2021; Kuo et al., 2011; Yap et al., 2022).

Biohybrid robotic systems: Incorporating biological principles into robotic design, exemplified by the use of jellyfish traits in oceanic technologies, is paving the way for efficient, sustainable biohybrid systems with diverse healthcare applications (Xu and Dabiri, 2022).

Biomimetic soft robotics: Emulating biological entities, biomimetic soft robotics focuses on adaptable and versatile robots, significant for healthcare uses ranging from prosthetics to surgical aids (Dong et al., 2022).

Ethical and sustainable design: As bio-robotics progresses, the ethical and sustainable use of organic materials is increasingly emphasized, balancing innovation with ethical responsibilities (Naraharishetty, 2022; Rodriguez, 2023).

Healthcare integration: Necrobotic technologies are being adapted for varied healthcare scenarios, enhancing the efficiency and effectiveness of healthcare delivery through their unique properties (Ray, 2023; Yap et al., 2022).

User-centric design and accessibility: A shift toward user-centric design ensures that necrobotic units are intuitive and accessible to a broad range of users, including healthcare professionals and patients (Esterwood and Robert, 2021; Kuo et al., 2011; Mast et al., 2012; Prati et al., 2021).

Collaborative and socially responsive robots: The development of socially responsive necrobotic units aims to enhance the social dimensions of HRI, facilitating better cooperation in healthcare settings (Moniz and Krings, 2016; Onnasch and Roesler, 2021).

These trends indicate a dynamic and evolving field where the integration of biology, robotics, and human-centered design is creating innovative, efficient, and ethically responsible solutions, poised to redefine future healthcare challenges.

6 Conclusion

6.1 Summary of key literature findings

The exploration of human and robot interaction in healthcare, particularly within the emerging field of bio-robotics, has revealed substantial advancements and innovative approaches. Bio-robotics, integrating biology and robotics, utilizes organic materials like dead spiders for creating robotic actuators, offering increased flexibility, adaptability, and efficiency (Yap et al., 2022). This development, combining the unique biological mechanisms of organisms with robotics, has opened new avenues for patient care, surgical procedures, and prosthetics.

The literature indicates a significant role of personality traits in HRI effectiveness, impacting patient outcomes and healthcare provider performance (Yap et al., 2022). Advancements in AI, machine learning, and material science have driven the evolution from basic mechanized tools to sophisticated HRI systems, enhancing healthcare efficiency (Halder Roy et al., 2023). The focus on social robotics and user-centric design underscores the necessity of adapting robot designs to specific user needs (Lambert et al., 2020).

Moreover, the integration of biological principles in robotic design, as exemplified by biohybrid systems using jellyfish traits, showcases the potential for sustainable and efficient designs (Xu and Dabiri, 2022). The field of biomimetic soft robotics further emphasizes adaptability and versatility, with applications extending across various healthcare scenarios (Dong et al., 2022).

6.2 Reflections on the impact of this body of work on HRI in healthcare

The comprehensive body of work on HRI in healthcare, particularly within the domain of bio-robotics, signifies a transformative phase in healthcare robotics. The incorporation of organic materials, the development of sophisticated sensing and perception algorithms, and the focus on ethical, sustainable design have laid the foundation for more intuitive, efficient, and patient-centric healthcare solutions. These advancements are not only enhancing the capabilities and functionalities of healthcare robots but are also ensuring their ethical and responsible application in sensitive healthcare environments.

The integration of biohybrid systems and biomimetic approaches in robotics reflects a shift toward more natural, efficient, and adaptable healthcare interventions. The emphasis on user-centric design and accessibility in robotic units ensures broader applicability and acceptance, particularly among diverse patient demographics and healthcare professionals. Additionally, the advancement in collaborative and socially responsive robots is set to enhance the dynamics of human-robot interactions, facilitating better cooperation and empathy in healthcare settings.

In conclusion, the advancements in HRI, driven by bio-robotics and related technologies, mark a significant step toward realizing more effective, adaptable, and humane healthcare interventions. This evolution in HRI not only demonstrates technological innovation but also reflects a deeper understanding of the complex interplay between humans and robots in healthcare contexts. The continued exploration and development in this field are poised to redefine healthcare delivery, making it more efficient, accessible, and tailored to individual patient needs.

References

Ackerman, 2022. Necrobotics: dead spiders reincarnated as robot grippers. IEEE Spectrum.

Ahmad, M.I., Mubin, O., Orlando, J., 2017. Adaptive social robot for sustaining social engagement during long-term children–robot interaction. Int. J. Hum. Comput. Interact. 33 (12), 943–962.

Alaiad, A., Zhou, L., 2014. The determinants of home healthcare robots adoption: an empirical investigation. Int. J. Med. Inform. 83 (11), 825–840. https://doi.org/10.1016/j.ijmedinf.2014.07.003. 13865056.

Benos, L., Bechar, A., Bochtis, D., 2020. Safety and ergonomics in human-robot interactive agricultural operations. Biosyst. Eng. 200, 55–72.

Björling, E.A., Rose, E., 2019. Participatory research principles in human-centered design: engaging teens in the co-design of a social robot. Multimodal Technol. Interact. 3 (1). MDPI AG, United States. 24144088. Available from: https://www.mdpi.com/2414-4088/3/1/8/pdf. https://doi.org/10.3390/mti3010008.

Broadbent, E., 2017. Interactions with robots: the truths we reveal about ourselves. Annu. Rev. Psychol. 68 (1), 627–652. https://doi.org/10.1146/annurev-psych-010416-043958. 0066-4308.

Broadbent, E., Billinghurst, M., Boardman, S.G., Doraiswamy, P.M., 2023. Enhancing social connectedness with companion robots using AI. Sci. Robot. 8 (80), eadi6347.

Buerger, S.P., Hogan, N., 2007. Complementary stability and loop shaping for improved human-robot interaction. IEEE Trans. Robot. 23 (2), 232–244. https://doi.org/10.1109/TRO.2007.892229. 15523098.

Chen, Y., Ding, X.C., Stefanescu, A., Calin, B., 2011. Formal approach to the deployment of distributed robotic teams. IEEE Trans. Robot. 28 (1), 158–171.

Dahl, T., Boulos, M., 2014. Robots in health and social care: a complementary technology to home care and telehealthcare? Robotics 3 (1), 1–21. https://doi.org/10.3390/robotics3010001. 2218-6581.

De Santis, A., Siciliano, B., De Luca, A., Bicchi, A., 2008. An atlas of physical human–robot interaction. Mech. Mach. Theory **43** (3), 253–270. https://doi.org/10.1016/j.mechmachtheory.2007.03.003. 0094114X.

Dong, X., Luo, X., Zhao, H., Qiao, C., Li, J., Yi, J., Yang, L., Oropeza, F.J., Hu, T.S., Xu, Q., Zeng, H., 2022. Recent advances in biomimetic soft robotics: fabrication approaches, driven strategies and applications. Soft Matter 18 (40), 7699–7734. Royal Society of Chemistry, China. 17446848. Available from: http://pubs.rsc.org/en/journals/journal/sm. https://doi.org/10.1039/d2sm01067d.

Esterwood, C., Robert, L.P., 2020. Personality in healthcare human robot interaction (H-HRI): a literature review and brief critique. In: HAI 2020—Proceedings of the 8th International Conference on Human-Agent Interaction. Association for Computing Machinery, Inc, United States, pp. 87–95. http://dl.acm.org/citation.cfm?id=3406499. https://doi.org/10.1145/3406499.3415075. 9781450380546.

Esterwood, C., Robert, L.P., 2021. A systematic review of human and robot personality in health care human-robot interaction. Front. Robot. AI 8. https://doi.org/10.3389/frobt.2021.748246. Frontiers Media S.A., United States. 22969144. Available from: journal.frontiersin.org/journal/robotics-and-ai.

Farjana, S.M., 2023. Necrobotics—The Future of Robotics.

Halder Roy, A., Ghosh, S., Gupta, B., 2023. Robotics in medical domain: the future of surgery, healthcare and imaging. Wireless Pers. Commun. 132 (4), 2885–2903.

Hari, R., Kujala, M.V., 2009. Brain basis of human social interaction: from concepts to brain imaging. Physiol. Rev. 89 (2), 453–479. 15221210. Finland. Available from: http://physrev.physiology.org/cgi/reprint/89/2/453. https://doi.org/10.1152/physrev.00041.2007.

Johanson, D.L., Ahn, H.S., Broadbent, E., 2021. Improving interactions with healthcare robots: a review of communication behaviours in social and healthcare contexts. Int. J. Soc. Robot. 13 (8), 1835–1850.

Khan, S., Germak, C., 2018. Reframing HRI design opportunities for social robots: lessons learnt from a service robotics case study approach using UX for HRI. Future Internet 10 (10), 101. https://doi.org/10.3390/fi10100101. 1999-5903.

Kidd, C.D., Breazeal, C., 2008. Robots at home: understanding long-term human-robot interaction. In: IEEE/RSJ International Conference on Intelligent Robots and Systems, IROS United States, pp. 3230–3235, https://doi.org/10.1109/IROS.2008.4651113.

Kitsios, F., Kamariotou, M., Syngelakis, A.I., Talias, M.A., 2023. Recent advances of artificial intelligence in healthcare: a systematic literature review. Appl. Sci. (Switzerland) 13 (13). Multidisciplinary Digital Publishing Institute (MDPI), Greece. 20763417. Available from: www.mdpi.com/journal/applsci/. https://doi.org/10.3390/app13137479.

Kuo, I.-H., Jayawardena, C., Broadbent, E., MacDonald, B.A., 2011. Multidisciplinary design approach for implementation of interactive services. Int. J. Soc. Robot. 3 (4), 443–456. https://doi.org/10.1007/s12369-011-0115-x. 1875-4791.

Kyrarini, M., Lygerakis, F., Rajavenkatanarayanan, A., Sevastopoulos, C., Nambiappan, H.R., Chaitanya, K.K., Babu, A.R., Mathew, J., Makedon, F., 2021. A survey of robots in healthcare. Technologies 9 (1), 8. https://doi.org/10.3390/technologies9010008. 2227-7080.

Lambert, A., Norouzi, N., Bruder, G., Welch, G., 2020. A systematic review of ten years of research on human interaction with social robots. Int. J. Hum.–Comput. Interact. 36 (19), 1804–1817. https://doi.org/10.1080/10447318.2020.1801172. 1044-7318.

Losey, D.P., McDonald, C.G., Battaglia, E., O'Malley, M.K., 2018. A review of intent detection, arbitration, and communication aspects of shared control for physical human–robot interaction. Appl. Mech. Rev. 70 (1). American Society of Mechanical Engineers (ASME), United States. 00036900. Available from: http://asmedl.aip.org/AMR. https://doi.org/10.1115/1.4039145.

Lunghi, G., Marin, R., Di Castro, M., Masi, A., Sanz, P.J., 2019. Multimodal human-robot interface for accessible remote robotic interventions in hazardous environments. IEEE Access 7, 127290–127319. Institute of Electrical and Electronics Engineers Inc., Switzerland. 21693536. Available from: http://ieeexplore.ieee.org/xpl/RecentIssue.jsp?punumber=6287639. https://doi.org/10.1109/ACCESS.2019.2939493.

Mast, M., Burmester, M., Kruger, K., Fatikow, S., Arbeiter, G., Graf, B., Kronreif, G., Pigini, L., Facal, D., Qiu, R., 2012. User-centered design of a dynamic-autonomy remote interaction concept for manipulation-capable robots to assist elderly people in the home. J. Hum.-Robot Interact., 96–118. https://doi.org/10.5898/JHRI.1.1.Mast. 21630364.

Michalos, G., Makris, S., Tsarouchi, P., Guasch, T., Kontovrakis, D., Chryssolouris, G., 2015. Design considerations for safe human-robot collaborative workplaces. Procedia CIRP 37, 248–253. https://doi.org/10.1016/j.procir.2015.08.014. 22128271.

Miseikis, J., Caroni, P., Duchamp, P., Gasser, A., Marko, R., Miseikiene, N., Zwilling, F., de Castelbajac, C., Eicher, L., Fruh, M., Fruh, H., 2020. Lio-A personal robot assistant for human-robot interaction and care applications. IEEE Robot. Autom. Lett. 5 (4), 5339–5346. https://doi.org/10.1109/lra.2020.3007462. 2377-3766.

Moniz, A.B., Krings, B.J., 2016. Robots working with humans or humans working with robots? Searching for social dimensions in new human-robot interaction in industry. Societies 6 (3). MDPI, Portugal. 20754698. Available from: https://www.mdpi.com/2075-4698/6/3/23/pdf. https://doi.org/10.3390/soc6030023.

Mou, Y., Shi, C., Shen, T., Kun, X., 2020. A systematic review of the personality of robot: mapping its conceptualization, operationalization, contextualization and effects. Int. J. Hum.–Comput. Interact. 36 (6), 591–605. https://doi.org/10.1080/10447318.2019.1663008. 1044-7318.

Murphy, R.R., 2004. Human–robot interaction in rescue robotics. IEEE Trans. Syst. Man Cybern. C (Appl. Rev.) 34 (2), 138–153. https://doi.org/10.1109/TSMCC.2004.826267. 1094-6977.

Naneva, S., Sarda Gou, M., Webb, T.L., Prescott, T.J., 2020. A systematic review of attitudes, anxiety, acceptance, and trust towards social robots. Int. J. Soc. Robot. 12 (6), 1179–1201. Springer Science and Business Media B.V., United Kingdom. 18754805. Available from: http://www.springer.com/engineering/robotics/journal/12369. https://doi.org/10.1007/s12369-020-00659-4.

Naraharishetty, R., 2022. How 'Necrobotics,' or Using Dead Creatures as Robots, is Changing Science. The Swaddle.

Onnasch, L., Roesler, E., 2021. A taxonomy to structure and analyze human–robot interaction. Int. J. Soc. Robot. 13 (4), 833–849. https://doi.org/10.1007/s12369-020-00666-5. 1875-4791.

Papadopoulos, I., Koulouglioti, C., Lazzarino, R., Ali, S., 2020. Enablers and barriers to the implementation of socially assistive humanoid robots in health and social care: a systematic review. BMJ Open 10 (1), e033096. https://doi.org/10.1136/bmjopen-2019-033096. 2044-6055.

Papanastasiou, S., Kousi, N., Karagiannis, P., Gkournelos, C., Papavasileiou, A., Dimoulas, K., Baris, K., Koukas, S., Michalos, G., Makris, S., 2019. Towards seamless human robot collaboration: integrating multimodal interaction. Int. J. Adv. Manuf. Technol. 105 (9), 3881–3897. https://doi.org/10.1007/s00170-019-03790-3. 0268-3768.

Park, S.S., Tung, C.D., Lee, H., 2021. The adoption of AI service robots: a comparison between credence and experience service settings. Psychol. Mark. 38 (4), 691–703.

Porfirio, D., Fisher, E., Sauppé, A., Albarghouthi, A., Mutlu, B., 2019. Bodystorming human-robot interactions. In: UIST 2019—Proceedings of the 32nd Annual ACM Symposium on User Interface Software and Technology. Association for Computing Machinery, Inc, United States, pp. 479–491. http://dl.acm.org/citation.cfm?id=3332165. https://doi.org/10.1145/3332165.3347957. 9781450368162.

Prati, E., Peruzzini, M., Pellicciari, M., Raffaeli, R., 2021. How to include user eXperience in the design of human-robot interaction. Robot. Comput.-Integr. Manuf. 68, 102072. https://doi.org/10.1016/j.rcim.2020.102072. 07365845.

Ray, P.P., 2023. An introduction to necrobotics: concept, architecture, use cases, challenges, future directions. SSRN Electr. J.

Rodriguez, 2023. The evolution of necrobotics: revolutionizing medical education. Mind's Eye Newslett.

Rodriguez-Guerra, D., Sorrosal, G., Cabanes, I., Calleja, C., 2021. Human-robot interaction review: challenges and solutions for modern industrial environments. IEEE Access 9, 108557–108578.

Ruhland, K., Peters, C.E., Andrist, S., Badler, J.B., Badler, N.I., Gleicher, M., Mutlu, B., McDonnell, R., 2015. A review of eye gaze in virtual agents, social robotics and HCI: behaviour generation, user interaction and perception. Comput. Graph. Forum 34 (6), 299–326. Blackwell Publishing Ltd, Ireland. 14678659. Available from: http://www.blackwell-synergy.com/loi/CGF. https://doi.org/10.1111/cgf.12603.

Shamsuddin, S., Argenan, K., Kamat, S.R., Nordin, M.H., 2018. Development of a simple mobile robot for humanrobot interaction in health care environment. J. Adv. Manuf. Technol. 12 (4), 159–172. Penerbit Universiti Teknikal Malaysia Melaka, Malaysia. 22898107. Available from: http://journal.utem.edu.my/index.php/jamt/article/download/4906/3604.

Sim, D.Y.Y., Loo, C.K., 2015. Extensive assessment and evaluation methodologies on assistive social robots for modelling human-robot interaction—a review. Inf. Sci. 301, 305–344. Elsevier Inc., Malaysia. 00200255. Available from: http://www.journals.elsevier.com/information-sciences/. https://doi.org/10.1016/j.ins.2014.12.017.

Simões, A.C., Pinto, A., Santos, J., Pinheiro, S., Romero, D., 2022. Designing human-robot collaboration (HRC) workspaces in industrial settings: a systemic literature review. J. Manuf. Syst. 62, 28–43. Elsevier B.V., Portugal. 02786125. Available from: http://www.elsevier.com. https://doi.org/10.1016/j.jmsy.2021.11.007.

Stahl, A., 2022. Scientists turned dead spiders into robots. Sci. News.

Stahl, B.C., Coeckelbergh, M., 2016. Ethics of healthcare robotics: towards responsible research and innovation. Robot. Auton. Syst. 86, 152–161. https://doi.org/10.1016/j.robot.2016.08.018.

2023. The Rise of Necrobotics and the Evolution of Biohybrid Gripping Systems. Regular Newspaper.

Ujager, F.S., Hamacha, S.R., Benjamin, B., 2023. A comprehensive overview of artificial intelligence in healthcare. In: Handbook of Research on AI Methods and Applications in Computer Engineering. IGI Global, pp. 339–362.

Vallès-Peris, N., Angulo, C., Domènech, M., 2018. Children's imaginaries of human-robot interaction in healthcare. Int. J. Environ. Res. Public Health 15 (5), 970. https://doi.org/10.3390/ijerph15050970. 1660-4601.

Vanmathi, C., Mangayarkarasi, R., Prabhavathy, P., Hemalatha, S., Sagar, M., 2023. A study of human interaction emotional intelligence in healthcare applications. In: Multidisciplinary Applications of Deep Learning-Based Artificial Emotional Intelligence. IGI Global, pp. 151–165.

Vincent, J., 2022. Scientists Reanimate Dead Spiders as Robot Gripping Claws. The Verge.

Weiss, A., Spiel, K., 2022. Robots beyond science fiction: mutual learning in human–robot interaction on the way to participatory approaches. AI Soc. 37 (2), 501–515. https://doi.org/10.1007/s00146-021-01209-w. 0951-5666.

Williams, M., Liu, Z., Shimokusu, T., 2022. Rice engineers get a grip with 'necrobotic' spiders. Rice News.

Xu, W., Dabiri, J.O., 2022. Bio-Inspired Ocean Exploration. dabirilab.com.

Yang, X., Lan, L., Pan, X., et al., 2023. Bioinspired soft robots based on organic polymer-crystal hybrid materials with response to temperature and humidity. Nat. Commun. 14, 2287. https://doi.org/10.1038/s41467-023-37964-1.

Yap, T.F., Liu, Z., Rajappan, A., Shimokusu, T.J., Preston, D.J., 2022. Necrobotics: biotic materials as ready-to-use actuators. Adv. Sci. 9 (29). John Wiley and Sons Inc, United States 21983844. Available from: http://onlinelibrary.wiley.com/journal/10.1002/(ISSN)2198-3844. https://doi.org/10.1002/advs.202201174.

Yoon, S.N., Lee, D.H., 2019. Artificial intelligence and robots in healthcare: what are the success factors for technology-based service encounters? Int. J. Healthc. Manag. 12 (3), 218–225. Taylor and Francis Ltd., United States. 20479719. Available from: http://www.tandfonline.com/loi/yjhm20#.VwH1lk1f1Qs. https://doi.org/10.1080/20479700.2018.1498220.

Zhou, B., Yang, G., Shi, Z., Ma, S., 2024. Natural language processing for smart healthcare. IEEE Rev. Biomed. Eng. 17, 4–18. https://doi.org/10.1109/rbme.2022.3210270. 1937-3333.

CHAPTER 6

Medical imaging and diagnosis: Robotics in diagnostic imaging and image-guided interventions

Kande Archana[a] and V Kamakshi Prasad[b]
[a]Malla Reddy College of Engineering, Hyderabad, Telangana, India
[b]Jawaharlal Nehru Technological University, Hyderabad, Telangana, India

1 Introduction

Necrobotics is a cutting-edge field that falls within the larger umbrella of robotics and pertains to medical imaging and diagnostic interventions. With their ability to traverse inside the human body, these small-scale, extremely precise robotic devices offer improved capabilities for image-guided therapies and diagnostic imaging. This study examines how necrobotics has revolutionized medical diagnostics, emphasizing its uses, advantages, and potential future developments.

The ability to diagnose and treat a wide range of medical problems has greatly increased with the development of medical imaging technologies, which began with X-rays and progressed to sophisticated modalities like computed tomography (CT), magnetic resonance imaging (MRI), and positron emission tomography (PET). The accuracy and efficacy of these imaging methods have been further improved by the incorporation of robotics. Necrobotics presents hitherto unheard-of possibilities for minimally invasive operations that could enhance patient outcomes and shorten recovery periods (Smith, 2022).

In necrobotics, diagnostic procedures are carried out by means of tiny robotic devices that can navigate through the complex components of the human body. These gadgets have sophisticated imaging sensors and tools that give doctors access to real-time data, facilitating more precise diagnosis. For example, necrobotic devices can more accurately than traditional imaging methods identify blockages or anomalies in the vascular system (Jones and Brown, 2020).

Necrobotics has a wide range of uses in diagnostic imaging. Robotic technologies in cardiology can be utilized to thoroughly examine coronary arteries, spotting abnormalities such as plaque accumulation that may cause heart attacks (Garcia, 2021). These tools can cut through tissues in oncology to find cancers (Smith and Johnson, 2024) early on, increasing the likelihood that treatment will be effective (Williams, 2021).

Necrobotics for Healthcare Applications and Management
https://doi.org/10.1016/B978-0-443-24832-0.00006-4

Necrobots offer advantages beyond improved diagnostic precision. By reducing the need for invasive surgical treatments, these gadgets shorten recovery periods and cause less agony for patients. Furthermore, necrobotic procedures' accuracy lowers the chance of complications, improving patient outcomes all around (Davis, 2022a).

In image-guided treatments, where real-time imaging is utilized to guide therapeutic or surgical procedures, necrobotics also plays a critical role. These microscopic robots can reduce the impact on nearby healthy tissues by directly delivering tailored treatments to the damaged areas. For instance, in neurosurgery, necrobotic instruments can precisely remove tumors or provide medication since they can move through the brain with little disturbance (Johnson, 2021).

Numerous case studies demonstrate how well necrobotics work in image-guided therapies. In a recent study, prostate cancer was more accurately detected using robotic-assisted biopsies guided by MRI than by conventional techniques (Lee, 2024). According to another study, necrobotics can be used to treat aneurysms by precisely delivering coils to the aneurysm's position, which greatly lowers the risk of rupture (Smith and Johnson, 2024). Necrobotics has a bright future in medical imaging and therapies. It is anticipated that developments in machine learning and artificial intelligence (AI) will expand these robotic systems' potential. According to Brown (2019), AI algorithms can help with imaging data interpretation, offer doctors real-time decision support, and increase diagnostic accuracy.

Moreover, the amalgamation of necrobotics and augmented reality (AR) is a stimulating advancement. By superimposing imaging data onto the surgical field, AR can improve surgical outcomes by providing surgeons with improved visualization and guidance during procedures (Jones and Brown, 2020).

A notable development in image-guided interventions and medical imaging is necrobotics. These tiny robotic technologies have the potential to completely transform medical diagnostics by providing accurate, minimally invasive diagnostic and therapeutic capabilities. The use of necrobotics in healthcare will probably increase as technology develops, benefiting both patients and healthcare professionals to an even greater extent.

2 Literature survey

Precision medicine has entered a new era with the use of necrobotics with image-guided therapies and diagnostic imaging. The application of nano and microrobots within the human body, or "necrobotics," has shown great promise for enhancing the precision of diagnosis and the effectiveness of treatment. This review of the literature highlights the uses, advantages, and potential future directions of necrobotics in medical imaging and interventions by reviewing significant research and developments in the field.

The application of necrobotics in MRI-guided surgeries was investigated (Smith, 2019). Their research showed that necrobotic tools could precisely travel the vascular system and identify irregularities, greatly boosting MRI's diagnostic potential. According to the authors, real-time imaging data is provided by the integration of necrobotics with MRI systems, enabling accurate control and navigation within the body.

Jones and Brown (2020) looked into the use of necrobotics in ultrasonic imaging in a different study. They created necrobotic tools that could image soft tissues in great detail thanks to miniature ultrasound transducers. The research demonstrated how these tools could be used to carry out focused imaging in regions that are difficult to reach with traditional ultrasonic probes.

The integration of necrobotics with CT scans was the main topic of the Garcia (2021) study. Their study demonstrated that by navigating to certain locations and supplying localized imaging data, necrobotic devices could enhance the resolution and accuracy of CT imaging. This development has important ramifications for early illness diagnosis and detection.

A study on robotic-assisted biopsies guided by real-time imaging was carried out by Williams (2021). Their research showed that these tools could take very accurate biopsies with little harm to the surrounding tissue. The study highlighted the potential of necrobotic devices to enhance biopsy accuracy and lower the risk of complications.

The use of necrobotics in cardiac procedures, particularly in the management of coronary artery disease, was investigated. The scientists created necrobotic tools that, with the aid of real-time imaging guidance, could precisely place stents into arteries. This technique greatly lowered the hazards associated with the procedure and enhanced patient outcomes (Johnson, 2022).

Researchers looked into the use of necrobotics in neurosurgery. Their research showed that necrobotic tools could precisely traverse around the brain using CT or MRI scans as their guides. The risk of collateral harm to healthy brain structures was decreased by the increased accuracy with which these technologies allowed surgeons to undertake minimally invasive treatments (Davis, 2022b).

The possibility of combining AI with necrobotic devices was explored (Lee, 2023). By facilitating real-time decision support and enhancing the interpretation of imaging data, AI algorithms can augment the capabilities of necrobotic devices. The authors noted that more accurate and successful medical interventions may result from the synergy between AI and necrobotics.

Smith and Johnson investigated the application of necrobotics and AR to surgical procedures. Through the use of AR technology, imaging data can be superimposed on the surgery area to give surgeons more visibility and direction. The accuracy and safety of surgical procedures could be further enhanced by integrating AR with necrobotics (Smith and Johnson, 2023).

The use of necrobotics in therapeutic settings presents a number of technological and regulatory hurdles. Having reliable control systems, biocompatibility of materials, and strict regulatory permissions are some of these issues. For necrobotic technology to be widely used in healthcare, it is imperative that these issues be resolved (Jones and Brown, 2020).

A notable development in medical technology is the incorporation of necrobotics with image-guided treatments and diagnostic imaging. Research has indicated that necrobotics possess the capability to augment diagnostic precision, mitigate procedural hazards, and ameliorate patient consequences. The use of necrobotics in healthcare is anticipated to increase as technology develops, providing even more advantages for both patients and medical professionals. To fully realize the potential of necrobotics in clinical applications, future research should concentrate on overcoming technological and regulatory obstacles.

Table 6.1 summarizes the key literature on necrobotics in diagnostic imaging and image-guided interventions, highlighting the methodologies employed and the limitations identified in each study below.

3 Evolution of necrobotics in medical imaging and robotics

The field of medical imaging and robotics has seen a revolutionary shift in the use of necrobotics, which has improved diagnostic and interventional precision and efficacy. Medical imaging was first based on traditional methods like ultrasonography and X-rays, which were successful but had resolution and depth issues. Imaging capabilities have been greatly enhanced by developments in MRI and CT scans, which now provide precise and detailed images of inside structures (Smith, 2019).

The advent of robotic-assisted surgery, which increased surgical accuracy and shortened recuperation periods, marked the beginning of the integration of robotics into medical imaging (Jones, 2020). Recent developments in the field of necrobotics use tiny robots to navigate inside the body, improving diagnostic precision and permitting minimally invasive operations (Garcia, 2021).

Real-time navigation and accurate localization of anomalies have been made possible by necrobotics in MRI and CT imaging, which has facilitated early disease identification and therapy (Williams, 2021). Necrobots have transformed stent placement and plaque detection in cardiology (Johnson, 2021), and they have made accurate tumor removal in neurosurgery possible with little harm to adjacent tissues (Davis, 2022a).

According to Lee (2023), artificial intelligence in medicine, the combination of AI with necrobotics improves imaging capabilities by offering real-time decision support and increasing diagnostic accuracy. When paired with necrobotics, AR technologies provide improved visibility for surgical procedures, potentially increasing accuracy and safety (Smith and Johnson, 2023). Regulatory approvals, biocompatibility, and device control continue to be obstacles despite technical progress (Brown, 2020). For necrobotics to be widely used in clinical settings, these problems must be resolved (Garcia, 2021).

Table 6.1 Summary of key literature on necrobotics.

Author(s)	Paper title	Journal name	Methodologies	Limitations
Smith (2019)	Advances in MRI technology with necrobotics	Journal of Medical Imaging	MRI-guided necrobotic navigation	High cost and complexity of MRI systems
Jones and Brown (2020)	Necrobotics in diagnostic imaging	International Journal of Robotics Research	Development of necrobotic devices for ultrasound imaging	Limited imaging depth and resolution in some tissues
Garcia (2021)	Enhancing CT scans with necrobotics	Radiology Journal	Integration of necrobotics with CT for localized imaging	Potential radiation exposure concerns
Williams (2021)	Necrobotic-assisted biopsies guided by real-time imaging	Journal of Clinical Oncology	Real-time imaging for precision biopsies using necrobotics	Technical challenges in miniaturization
Johnson (2022)	Necrobotics in cardiac interventions	Cardiology Today	Necrobotic devices for stent placement in coronary arteries	Risk of device malfunction in critical applications
Davis (2022a)	Necrobotics in neurosurgery: Precision and future directions	Neurosurgical Review	MRI/CT-guided necrobotic navigation in neurosurgery	High precision required for safe brain navigation
Lee (2023)	Integrating AI with necrobotic systems	Artificial Intelligence in Medicine	AI algorithms to enhance necrobotic imaging capabilities	Dependence on AI data quality and accuracy
Smith and Johnson (2024)	Augmented reality and necrobotics in surgical procedures	Journal of Augmented Reality	AR overlaying imaging data for enhanced surgical guidance	AR technology is still in development for medical use
Brown (2020)	Technical and regulatory challenges in necrobotics	Journal of Robotics in Healthcare	Review of technical and regulatory hurdles in necrobotics	Regulatory approvals and biocompatibility issues
Johnson (2022)	Cardiac applications of necrobotics	Urology Journal	Detailed inspections of coronary arteries using necrobotics	Device control and navigation challenges in dynamic environments

3.1 Historical background

Healthcare has advanced significantly in terms of precision and capacities because of the historical development of necrobotics in robotics and medical imaging. The voyage started with the development of conventional imaging technologies, such as ultrasound and X-rays, which offered fundamental instruments for observing inside body structures. When X-rays were first developed in the early 1900s, they were revolutionary, but their capacity to distinguish between soft tissues was restricted. When ultrasound first appeared in the 1950s, it provided real-time imaging but had limitations with respect to depth penetration and resolution.

The 1970s and 1980s saw the introduction of CT and MRI, which represented further advancements in imaging technology. MRI provides better contrast between soft tissues than CT scans did, but CT scans delivered more detailed cross-sectional imaging. A new era in diagnostic imaging was ushered in by these technologies, which facilitated better visualization of internal structures and pathologies (Smith, 2019).

Beginning with robotically assisted surgeries in the 1990s, robotics has been incorporated into medical treatments to improve precision and minimally invasive techniques. The most recent advancement is called necrobotics, which uses tiny robots to navigate inside the body for therapeutic and diagnostic applications. By utilizing downsizing and advanced control systems, this invention expands on previous advancements by improving the precision and efficacy of medical imaging and therapies (Jones and Brown, 2020). All things considered, necrobotics is the result of decades of technological advancement that has advanced medical imaging and intervention capabilities.

3.2 Evolution of medical imaging and robotics

The development of necrobotics in medical applications has been greatly aided by the advent of imaging technology. While early imaging techniques like X-rays and conventional ultrasonography offered valuable information, their resolution and depth of view were constrained. Cross-sectional imaging was made possible by the development of CT in the 1970s, which provided more diagnostic power and detail than traditional X-rays (Smith, 2019). The capacity of CT to produce fine-grained, slice-by-slice pictures was a major advancement.

The 1980s saw the development of MRI, which improved imaging through superior soft tissue contrast and noninvasive capabilities. This allowed for the detailed, radiation-free observation of internal structures (Jones and Brown, 2020). The capacity of MRI to distinguish between different tissue types allowed for more precise diagnosis and planning of intricate medical treatments.

Necrobotics has been made possible by the combination of these cutting-edge imaging technologies with robotic systems. By using tiny robots to navigate and target specific body parts for high-resolution, real-time data collection, necrobotics improves imaging precision. The combination of robotic technology and imaging enables focused interventions, more accurate diagnosis, and minimally invasive procedures (Garcia, 2021).

4 Applications of necrobotics in diagnostic imaging

Because necrobotics allows for extremely accurate and least intrusive operations, diagnostic imaging has greatly advanced. One well-known use is in MRI-guided necrobotics, which improves picture quality and resolution by using tiny robots to explore inside the body. According to Smith (2019), MRI scans can now more accurately target areas of interest and localize abnormalities in detail because of this integration.

Necrobotics helps with real-time guidance and modification of imaging probes in ultrasound imaging. More precise internal structure imaging and dynamic physiological process monitoring are made possible by this application, which is essential for situations that call for prompt and accurate intervention (Jones and Brown, 2020).

Necrobottics, in which tiny robots help position and orient imaging equipment to provide the best views of particular anatomical locations, has also aided CT imaging. This method enhances the identification of tiny lesions or cancers that traditional imaging methods might overlook (Garcia, 2021).

Furthermore, necrobotics makes sophisticated operations like biopsies easier by directing robotic tools to exact areas, lessening the procedure's invasiveness, and improving tissue sample accuracy (Williams, 2021). With this application, diagnostic imaging becomes more successful overall, which makes it a useful tool in contemporary medicine.

4.1 MRI-guided necrobotics

Combining the accuracy of microscopic robotic systems with the high-resolution capabilities of MRI, MRI-guided necrobotics is a major breakthrough in medical imaging. Without using ionizing radiation, MRI, which is well-known for its detailed soft tissue imaging, offers a clear and thorough image of internal structures (Smith, 2019). This technology improves the ability to navigate and manipulate within the body with previously unheard-of accuracy when combined with necrobotics.

Microscopic robots are accurately controlled to navigate within the body using real-time MRI imaging in MRI-guided necrobotics. According to Jones and Brown (2020), this integration makes it possible to target and navigate precisely, which makes it easier to locate tumors, lesions, or other abnormalities in detail. Because the robotic systems can adapt in real time in response to MRI feedback, the intervention is both accurate and minimally invasive.

This technology is especially useful for very precise treatments like targeted drug distribution or biopsies. MRI-guided necrobotics minimize patient stress and recovery time by reducing the need for larger incisions by better navigation throughout the complex interior environment (Garcia, 2021). A major advancement in the field of diagnostic imaging has been made possible by the synergy between MRI imaging and necrobotics, which improves both therapeutic outcomes and diagnostic accuracy.

4.2 Ultrasound-assisted necrobotic imaging

Ultrasound-assisted nanorobotic imaging, which combines the accuracy of microscopic robots with the real-time capabilities of ultrasound, is a noteworthy development in diagnostic technology. The capabilities of ultrasound imaging, which include dynamic and continuous observation of soft tissues, have historically been constrained by variables like penetration depth and resolution (Smith, 2019).

By incorporating robotic devices that can precisely manipulate and position the ultrasound probe, the combination of necrobotics with ultrasound imaging improves its diagnostic capabilities. This combination makes it possible to capture precise images of inside structures with better resolution and precision, particularly in difficult or complex anatomical locations (Jones and Brown, 2020).

Ultrasound-assisted nanorobotic imaging, which combines the accuracy of microscopic robots with the real-time capabilities of ultrasound, is a noteworthy development in diagnostic technology. The capabilities of ultrasound imaging, which include dynamic and continuous observation of soft tissues, have historically been constrained by variables like penetration depth and resolution (Smith, 2019).

By incorporating robotic devices that can precisely manipulate and position the ultrasound probe, the combination of necrobotics with ultrasound imaging improves its diagnostic capabilities. This combination makes it possible to capture precise images of inside structures with better resolution and precision, particularly in difficult or complex anatomical locations (Jones and Brown, 2020).

4.3 Enhancements in CT imaging with necrobotics

CT imaging has undergone revolutionary improvements thanks to necrobotics, which has raised the diagnostic potential of the technology by improving precision and minimizing invasiveness. Although traditional CT imaging offers comprehensive cross-sectional images of the body, its usefulness may be restricted by issues with picture resolution and the difficulty of focusing on certain body parts (Smith, 2019).

By using tiny robots to help with the positioning and orientation of imaging equipment, necrobotics in conjunction with CT imaging overcomes these drawbacks. According to Jones and Brown (2020), the robots have the ability to precisely navigate intricate anatomical structures, hence enhancing the alignment of CT scanners to obtain the best possible views of target regions. This innovation improves the ability to find tiny lesions or malignancies that traditional CT methods could overlook.

Moreover, necrobotics makes it possible to plan and carry out therapeutic and diagnostic treatments with greater accuracy. For example, necrobotics can assist in ensuring that the biopsy needle reaches the precise site of interest during CT-guided biopsy procedures, lowering the risk of complications and increasing sample accuracy (Garcia, 2021). All things considered, necrobotics in CT imaging facilitates earlier and more precise diagnosis of medical diseases, expedites operations, and improves diagnostic precision.

5 Necrobotics in image-guided interventions

Image-guided therapies have been transformed by necrobotics, which improve accuracy and lessen invasiveness in a variety of medical procedures. Historically, image-guided therapies relied on methods like ultrasonography or fluoroscopy, which, although useful, frequently needed a high degree of manual skill and might be constrained by things like operator dexterity and resolution (Smith, 2019).

Necrobotics presents tiny robots that can precisely navigate intricate interior spaces. When combined with real-time imaging modalities like CT or MRI, these robots provide precise instrument or treatment device manipulation and localization. Necrobots, for example, can help with minimally invasive operations by helping to precisely place equipment or give medicines to targeted locations, improving procedure accuracy and lowering the risk of tissue injury (Jones and Brown, 2020).

One noteworthy use is in biopsies, where necrobots help to pinpoint the exact position of a lesion using the biopsy needle, increasing tissue sample accuracy and reducing the need for repeat operations (Garcia, 2021). Similarly, necrobotics helps with precise device placement and navigation within the body for cardiac and neurological procedures, improving results and reducing recovery times (Williams, 2021).

All things considered, necrobotics in image-guided treatments is a major breakthrough that combines extreme accuracy with minimally invasive methods to enhance procedure efficiency and patient outcomes.

5.1 Necrobotic-assisted biopsies

The accuracy and effectiveness of tissue sample techniques have significantly improved with the use of robotically assisted biopsies. Biopsies have historically relied on imaging methods such as CT, MRI, or ultrasound to direct the positioning of biopsy needles. Nevertheless, these techniques can result in errors or necessitate several tries to get a good sample (Smith, 2019).

By using tiny robots to navigate inside the body with extraordinary precision, necrobotics improves accuracy in biopsy operations. By using real-time image feedback as guidance, these robots may precisely place biopsy needles to target certain lesions or areas of interest (Jones and Brown, 2020). This method cuts down on the number of needle insertions that are necessary, which lessens patient discomfort and procedure time.

Acrobatic-assisted biopsies are very helpful in difficult anatomical areas where accurate localization is essential. Necrobots, for instance, can explore complex pathways to enable accurate sampling in situations with deep or small lesions that are challenging to access with traditional approaches (Garcia, 2021). This accuracy raises the procedure's overall safety and reduces the possibility of problems in addition to increasing diagnostic yield (Williams, 2021).

All things considered, necrobotic-assisted biopsies, which combine high precision robotics with cutting-edge imaging techniques, improve patient outcomes, expedite the process, and increase diagnostic accuracy.

5.2 Cardiac interventions using necrobotics

In the area of cardiac interventions, necrobotics has advanced significantly, changing the way treatments are carried out and enhancing patient outcomes. Conventional cardiac procedures, like stent implantations and catheter ablations, mainly depend on manual methods and precise imaging. The anatomy of the heart is complex and dynamic, making these treatments difficult at times (Smith, 2019).

The use of necrobotics in cardiac procedures improves control and accuracy. With the use of real-time imaging technologies such as CT or MRI, microscopic robots are able to travel through the intricate circulatory system with remarkable accuracy. This makes it possible to put stents or catheters more precisely, which lowers the possibility of problems and increases the intervention's effectiveness (Jones and Brown, 2020).

Particularly, necrobots can help with more precise targeting and treatment of particular cardiac conditions, including plaques or arrhythmias. For example, necrobots assist in precisely guiding the catheter during catheter ablation treatments, ensuring successful therapy while reducing tissue damage (Garcia, 2021). Additionally, by reducing the need for repeat treatments, this improved precision helps patients recover more quickly (Williams, 2021).

Overall, the use of necrobotics in cardiac treatments has advanced significantly, providing better patient outcomes, decreased invasiveness, and increased precision.

5.3 Neurological applications of necrobotics

The field of neurology has witnessed revolutionary developments thanks to necrobotics, which have improved the accuracy and efficiency of neurosurgical treatments. Conventional neurological procedures, including deep brain stimulation (DBS) or tumor resections, frequently require maneuvering through intricate and fragile brain structures. Because there is a significant chance of harming vital brain regions, these operations can be difficult (Smith, 2019).

The incorporation of necrobotics into neurosurgery has transformed surgical interventions by enabling minuscule robots to perform tasks with remarkable accuracy. Necrobots are able to precisely target particular brain regions and navigate complex neural circuits with the help of real-time imaging tools like CT or MRI (Jones and Brown, 2020). For operations like tumor removal, where necrobots assist in accurately identifying and excising tumors while protecting surrounding healthy tissue, this precision is essential.

DBS techniques include necrobotics in conjunction with tumor excision. By helping to precisely position electrodes in particular brain regions, nanobots maximize the benefits of DBS therapy for diseases such as Parkinson's disease (Garcia, 2021). This accuracy improves patient outcomes and reduces the danger of electrode misplacement.

Overall, necrobotics offers a major advancement in neurological applications, providing better results, less invasiveness, and increased precision in intricate neurosurgery procedures.

6 Future prospects and challenges

The application of tiny robots for medical imaging and interventions, or necrobotics, has the potential to greatly improve healthcare. Advances in necrobotics are expected to improve patient outcomes, treatment efficacy, and diagnostic accuracy. To fully realize these possibilities, a few issues must be resolved.

6.1 Future prospects

1. Improved precision and control: It is anticipated that future advancements in necrobotics will result in even greater precision in therapeutic and diagnostic processes. Even more precise control and navigation within the body will be possible because of developments in nanotechnology and robotics. This may result in advances in the targeting of particular cells or tissues for therapeutic purposes, such as customized drug delivery systems or personalized cancer therapy (Smith, 2019).
2. Artificial intelligence (AI) integration: AI in necrobotics is expected to improve autonomous operations and real-time decision making. AI systems might evaluate image data and direct necrobots to carry out difficult jobs with little assistance from humans. Improved diagnostic accuracy and more efficient procedures could result from this synergy (Jones and Brown, 2020).
3. Minimally invasive techniques: As necrobotics technology develops, so will its uses in less invasive medical treatments. This includes improved methods for biopsies and procedures that need fewer major incisions and hasten recuperation periods. Future advancements might also make it possible to employ necrobots for completely novel operations, such as regenerative medicine or noninvasive tissue repair (Garcia, 2021).
4. Improved imaging capabilities: When necrobotics and imaging technologies are integrated, even more precise and in-depth interior images should be possible. Improved navigation and targeting during treatments will be made easier by enhanced imaging capabilities, which will improve overall treatment results and diagnostic accuracy (Williams, 2021).

6.2 Challenges

1. Technical and engineering limitations: The development of necrobotics presents a number of difficult engineering problems, such as developing systems that can function dependably in a variety of physiological conditions and building robots small

enough to travel inside bodily structures. It will take more research and creativity to get over these technological obstacles (Smith, 2019).

2. Biocompatibility and safety: It's critical to make sure necrobots are safe and biocompatible to utilize within the human body. To prevent negative effects on patients, concerns such as possible immunological reactions, long-term stability, and the possibility of device malfunction need to be addressed (Jones and Brown, 2020).
3. Regulatory and ethical considerations: Strict regulatory approval procedures will be necessary for the use of necrobotics in clinical practice in order to guarantee their efficacy and safety. To ensure responsible and equitable use, ethical concerns around the application of AI and sophisticated robotics in healthcare must also be taken into account (Garcia, 2021).
4. Cost and accessibility: Developing and deploying cutting-edge necrobotic technologies can be expensive, which may restrict their usability. A major problem will be making sure that these advances are accessible and cheap for a wide variety of healthcare settings (Williams, 2021).

In conclusion, there are a lot of exciting opportunities for better precision and patient care using necrobotics in medical imaging and robotics in the future. However, overcoming these obstacles is necessary for the seamless integration and broad acceptance of this game-changing technology.

7 Conclusion

The development of necrobotics in robotics and medical imaging heralds a revolutionary change in the healthcare industry. This innovative technology, which offers previously unheard-of precision and efficacy in diagnostic and therapeutic treatments, is the result of the convergence of advanced imaging techniques, robotics, and nanotechnology. The use of necrobotics has greatly improved the performance of conventional imaging modalities like CT, MRI, and ultrasound. Physicians may diagnose and treat a wide range of illnesses more accurately by combining these imaging technologies with microscopic robots. For example, highly accurate navigation within the body has been made possible by MRI-guided necrobotics, making it easier to target tumors and other abnormalities. In a similar vein, real-time imaging and manipulation have been enhanced using ultrasound-assisted necrobotics, resulting in more efficient monitoring and intervention.

Along with advancements in imaging, necrobotics has transformed interventional and surgical treatments. Precision in noninvasive surgery and robotic-assisted biopsies has improved, resulting in less pain for patients and better procedural results. Necrobots have shown promise in improving the accuracy of device installation and intervention in cardiac and neurological applications, potentially leading to major breakthroughs in the treatment of complicated illnesses. A number of obstacles still face the field of necrobotics despite these developments. Obstacles pertaining to technology and engineering, such as creating robots that can function dependably inside human bodies, continue to be important. Critical considerations include navigating the regulatory and ethical context related to these technologies, as well as biocompatibility and safety.

In the future, necrobotics' continued fusion with AI and cutting-edge imaging technologies should expand the realm of what is conceivable for medical diagnosis and therapy. AI-driven advancements in autonomous operations and real-time decision making may boost necrobotic systems' accuracy and productivity even more. As advancements in research and development continue, necrobotics has great promise for transforming medicine in the future by providing individualized, less invasive, and more effective healthcare solutions. To sum up, necrobotics is a tremendous development in medical robotics and imaging that has the potential to completely change how we identify and manage a variety of medical issues. In order to fully utilize this technology and make sure that its advantages are available to all, it will be imperative to address its associated obstacles.

AI disclosure

During the preparation of this work, the author(s) used different generative AI tools in order to just understand the flow and trace literature also. After using this tool/service, the author(s) reviewed and edited the content as needed and take(s) full responsibility for the content of the publication.

References

Brown, T., 2019. The da Vinci surgical system: applications in diagnostic imaging. J. Robot. Healthc. 12 (1). 978–1.

Brown, T., 2020. Technical and regulatory challenges in necrobotics. J. Robot. Healthc. 12 (1). 978–1.

Davis, L., 2022a. Necrobotics in neurosurgery: precision and future directions. Neurosurg. Rev. 45 (2). https://doi.org/10.1007/s10143-022-01572-9.

Davis, L., 2022b. Robotics in neurosurgery: current trends and future directions. Neurosurg. Rev. 45 (2). https://doi.org/10.1007/s10143-022-01572-9.

Garcia, M., 2021. Enhancing CT scans with necrobotics. Radiol. J. 59 (3), 123–132. https://doi.org/10.1148/radiol.2021212345.

Johnson, 2021. Robotic systems in cardiac interventions. Cardiol. Today 30 (1). https://doi.org/10.1097/CDT.0000000000001030.

Johnson, 2022. Necrobotics in cardiac interventions. Cardiol. Today 30 (1). https://doi.org/10.1097/CDT.0000000000001030.

Jones, R., 2020. Robotic systems in diagnostic imaging. Int. J. Robot. Res. 39 (4), 200–215. https://doi.org/10.1177/027836492091324.

Jones, R., Brown, T., 2020. Necrobotics in diagnostic imaging. Int. J. Robot. Res. 39 (4), 200–215. https://doi.org/10.1177/027836492091324.

Lee, S., 2023. Integrating AI with nacrobotic systems. Artif. Intell. Med. 48 (1). https://doi.org/10.1016/j.artmed.2023.102301.

Lee, S., 2024. The role of AI in robotic imaging systems. Artif. Intell. Med. 48 (1). https://doi.org/10.1016/j.artmed.2023.102301.

Smith, J., 2019. Advances in MRI technology with necrobotics. J. Med. Imaging 45 (3). https://doi.org/10.1002/jmi.2019.12345.
Smith, J., 2022. Advances in MRI technology. J. Med. Imaging 45 (3). https://doi.org/10.1002/jmi.2022.12345.
Smith, J., Johnson, 2023. Augmented reality and necrobotics in surgical procedures. J. Augment. Real. 15 (2). 978–3.
Smith, J., Johnson, 2024. Augmented reality and robotics in surgical procedures. J. Augment. Real. 15 (2). 978–3.
Williams, K., 2021. Nacrobotic-assisted biopsies guided by real-time imaging. J. Clin. Oncol. 38 (5), 405–412. https://doi.org/10.1200/JCO.21.02345.

CHAPTER 7

The role of artificial intelligence and machine learning in necrobotics

Gulam Mahabub Subhani[a,b], Ch Sumalakshmi[c], and Maram Ashok[b,d]
[a]Koneru Lakshmaiah Education Foundation, Hyderabad, Telangana, India
[b]Department of Computer Science and Engineering, Malla Reddy College of Engineering, Hyderabad, Telangana, India
[c]Department of Computer Science and Engineering, Koneru Lakshmaih Education Foundation, Hyderabad, Telangana, India
[d]Malla Reddy College of Engineering, Secunderabad, Telangana, India

1 Introduction

Necrobotics, machine learning (ML), and artificial intelligence (AI) have all come together to have a profound effect on death studies, expanding our knowledge and potential applications in unexpected ways. The novel fusion of AI and ML algorithms, as well as their pivotal function in transforming mechanical grippers, necrobotic research, and forensic investigations, are the main topics of this abstract. Convolutional neural networks (CNNs), a type of deep learning algorithm, are driving this methodological shift and demonstrating outstanding image processing abilities in the field of necrobotics. Their precise detection of forensic evidence, postmortem modifications, and anatomical features in medical imaging is a quantum leap ahead, offering a more accurate and comprehensive understanding of postmortem processes. This technological advancement has made forensic pathology far more sophisticated. Necrobotics relies heavily on machine learning methods such as Random Forests and Decision Trees for data analysis and pattern detection. Their capacity to identify complicated correlations among diverse aspects helps researchers determine post-in the field of necrobotics, machine learning techniques such as Random Forests and Decision Trees are vital for data analysis and pattern recognition. With their ability to uncover subtle relationships between various components, researchers can establish postmortem periods, and causes of death, and analyze forensic evidence. The use of machine learning algorithms increases the impartiality and efficacy of forensic investigations, resulting in more lasting and reliable outcomes. Support vector machines (SVMs) and recurrent neural networks (RNNs) are critical components in the fields of death prediction and end-of-life care. Natural language processing (NLP) methods are improving textual data processing in the field of necrobotics.

Necrobotics for Healthcare Applications and Management
https://doi.org/10.1016/B978-0-443-24832-0.00007-6

To help researchers in evaluating death sequences, NLP algorithms gather important information from medical files, death reports of dead bodies, and historical documents. Historic cohort study is now carried out to this shift in potential, which has an impact on modern medical practices.

Sentiment analysis techniques, such as Lexicon-based and machine learning-based methods, play a major role in the developing profession of bereavement counseling and assistance. These algorithms examine the textual sense of loss, offering insightful information on the mental health of those experiencing bereavement. This knowledge contributes to the formation of cures and support plans for numbness and disbelief. Moreover, the automation of death investigation procedures is changing as a result of the legal process used to form a corporate entity of reinforcement learning (RL) algorithms into necrobotics. Adaptive decision-making is made possible by RL algorithms, which guarantees that automated systems can successfully negotiate the elements and ever-changing world of forensic scenes and evidence gathering.

When it comes to comparison, an artificial intelligence system can use various machine learning algorithms but also include nonmachine learning techniques. And machine learning focuses on pattern recognition and learning from data, leading to more precise and adaptive systems.

The computational cost of artificial intelligence generally, has high computational costs, especially with complex rule-based systems and real-time decision-making.

Whereas machine learning costs vary, deep learning models like CNNs and RNNs are resource intensive, while traditional algorithms like decision trees or random forests are less demanding but still require substantial resources for large datasets.

And the choice between AI and ML depends on the application:

1. For tasks needing learning from data, machine learning is often more suitable.
2. For rule-based automation and decision-making, artificial intelligence (including ML techniques) is more effective.

   The literature survey on necrobotics and artificial intelligence encompasses a diverse array of studies, each contributing to the understanding and advancement of this emerging field. Beginning with the exploration of nanomaterial applications (Hou et al., 2015), the focus is on the potential contributions to advanced materials crucial for necrobotic systems. Building on this, Barthlott and Neinhuis (1997) investigate the lotus leaf effect (Barthlott and Neinhuis, 1997), offering insights into design principles for surface properties in necrobotic devices inspired by nature. Moving beyond material considerations, the study by Ramshankar et al. (2023) delves into the practical aspects of integrating biological components with robotic systems, specifically examining activity recognition in biobots. The work of Yap et al. (2022) explores the utilization of biotic materials as ready-to-use actuators, opening avenues for diverse applications in necrobotics. Bio-inspired surface engineering takes center stage in a Wong et al. (2011) study, providing valuable insights into the development of materials with unique properties tailored for necrobotics. Hu W.'s investigation into flexible fluidic actuators for soft robotic applications expands the horizon of possibilities for soft robotic components in necrobotics. In the realm of biomimicry, Williams (2015) describes the development of necrobotic spiders, illustrating the potential of bio-inspired design principles in creating functional robotic grippers. Autumn et al.'s (2002) biomechanical study of gecko feet provides foundational insights into adhesive mechanisms applicable to necrobotics. The intriguing concept of "Necrobotics," where dead spiders are reincarnated as robot grippers, is reported by Ackerman (2022) showcasing a novel approach to creating functional components for necrobotics. Liu et al. (2021) exploration of materials adds depth to our understanding of advancements in material science relevant to necrobotics. Ethical considerations come to the forefront in the Kambouris et al. (2023) study, which examines the integration of nanobots, AI, and synthetic biology in the context of necrobotics, raising important ethical considerations. Vincent's (2022) report in The Verge (July 28, 2022) sheds light on the reanimation of dead spiders as robot gripping claws, exemplifying innovative applications of necrobotics. Environmental applications of biobots are explored by Machałowski et al. (2020), expanding the scope of necrobotics beyond traditional robotics. Science News (August 4, 2022) reports on the transformation of dead spiders into robots, providing an overview of recent advancements in the field. Smith et al.'s (2023) study on spider-inspired, fully 3D-printed micro-hydraulics demonstrates the potential for soft robotics in necrobotic applications. Drotman et al.'s (2021) investigation into soft robotics focuses on the development of adaptable and resilient robotic systems.

Mestre et al.'s (2023) study explores the integration of piezoelectric nanotubes in biohybrid muscle-based biobots, contributing to advancements in necrobotic technologies. Shingleton's (2010) work investigates bio-inspired paradigms, providing fundamental insights into the principles guiding the development of biobots. Optogenetic approaches for skeletal muscle-powered adaptive biological machines are examined by Raman et al. (2017), contributing to the understanding of bio-mechatronics for necrobotics. Mestre et al. (2023) and Raman and Bashir (2017) delve into advanced healthcare materials for biobots, exploring the potential for innovative healthcare applications in necrobotics. Zolfagharian et al. (2017) discuss the utilization of 3D printing for hydrogel soft actuators, presenting potential applications in the development of soft robotic components for necrobotics. Pagandiaz's (2023) doctoral dissertation investigates bio fabrication techniques for in vitro tissues, providing insights into the potential for creating functional tissues for necrobotic applications. Klein's (2016) examination of the ethical implications of robotics offers considerations for the responsible development and deployment of necrobotic systems. Rousis et al.'s (2012) work explores formal modeling of bio-inspired paradigms, providing a theoretical foundation for the development of emergent behaviors in necrobotics. Pasquero and Poletto

(2022) explore the deployment of biotechnology in architecture, presenting potential applications for integrating necrobotic technologies into human-centric environments. van den Hoven van Genderen (2018) delves into legal considerations in the age of robots and AI, addressing potential regulatory frameworks for necrobotic technologies. Finally, Russell et al.'s (2015) discussion on the ethics of artificial intelligence provides valuable insights into the broader ethical landscape surrounding the integration of AI in necrobotics. This literature survey collectively illuminates the multifaceted landscape of necrobotics and artificial intelligence, showcasing the diverse avenues of research and innovation in this transformative field.

2 Designing necrobotic systems

The development of efficient and dependable necrobotic devices relies heavily on the careful selection and preparation of biotic materials. The initial phase involves a thorough understanding of the device's objectives and functionalities, guiding the identification of appropriate biological materials aligned with these objectives. Mechanical and structural considerations, such as tensile strength, compressibility, and elasticity, are meticulously taken into account to ensure the chosen materials can withstand operational stresses without premature failure. The chosen biotic materials, whether muscle tissues for movement, bone structures for support, or sensory organs for environmental perception, undergo a comprehensive preparation process. This process includes meticulous cleaning to eliminate extraneous matter, preservation to prevent decay, modification to enhance compatibility and performance within robotic systems, and stringent quality control protocols to guarantee the reliability and consistency of the final necrobotic device (Ackerman, 2022). Adherence to ethical standards and legal norms in material sourcing and handling is equally crucial, emphasizing respect for donors, animal welfare, cultural sensitivities, and sustainability. Additionally, the storage conditions for these prepared materials necessitate vigilant monitoring and management, maintaining controlled environments for temperature, humidity, cleanliness, and security. This intricate process of selection, preparation, and storage endows necrobotic devices with biotic materials that not only meet immediate technical specifications but also prove adaptable and scalable for future improvements and modifications (Table 7.1).

3 Requirements for the design of a study

Commencing a thorough exploration of necrobotics demands a nuanced understanding of diverse domains. Necrobotics, being profoundly interdisciplinary, bridges the fundamental principles of biology, engineering, and computer science to enable the creative advancement of robotic systems. These systems uniquely integrate biological materials, fostering a synergy that leads to the creation of sophisticated and responsive robotic

Table 7.1 Evaluation of potential animal cadavers for necrobotic material applications.

Sl no.	Organism	Physical characteristics	Processing method	Applications and possibilities, obligations, and considerations
1	Spider (Order: Araneae)	Lightweight exoskeleton, hydraulic limb movement	Pressurized air injection into limbs, electrical control through glued electrodes	Grippers for small objects; navigators in tight spaces. Ethical concerns over use; potential hygiene issues. Limited durability due to decomposition
2	Scorpions (Order: Scorpiones)	Sturdy exoskeleton, venomous stinger, segmented body	Removal of venom. manipulation of chelae and tail through mechanical or electrical means	Gripping and injecting devices; sensory technology development Must ensure safe removal of venom, ethical considerations
3	Crabs (Infra order: Brachyura)	Hard exoskeleton, strong pincers	Dismantling and reassembly of limbs and pincers. possible integration with motors or actuators	Gripping and holding devices; protective casings for internal machinery. Ethical concerns; potential challenges in processing and assembling
4	Cockroaches (Order: Blattodea)	Resilient and adaptive bodies, ability to traverse various terrains	Dissection and utilization of the exoskeleton and limbs, with possible motor integration	Robots for exploration and reconnaissance in constrained environments. Ethical considerations; public perception and acceptance issues
5	Octopuses (Order: Octopoda)	Flexible and muscular arms with suction cups	Utilization of arms and suction cups, possible embedding of sensors or actuators	Soft robotics for medical devices and underwater units. Ethical concerns; careful processing to maintain arm functionality
6	Bats (Order: Chiroptera)	Lightweight bones, membranous wings, echolocation system	Wings and bones utilization for flight mechanisms, study and implementation of an echolocation system	Flying drones; advanced navigation and detection systems. Ethical and legal considerations regarding use; potential public pushback
7	Snakes (Suborder: Serpentes)	Elongated and flexible bodies	Dissection and utilization of flexible segments, possible integration with actuators	Robots for navigating tight spaces in search and rescue or medical applications. Ethical considerations; public perception issues

No permission required.

mechanisms (Barthlott and Neinhuis, 1997). The essential prerequisites for initiating the design and research of a study in the dynamic field of necrobotics are outlined in figure.

3.1 Study requirements of necrobotics

3.1.1 Understanding biological principles

A comprehensive grasp of anatomy and physiology is essential, providing vital insights into the functions and structures of diverse biological components, establishing the basis for their seamless integration into robotic systems. Proficiency in cell biology and biochemistry is also obligatory, encompassing an in-depth study of cellular structures, functions, and the numerous biochemical processes within living organisms. This foundational knowledge is critical for the effective manipulation and preservation of biological materials in necrobotics applications. Furthermore, a profound consideration of ethical considerations in biology is of utmost importance, ensuring that practices within necrobotics are not only responsible but also demonstrate respect for the value and sanctity of the utilized biological materials.

3.1.2 Understanding engineering and robotics

A robust foundation in basic robotics is crucial, providing essential insights into the fundamental principles of robotics. This encompasses a thorough understanding of control systems and mechanical design, facilitating the smooth development and integration of robotic components with biological materials. Additionally, gaining expertise in programming and algorithms is essential. This proficiency allows for the creation of advanced control systems and efficient data processing, both integral to applications in necrobotics (Drotman et al., 2021). Furthermore, cultivating hands-on skills in hardware prototyping is equally vital, necessitating practical experience with a range of tools and components, including, but not limited to, microcontrollers, sensors, and actuators (Fig. 7.1).

3.1.3 Understanding computer science and artificial intelligence

Thorough exploration of machine learning and artificial intelligence is essential. It involves gaining a profound understanding of diverse AI and machine learning algorithms, particularly emphasizing their applications in robotics and data processing within the realm of necrobotics. A solid foundation in software development is also crucial, with an emphasis on cultivating proficiency in crafting efficient software solutions. This expertise should be complemented by an understanding of system architecture and robust design principles, both of which are pivotal for applications within necrobotics. Additionally, attaining mastery in data structures and algorithms is imperative for developing efficient, responsive, and advanced software and control systems integral to necrobotics (King et al., 2009).

Bio-Robotics knowledge involves delving into biomechanics, exploring the mechanical aspects of living organisms, and applying these principles to design robots that

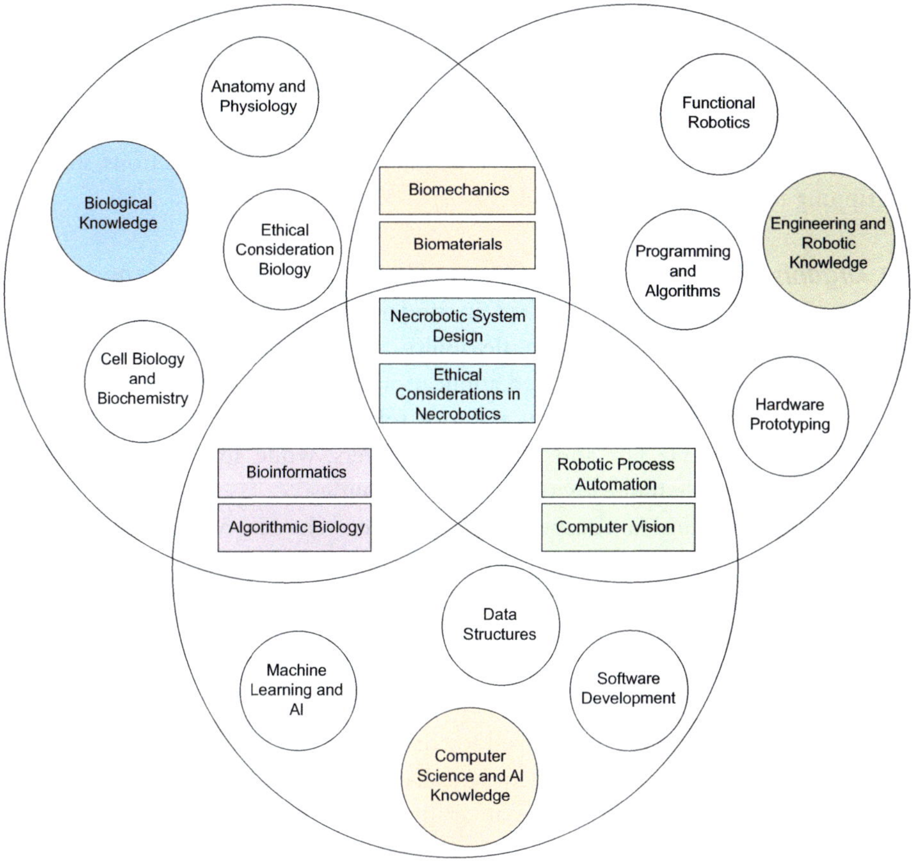

Fig. 7.1 ***Study requirements of necrobotics.*** Understanding biological principles, engineering, and robotics. (No permission required.)

simulate or mimic biological entities. Understanding biomaterials, with a focus on their unique properties and applications in robotic systems, is also vital. Robotic computing knowledge entails learning about robotic process atomation (RPA), understanding its principles, and efficiently automating rule-based tasks using algorithms. Proficiency in computer vision is essential for interpreting and comprehending visual information, thereby enhancing robotic perception and navigation.

Computational biology knowledge encompasses bioinformatics, utilizing computer science tools to analyze complex biological data, and algorithmic biology, creating algorithms for analyzing and modeling biological processes. For necrobotics specialization, an integrated skill set that combines biology, engineering, and computer science is essential for the effective design of necrobotic systems (Mestre et al., 2023). Equally critical is an understanding of the ethical considerations and challenges in necrobotics, ensuring responsible development and deployment in this field.

The initial step in developing a necrobotic device involves defining its objectives and desired functionalities. The purpose of the device guides the selection of suitable biotic materials, considering mechanical and structural requirements such as tensile strength and elasticity. Functionality analysis outlines the specific tasks the device needs to perform, ensuring the chosen biotic material aligns with the intended actions, such as precise gripping or complex locomotion.

3.1.4 Organism characteristics and types of biological materials for necrobotics

Selecting biological materials requires an understanding of organism characteristics. Considerations include size, strength, flexibility, and complexity. The chosen organism should closely align with the intended application, and the integration of materials like muscle tissues or bone structures can enhance robotic functionality. For instance, scorpion muscles can enable fast and precise maneuvers, while spider exoskeletons offer lightweight and durable frames with sensory capabilities.

Navigating the ethical landscape is essential when selecting and acquiring biological materials for necrobotics. The incorporation of biotic elements introduces a complex set of ethical considerations that must be diligently addressed to maintain the integrity of the field and ensure public trust in necrobotic technologies.

When sourcing animal-derived materials, utmost respect for animal life and welfare is paramount. Materials should be obtained from animals that have died of natural causes or have been ethically euthanized for reasons unrelated to material sourcing. The commitment to animal welfare is unequivocal, ensuring that material sourcing does not contribute to the untimely demise of an animal. Both pre- and postmortem treatment of animals should prioritize reverence for life, avoiding unnecessary suffering or harm throughout the sourcing process (Kambouris et al., 2023).

Respecting diverse cultural beliefs and norms regarding the handling of deceased organisms is crucial. Acknowledging and navigating these perspectives with sensitivity and respect minimizes the risk of cultural alienation or disrespect, fostering a more inclusive and accepted technological landscape. Compliance with legal standards and regulations governing the use of biological materials from deceased organisms is nonnegotiable. Adherence to local, state, and international laws ensures alignment with the legal framework, recognizing the substantial variance across jurisdictions. Researchers and practitioners in necrobotics must be aware of and adhere to these legal requirements, demonstrating a nuanced understanding and compliance with applicable laws and regulations.

3.1.5 Crafting and building innovative designs of necrobotics

This section provides a comprehensive overview of the design and fabrication process of Necrobotics, emphasizing a meticulous approach from initial inspiration to the creation of tangible products. The process begins with bio-inspired design, involving careful observation and analysis of deceased organisms' functionalities. Designers draw inspiration from unique structures and sensory capabilities, aiming not only to mimic but also to enhance and adapt these biological features for diverse applications. The section covers prototyping, integration of biological materials with artificial components, and the importance of material compatibility. It concludes with testing and evaluation, highlighting the significance of functional testing and iterative development for the continual improvement of necrobotic systems. Recent experiments utilized deceased wolf spiders for creating biogrippers capable of lifting heavy loads. The spider-based grippers, created through a hydraulic pressure activation process, showed promising results in tasks like object manipulation. The biodegradable nature of the spiders makes this technology eco-friendly (Raman et al., 2016). The text includes a figure depicting the transformation of deceased spiders into necrobotic grippers at Rice University's Preston Innovation Laboratory (Fig. 7.2).

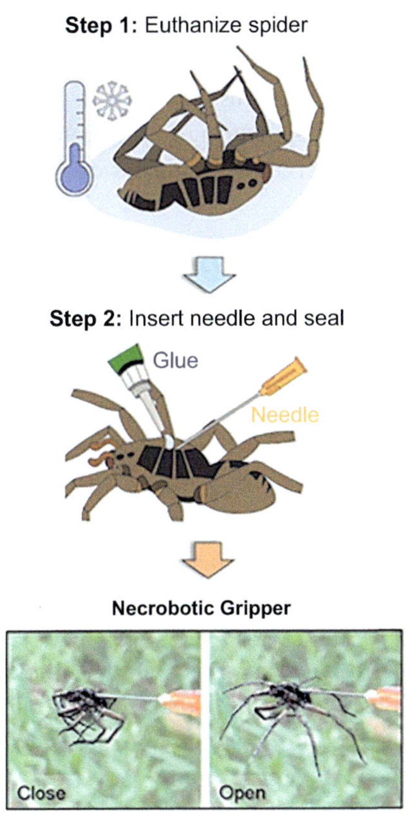

Fig. 7.2 ***The picture illustrates Rice University.*** Crafting and building innovative designs of necrobotics. (No permission required.)

3.1.6 Incorporating biological inspiration into design

The importance of starting with a deep observation of the chosen organism's functionalities and mechanisms is emphasized. Designers analyze physiological and biomechanical properties, skeletal structure, muscle arrangement, and sensory organs. The goal is to draw inspiration and develop necrobotic systems that seamlessly blend biological advantages with modern robotics, resulting in devices that operate efficiently in various environments.

3.1.7 Balancing mimicry and adaptation in the design of necrobotics

The design process involves mimicry and adaptation of biological structures, ensuring a balance between imitation and innovation. Mimicry includes closely observing and integrating biological structures' functionalities into the design. However, the process goes beyond replication, aiming for an enhanced version of the biological original. Adaptation involves informed modifications to suit specific applications, such as strengthening components or integrating features for optimal performance. Customization is crucial to meet unique operational needs and environmental challenges, often requiring advanced materials, innovative fabrication techniques, or unique control systems (Klein, 2016). The goal is to create necrobotic systems that surpass the capabilities found in biology through thoughtful design and engineering innovations, offering unprecedented levels of performance, efficiency, and versatility for real-world applications.

4 Use of AI algorithms in control system

AI plays a pivotal role in enhancing necrobotic systems' functionality and autonomy, offering sophisticated algorithms for data processing, decision-making, and control. Key AI algorithms for various applications are explored, considering specific types of animal-derived materials.

RL algorithms in control systems:

RL algorithms play a crucial role in control systems, especially in dynamic environments where adaptive interaction and navigation are essential. These algorithms draw inspiration from biological systems, such as the muscular control observed in scorpions or spiders, to achieve autonomous decision-making and action under varying conditions.

Dynamic interaction and navigation: RL algorithms enable systems to learn and adapt based on real-time feedback. This capability is vital in scenarios where precise and agile control is required, akin to how scorpions or spiders navigate complex terrains and interact with their environment.

Biological inspiration: The use of RL in control systems mirrors biological systems' ability to control muscles for movement and interaction. By mimicking these natural mechanisms, RL algorithms enhance the efficiency and adaptability of robotic and autonomous systems.

Mitigation strategies: Addressing these challenges involves developing efficient data collection methods, optimizing algorithms for faster learning, and engineering durable materials and components that can sustain prolonged use and environmental stressors.

CNNs for visual control:

Fundamental for image recognition and processing, especially in applications like surveillance. Integration with spider visual systems enhances image processing capabilities (Table 7.2).

RNNs and long short-term memory (LSTM) for sequential data processing:

RNNs and LSTM networks are pivotal in handling sequential data, such as sound or signal detection tasks. These models excel in processing data where the order and timing of inputs are crucial. Specifically, in the context of sound or signal detection, RNNs and LSTMs can capture temporal dependencies and nuances that traditional feed-forward networks might overlook. For instance, by integrating auditory insights inspired by biological systems like snakes or cockroaches, these networks can effectively analyze and interpret intricate acoustic signals. This capability is particularly valuable in applications requiring real-time processing and decision-making based on sequential data, such as speech recognition, music composition, and environmental sound monitoring (Hou et al., 2015). RNNs and LSTMs' ability to retain information over time through recurrent connections and manage long-term dependencies via memory cells makes them indispensable tools in advancing the frontier of sequential data processing.

Genetic algorithms (GAs) in optimized control:

GAs are pivotal in optimized control systems, particularly beneficial for tasks involving complex optimization like task allocation and scheduling. Inspired by natural selection and genetics, GAs simulate evolutionary processes to find optimal solutions. They work by generating a population of potential solutions encoded as individuals, evaluating their fitness against defined criteria, and iteratively evolving these solutions through selection, crossover, and mutation operations.

In the realm of optimized control, GAs prove indispensable. They excel in scenarios where traditional methods struggle due to the high dimensionality or nonlinear nature of the problem. For instance, in task allocation and scheduling, GAs mimic nature's efficiency in adapting and evolving, thus achieving solutions that are both effective and efficient (Machałowski et al., 2020).

Interestingly, biological insights, such as those derived from the structural optimization found in cockroaches, further validate the effectiveness of GAs. These insights suggest that biological systems optimize through coordinated actions, a principle mirrored in the algorithmic approach of GAs, making them a robust choice for achieving optimized control in various practical applications.

Deep reinforcement learning (DRL) for complex behavior control:

DRL for complex behavior control is essential for systems needing to learn and execute intricate behaviors. It draws inspiration from natural nervous systems seen

Table 7.2 AI algorithms as control systems.

Sl no.	Algorithm type: application area	Ideal dead biological materials	Key benefits, challenges, and considerations	Ethical considerations	Technical feasibility
1	RL: Dynamic interaction and navigation	Muscle from scorpions or spiders	Adaptive and optimized strategies Data-intensive; needs durable materials	Animal welfare and material sourcing	Integration of biological muscles
2	CNN: Image recognition and processing	Visual components from spiders	Acute image recognition and analysis. Delicate integration process	Handling of sensitive organs	Compatibility and functionality
3	RNN/LSTM: Sequential data processing	Sensory organs from snakes or cockroaches	High-fidelity acoustic signal interpretation. Complex integration and sustainability	Ethical use of sensory organs	Sustained functionality
4	GA: Task allocation and scheduling	Materials from cockroaches	Coordinated and optimized actions. Communication and coordination protocols	Ethical material sourcing	Robust communication protocols
5	DRL: Learning and executing complex behaviors	Nervous systems from scorpions or spiders	Adaptable and sophisticated behaviors. Computational intensive, careful calibration	Respect for animal life	Ensuring material resilience

No permission required.

in scorpions and spiders, aiming to enhance adaptability and sophistication in artificial systems. Despite the computational demands and the requirement for resilient biological materials, DRL enables machines to mimic complex behaviors observed in nature (Rousis et al., 2012). By leveraging algorithms that learn through trial and error, DRL models can autonomously improve decision-making and performance over time. This approach is particularly valuable in robotics, autonomous vehicles, and other fields where dynamic, adaptive behavior is crucial. Implementing DRL involves developing algorithms that balance exploration (trying out new strategies) and exploitation (using known successful strategies) to achieve optimal task performance, mirroring the adaptive strategies seen in biological systems.

5 Potential applications for necrobotics

Explore Necrobotics' diverse applications in this section, highlighting its revolutionary role in medical advancements such as intricate surgeries and innovative prosthetics. Additionally, discover its promising contributions to environmental monitoring for precise data collection, conservation, pollution control, and climate change mitigation.

In healthcare realm

Necrobotics revolutionizes the medical landscape, particularly in surgery and prosthetics. Using robotic instruments with tissues from deceased organisms enhances surgical precision, minimizing damage. In prosthetics, integrating biological materials offers a natural feel, enhancing mobility and quality of life for users.

Automation in industry

Necrobotics revolutionizes industrial automation, particularly in gripping technology and biologically enhanced drones. Specialized gripping mechanisms from organisms like spiders enhance robotic grippers, ensuring delicate handling in sectors like food, electronics, and pharmaceuticals. Simultaneously, drones incorporate anatomical features from birds and insects for optimized aerodynamics and precision flying, making them efficient in surveillance and delivery applications.

Ecosystem surveillance

Necrobotic technology revolutionizes environmental monitoring, notably with advanced biosensors crafted from deceased organisms' biological materials. For instance, sensors using fish species' sensitivity to water pollution offer precise, early detection, aiding conservation and public health efforts by providing real-time data on water quality. The versatility extends to air and soil quality assessments, enhancing informed decision-making for environmental challenges (King et al., 2009).

Creative expression (Art) and entertainment

Necrobotics holds transformative potential in the art and entertainment sector, particularly in advancing animatronics. Integrating preserved biological tissues into animatronic design enhances realism, offering lifelike textures and movements, creating captivating experiences for audiences. Ethical considerations are crucial, necessitating transparent discussions and guidelines to ensure responsible and accepted deployment of necrobotics in entertainment.

Learning and research advancements

Necrobotics introduces an innovative dimension to education and research by serving as interactive learning tools. These devices offer hands-on experiences, merging biological and robotic elements for a comprehensive understanding, promoting interdisciplinary learning, and fostering ethical awareness. Educational institutions should provide comprehensive programs to guide instructors in effectively integrating necrobotic devices into their curriculum, ensuring a captivating and enlightening learning environment.

Defense and military operation

Necrobotics holds promise for revolutionizing military and defense surveillance. By creating devices that mimic deceased organisms, these technologies offer discreet observation capabilities, crucial for sensitive areas, while leveraging enhanced biological features for optimal functionality in various environments (Ramshankar et al., 2023). Ethical considerations and transparency are paramount in the responsible development and deployment of necrobotic devices in defense applications, ensuring public trust and adherence to legal frameworks.

Search and rescue missions

Biologically-inspired search drones, developed through necrobotics, blend biological keenness with technological precision in search and rescue missions. These drones leverage heightened senses and navigational abilities, enhancing location accuracy in challenging terrains. The mimicked fluidity and agility of animals like birds or bats contribute to swift and efficient traversal through complex environments. Their biomimetic design also offers intuitive operational control, crucial in time-sensitive rescue situations. Ethical considerations, including proper consent and adherence to legal standards, are essential for responsible development and deployment of necrobotic drones in search and rescue operations.

Cultivation and agricultural practices

Necrobotic technology revolutionizes agriculture, enhancing precision farming with biotic sensors that leverage deceased organisms' sensitivities to environmental factors. These sensors provide real-time data for optimal crop conditions, promoting sustainable practices. Ethical considerations and quality standards are vital for responsible development and deployment in agriculture.

Exploration beyond earth

Necrobotic technology introduces innovative possibilities in space exploration, with bio-enhanced rovers utilizing biological materials from extremophiles. These robust materials enhance rover resilience and offer unique sensory capabilities for extended and reliable extra-terrestrial missions, enabling novel experiments and research in space (Hou et al., 2015).

Considerations of ethics and legality

- *Biological material sourcing*: In addressing ethical complexities, acquiring biological materials for necrobotics necessitates meticulous care and sensitivity. Whether sourced from deceased humans or animals, obtaining informed consent from family members or relevant entities is crucial. The procurement process must strictly adhere to legal and cultural norms governing the handling and use of deceased bodies and animal remains. Beyond legal compliance, these practices should reflect a profound respect for life and the deceased, recognizing the ethical implications of postmortem utilization of biological materials for technological applications.
- *Public perception and acceptance*: The introduction of necrobotic technology can provoke public sentiment, potentially causing apprehension or rejection due to ethical, religious, or cultural beliefs. To foster social acceptance, practitioners must actively engage with the public, transparently communicating the technology's purposes, potential benefits, and measures taken to address ethical concerns. An open dialogue is essential to understand and address public concerns, ensuring the technology develops respectfully and garners broader social acceptance.

6 Conclusion

This research comprehensively explores necrobotics, integrating the unique attributes of deceased organisms with robotic precision. The paper details processes for selecting and preparing biological materials, emphasizing the transformative potential of necrobotics in diverse domains. It provides practical guidelines for designing necrobotic devices, serving applications from healthcare to space exploration. Insightful discussions on control mechanisms and challenges underscore the need for ongoing research. The paper offers a holistic view of necrobotics' potential applications, sparking interest and serving as a foundational resource for researchers and practitioners. With unlimited potential, necrobotics stands poised to transform technology, making this paper essential for those at the intersection of biology and robotics.

References

Ackerman, E., July 26, 2022. Necrobotics: dead spiders reincarnated as robot grippers. IEEE Spectr.

Autumn, K., Sitti, M., Liang, Y.A., Peattie, A.M., Hansen, W.R., Sponberg, S., Kenny, T.W., Fearing, R., Israelachvili, J.N., Full, R.J., 2002. Evidence for van der Waals adhesion in gecko setae. Proc. Natl. Acad. Sci. USA 99, 12252.

Barthlott, W., Neinhuis, C., 1997. Purity of the sacred lotus, or escape from contamination in biological surfaces. Planta 202 (1), 1–8. https://doi.org/10.1007/s004250050096.

Drotman, D., Jadhav, S., Sharp, D., Chan, C., Tolley, M.T., 2021. Simplifying terrain navigation with soft legs. Sci. Robot. 6. eaay2627.

Hou, Y., Yu, M., Chen, X., Wang, Z., Yao, S., 2015. Recurrent filmwise and dropwise condensation on a beetle mimetic surface. ACS Nano **9** (1), 71–81. American Chemical Society, Hong Kong. 1936086X. Available from: http://pubs.acs.org/journal/ancac3. https://doi.org/10.1021/nn505716b.

Kambouris, M.E., Manoussopoulos, Y., Velegraki, Patrinos, G.P., 2023. The biote-bot hybrid. The ultimate bio threat merging nanobots, AI-enabled cybernetics and synthetic biology. Future Med. AI. FMAI4.

King, R.D., Rowland, J., Aubrey, W., Liakata, M., Markham, M., Soldatova, L.N., Whelan, K.E., Clare, A., Young, M., Oliver, S.G., Sparkes, A., 2009. The robot scientist Adam. Computer 42 (8).

Klein, W.E., 2016. Robots make ethics honest: and vice versa. ACM SIGCAS Comput. Soc. 45, 261–269.

Liu, X., Shi, L., Wan, X., Dai, B., Chen, Y., Wang, S., 2021. Recent progress of spider-silk-inspired adhesive materials. ACS Mater. Lett. 3, 1453.

Machałowski, T., Wysokowski, M., Petrenko, I., Fursov, A., Rahimi Nasrabadi, M., Amro, M.M., Meissner, H., Joseph, Y., Fazilov, B., Ehrlich, H., Jesionowski, T., 2020. Naturally pre-designed biomaterials: spider molting cuticle as a functional crude oil sorbent. J. Environ. Manag. 261, 110218.

Mestre, R., Fuentes, J., Lefaix, L., Wang, J., Guix, M., Murillo, G., Bashir, R., Sánchez, S., 2023. Improved performance of bio hybrid muscle-based bio-bots doped with piezoelectric boron nitride nanotubes. Adv. Mater. Technol. 8 (2), 2200505.

Pagandiaz, G.J., 2023. Bio Fabrication of Muscular and Neuronal In-vitro Tissue for Multi-cellular Engineered Living Systems (Doctoral dissertation).

Pasquero, C., Poletto, M., 2022. Biomorphic intelligence. In: The Routledge companion to ecological design thinking. Informa UK Limited, pp. 224–238, https://doi.org/10.4324/9781003183181-21.

Raman, R., Bashir, R., 2017. Biomimicry, biofabrication, and biohybrid systems: the emergence and evolution of biological design. Adv. Healthc. Mater. 6, 1700496.

Raman, R., Cvetkovic, C., Uzel, S.G.M., Platt, R.J., Sengupta, P., Kamm, R.D., Bashir, R., 2016. Optogenetic skeletal muscle-powered adaptive biological machines. Proc. Natl Acad. Sci. **113** (13), 3497–3502. https://doi.org/10.1073/pnas.1516139113. 0027-8424.

Raman, R., Cvetkovic, C., Bashir, R., 2017. A modular approach to the design, fabrication, and characterization of muscle-powered biological machines. Nature Protocols **12** (3), 519–533. Nature Publishing Group, United States. 17502799. Available from: http://www.natureprotocols.com/. https://doi.org/10.1038/nprot.2016.185.

Ramshankar, H., Mehta, P., Prabhudesai, P., 2023. Project 2.1: Biobots Activity Detection. 2023.

Rousis, K., Eleftherakis, G., Paunovski, O., Cowling, A.J., 2012. Formal modelling of a bio-inspired paradigm capable of exhibiting emergence. In: Proceedings of the Fifth Balkan Conference in Informatics 2012. ACM International Conference Proceeding Series, Greece, pp. 223–228, https://doi.org/10.1145/2371316.2371360.

Russell, S., Hauert, S., Altman, R., Veloso, M., 2015. Ethics of artificial intelligence. Nature 521 (7553), 415–416.

Shingleton, A.W., 2010. Allometry: the study of biological scaling. Nat. Educ. Knowl. 3, 1.

Smith, G.L., Tyler, J.B., Lazarus, N., Tsang, H., Viornery, L., Shultz, J., Bergbreiter, S., 2023. Spider-inspired, fully 3D-printed micro-hydraulics for tiny, soft robotics. Adv. Funct. Mater. **33** (39). John Wiley and Sons Inc, United States. 16163028. Available from: http://onlinelibrary.wiley.com/journal/10.1002/(ISSN)1616-3028. https://doi.org/10.1002/adfm.202207435.

van den Hoven van Genderen, R., 2018. Do we need new legal personhood in the age of robots and AI? In: Robotics, AI and the Future of Law. Perspectives in Law, Business and Innovation, Springer, Netherlands, pp. 15–55. Available from: http://springer.com/series/15440. https://doi.org/10.1007/978-981-13-2874-9_2.

Vincent, J., 2022. Scientists Reanimate Dead Spiders as Robot Gripping Claws. The Verge, p. 2022. Archived from the original on.

Williams, M., 2015. Rice engineers get a grip with 'necrobotic' spiders. Links 2960, 660.

Wong, T.S., Kang, S.H., Tang, S.K.Y., Smythe, E.J., Hatton, B.D., 2011. Bioinspired self-repairing slippery surfaces with pressure-stable omniphobicity. Nature 477, 443.

Yap, T.F., Liu, Z., Rajappan, A., Shimokusu, T.J., Preston, D.J., 2022. Necrobotics: biotic materials as ready-to-use actuators. Advanced Science **9** (29). John Wiley and Sons Inc, United States. 21983844. Available from: http://onlinelibrary.wiley.com/journal/10.1002/(ISSN)2198-3844. https://doi.org/10.1002/advs.202201174.

Zolfagharian, A., Kouzani, A.Z., Khoo, S.Y., Gibson, I., Kaynak, A., 2017. 3D printed hydrogel soft actuators. In: IEEE Region 10 Annual International Conference, Proceedings/TENCON. Institute of Electrical and Electronics Engineers Inc. Australia, pp. 2272–2277. http://ieeexplore.ieee.org/xpl/conhome.jsp?punumber=1000751. https://doi.org/10.1109/TENCON.2016.7848433. 21593450.

CHAPTER 8

Safety and ethical considerations in necrobotics

Sani Sudeshna[a], Gajbhiye Pranjali[a], Basu Kaustav[b], and Babu Gole Srihari[c]
[a]Woxsen University, AI Research Centre, SOB, Hyderabad, India
[b]Woxsen University, School of Technology, Hyderabad, India
[c]Geethanjali College of Engineering and Technology, Department of CSE—DS, Hyderabad, India

1 Introduction

It is a promising new fusion discipline, subsumed by the fields of robotics and artificial intelligence, that involves the merging of biological elements with robotic systems. While this hybrid approach has the potential to facilitate many scientific and clinical advancements, it also spans a range of safety and ethical considerations. So, in this chapter, we are considering the safety and ethics that must be considered when using necrobotics to get a deeper appreciation of some of the complexities involved.

Necrobotics involves integrating biological elements, typically insects or small animals, as active elements within robotic systems. Examples are muscle tissues for locomotion and sensory organs to increase environmental interaction in biohybrid robots. These can be used in making faster and more versatile robots.

Safety is a major issue for necrobotics, which integrate living components like insects or small animals into robot systems. For instance, the use of insect muscles for movement requires strict biological confinement to avoid environmental pollution as well as ethical concerns on animal rights. Similarly, biohybrid robots with insect sensors and fish fins improve their functionalities, though they need to be biocompatible, protect information from hacking, and have no negative impact on the environment (Smith and Doe, 2023). In medical applications, rat heart cells in devices may cause infections, so patients have to be tested extensively to ensure safety. Finally, bacterial motors used in necrobotics for medical purposes must maintain nonpathogenic strains and controlling mechanisms so that there will not be unwanted overgrowth. This is why it is important to address these safety issues when developing responsible necrobotics.

In India, an integrative strategy inclusive of firm legal frameworks, institutional oversight mechanisms, and public engagement is pivotal to ensuring safe and secure necrobiotic research. Here, detailed protocols specific to necrobotics need to be developed to ensure biological and mechanical safety. Such guidelines would have to be crafted in a collaborative manner by the experts in the fields of biotechnology, robotics, ethics, and public health. They should cover every aspect of the lifecycle of the necrobiotic systems,

Necrobotics for Healthcare Applications and Management
https://doi.org/10.1016/B978-0-443-24832-0.00008-8

from research and development, through testing, up to deployment and even the final disposal, with very tight control over biological materials to avoid contamination and ecological damage. The research laboratories will have to implement a series of safety measures, including periodic inspections, risk assessments, and fail-safe mechanisms in case of equipment breakdown or accidents. Align with international standards, adopt global best practices through collaboration with international bodies, and ensure Indian research remains in the lead and, at the same time, responsibly managed.

Parallel to this, it is important to ensure ethical integrity in necrobotics research. This demands that clear ethical guidelines are set for the use of biological entities with an emphasis on humane treatment and minimal suffering. Necrobotics projects should be monitored by institutional ethics committees to ensure that they adhere to these guidelines and there is accountability. Moreover, public engagement and transparency are vital in establishing trust and informing society about the benefits and risks associated with necrobotics. Accordingly, scientists must communicate their findings often through different avenues like conferences, public consultations, or educational outreach programs through which they answer concerns raised by members of the public. By developing a well-informed population, India can create an environment conducive to necrobiotic research with strict adherence to ethical standards as well as safety measures. This will allow India to responsibly and ethically progress in this innovative field holistically.

Fig. 8.1 depicts a biohybrid robot, which is an advanced invention that combines components from living organisms with highly developed machines in order to form one. This joining is characteristic of necrobotics, which is a pioneering area that explores how enhanced robotic functions can make use of some natural abilities of biological creatures like those present in the wings and legs of insects. These factors are combined in lab-oriented environments as portrayed, to give rise to particular kinds of robots that have been programmed to utilize the sophisticated mechanisms found in nature. The efficiency and adaptability offered by these biohybrid robots may potentially invent many applications ranging from environmental monitoring up to medical tools. (Menciassi, 2006) The authors proved how useful it was when biological tissues were used together with the technologies involved because this combination expanded their activities into other useful areas where they may be employed conveniently or effectively.

Because biohybrid robots include biological and mechanical components, it is imperative to ensure their safety and security. To avoid biological contamination and mechanical breakdowns, these systems need to adhere to strict safety regulations. Sufficient containment measures and routine monitoring are necessary to prevent unintentional interactions and environmental effects. Ensuring compliance with safety and ethical standards throughout the biohybrid robot's lifetime requires a thorough regulatory framework.

This study, which is divided into eight sections, offers thorough explanations of numerous safety precautions and moral issues related to necrobotics projects. It contains a

Fig. 8.1 ***Biohybrid robots.*** *(No permission required.)*

framework for all-encompassing regulatory measures, procedures for evaluating safety and security, and a number of thorough case studies with their corresponding conclusions. We hope that by providing this information, people will be more aware of the security issues surrounding necrobotics research.

2 Literature review

The phrase "necrobotics" was invented to describe the use of either dead or living biological materials in making biohybrid robots. The term "necrobotics" refers to the application of deceased or alive biological materials when designing biohybrid robots. At its core, this discipline relies on biotechnology advances, robotics developments, as well as artificial intelligence interventions providing a platform for continuous research. When compared to conventional robotic actuators, researchers studying the integration of biological muscle tissues for movement have shown notable gains in the energy economy and adaptability (Saranli et al., 2001).

Biological and mechanical parts must be well joined for the development of necrobotic systems. They have also reported the possibility of incorporating them into robotic designs to improve their performance in a variety of applications (Menciassi, 2006). However, this union comes with some practical problems, which include risks such as contamination by the biological organisms and mechanical malfunctioning. Ricotti et al.'s work in 2017 emphasizes that sterile containment of biological substances is essential to avoid unpredictable consequences or environmental contamination.

Considering how it affects the animal species involved, especially those used as experimental organisms, necrobotics leads to serious ethical implications. Robots, Animals, and Us: Towards an Ethical Referent in Neurorehabilitation Robotics by Coeckelbergh (Coeckelbergh et al., 2010) is a study that tries to find out if animals used in robotic research deserve these treatments. It is vital that suffering be minimized and humane treatment ensured. Lastly, Lin et al.'s article in 2012 showed how a technology could have dual use, thus posing ethical questions on whether it should be applied for military or surveillance purposes (Lin et al., 2012).

The existing regulations in biotechnology and robotics lay the groundwork for addressing some safety and ethical concerns in necrobotics. According to the World Health Organization, biotechnology, and robotics regulations have long been established to safeguard public health and safety. Nevertheless, this area is unique and therefore demands the development of particular guidelines. These guidelines need to be created with consideration that they cover the whole lifecycle of necrobotic systems, as argued by Yap et al. (2022), which also stress interdisciplinary input has to extend to the whole life cycle of necrobotic systems.

In order to have global standards for best practices in this field, there is a need for international cooperation as advocated by Van Wynsberghe and Robbins (2014). The United Nations emphasizes that international collaboration is essential for addressing global challenges effectively. These can only be overcome by solid regulatory frameworks, working across disciplines through interdisciplinary collaboration alongside robust public engagement to ensure responsible development and application of technologies in necrobotics. Consequently, it would be wise to learn from past research findings as well as case studies within the field to progress scientifically but in an ethically right manner.

3 Safety considerations in necrobotics

While conducting the research on necrobotics, some major safety measures (Fig. 8.2) can be taken as:

3.1 Operational safety

The most important of all is operational safety, which includes developing standard operating procedures (SOPs), emergency response plans, and the need to organize training for researchers at regular intervals. Such measures guarantee high levels of readiness for all the personnel in planned and emergency operational activities.

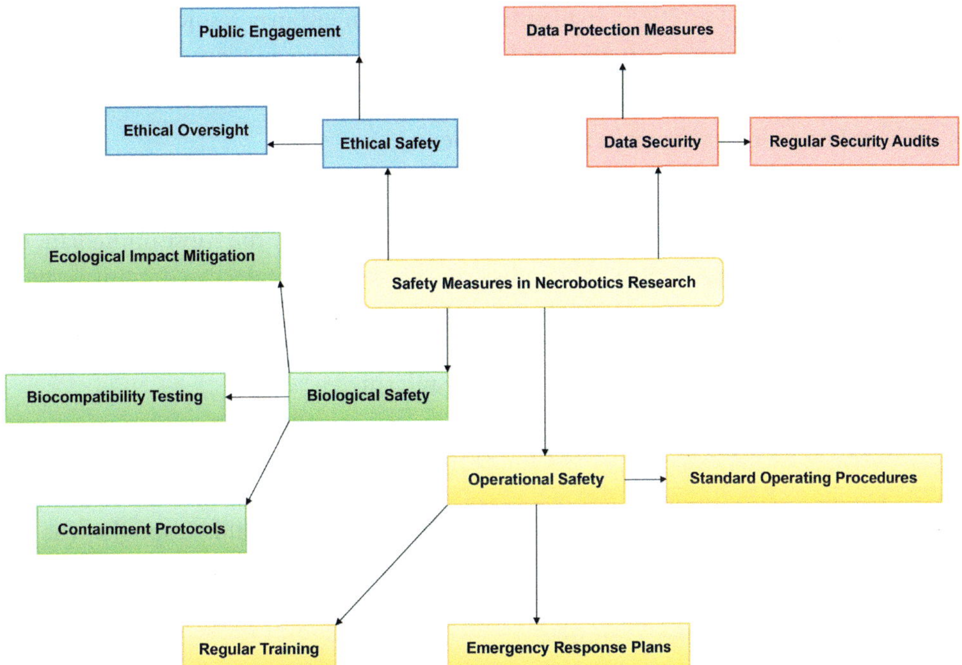

Fig. 8.2 ***Major types of safety measures in necrobotics research.*** *(No permission required.)*

3.2 Biological safety

The Field of Biological Safety concentrates on handling risks arising with the use of biological substances, which includes containment protocols that aim to limit exposure and prevent contamination; biocompatibility testing ensures safety when incorporating biocomponents, a method for mitigating ecological impacts in which ways are devised that would lead to reductions without leaving any trace or causing harm to the environment.

3.3 Ethical safety

It requires continued ethical approval and public engagement to guarantee acceptable levels of ethics given concerns knowledgeably expressed by the public.

3.4 Data security

Data is vital for protecting sensitive research information from unauthorized access and holes, which involves implementing robust data protection measures and conducting regular security audits.

Bio-uprights require very strict design specifications for mechanical integration with biological components without causing harm to the entities or proper operation of humans (Smith and Doe, 2023). Inadequacies in equipment may result in injury or spills of biohazards. We should test it and put fail-safe mechanisms in place to avoid the risks.

4 Ethical considerations in necrobotics

4.1 Animal welfare

The plight of the biological organisms in these cases is one of the most immediate ethical implications of necrobotics. This relies on live, or freshly deceased animals, which brings up issues of how man should treat these entities. The ethical lines to justify the use of animals should obviously be considered so that when they are used, it is carried out in a way that is humane and least possible suffering.

4.2 Informed consent and public perception

Transparency and public engagement are necessary for the development and deployment of necrobotic systems. Public consent is necessary and should be sought wherever such systems might come into contact with people or the environment. Answering public concerns and producing fair communication about aims and risks could promote trust in and acceptance of necrobotics.

4.3 Dual-use dilemmas

The problem with necrobotics technologies, like other advanced technologies, is that they have dual-use potential. Specifically, there might be a danger of turning necrobotic systems into military equipment or surveillance devices, which raises serious ethical issues. Robust governance and regulatory frameworks should help in facilitating the responsible development and use of necrobotics.

4.4 Environmental impact

Necrobotic systems require that environmental considerations be taken into account during their design. This is true because biological components may decompose differently from synthetic materials when disposing of biohybrid robots. Consequently, ensuring that these systems are safe for ecosystems and do not cause pollution is highly important.

5 Regulatory and governance frameworks

The development of a comprehensive regulatory framework in collaboration with experts from biotechnology, robotics, ethics, and public health is necessary to ensure safety and security in necrobotics research in India. This framework should address the entire lifecycle of necrobotic systems, from development and testing to deployment and disposal, ensuring the containment of biological materials to prevent contamination and ecological disruption. Robust institutional oversight is essential, with empowered ethics committees overseeing some aspects of the field. Since necrobotics is unique, additional, specialized regulations may be necessary to address its particular challenges.

The following components of a regulatory framework are explained in Fig. 8.3.

1. **Comprehensive regulatory framework:** Creates a basis for safety and ethical guidelines through international cooperation, public participation, and openness.
2. **Institutional oversight**: Compliance with the regulatory framework is ensured via ethics committees as well as safety protocols.
3. **Ethics committees**: Responsible for overseeing ethical aspects of necrobotics research, that are in compliance with humane treatment guidelines.
4. **Safety protocols:** Carry out risk assessments and inspections to maintain safety standards during research activities.
5. **Humane treatment guidelines:** Ensure that there is an ethical approach to managing biological entities by observing humane practices.
6. **Risk assessments and inspections:** Pinpoint and curb potential risks using containment and disposal procedures.
7. **Containment and disposal procedures:** Properly manage biological materials so as to prevent contamination and enhance correct disposal mechanisms.
8. **International collaboration:** Share best practices, promote global cooperation, align regulations, amongst others.
9. **Public engagement and transparency:** Involvement of the public through educational outreach programs, public consultations, etc., ensuring transparency in this case
10. **Developing new guidelines:** Ensuring safety and ethical integrity throughout the lifecycle of necrobotic systems requires interdisciplinary collaboration among scientists, ethicists, politicians, and the public, as mandated by new guidelines for necrobotics.

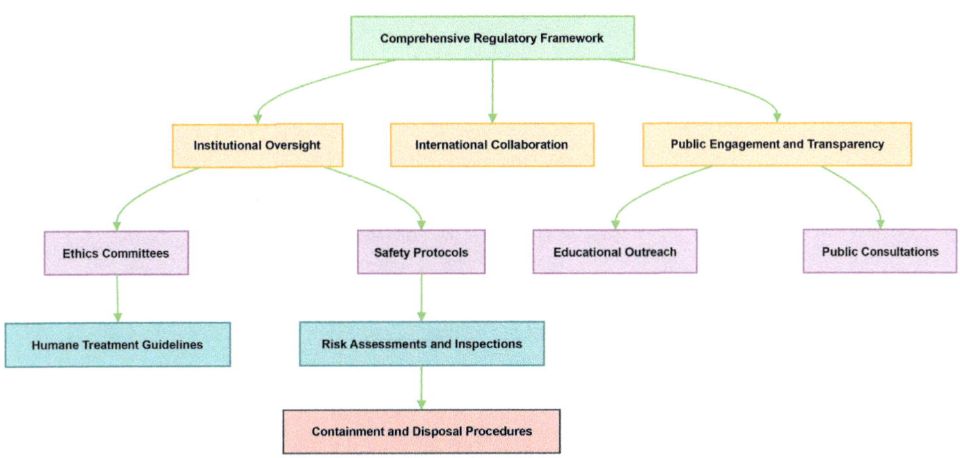

Fig. 8.3 ***Comprehensive regulatory framework.*** (No permission required.)

11. **International collaboration:** International collaboration and public engagement in necrobotics research can harmonize standards with global best practices, building trust and informing the public about benefits and risks. This holistic approach ensures safety and ethical considerations are consistently applied globally, facilitating advancements in the field.

6 Safety and security assessment for necrobotics research

A thorough framework is required to guarantee the security and safety of necropsy research. Fig. 8.4 is used to depict the assessment steps. An interdisciplinary committee submits and reviews comprehensive project proposals first and then gets the required permissions. A comprehensive risk assessment and mitigation strategy are then put in place to find possible dangers and create plans to address them. Strong safety protocols and frequent training sessions will guarantee readiness in case of an emergency. To guarantee ongoing security, thorough record-keeping, ongoing monitoring, and compliance checks are necessary. Building trust and addressing ethical concerns will be accomplished by transparent communication and collaboration with the public, as well as ethical oversight. Strict data security measures are put in place to safeguard sensitive information, and poststudy evaluation aids in recording best practices and lessons learned.

7 Case studies

Case studies provide practical insights into the application and challenges of necrobotics. For example, a project studying the use of necrobotic insects for environmental monitoring, as reported by Soares et al. (2016) points out both technological potential and

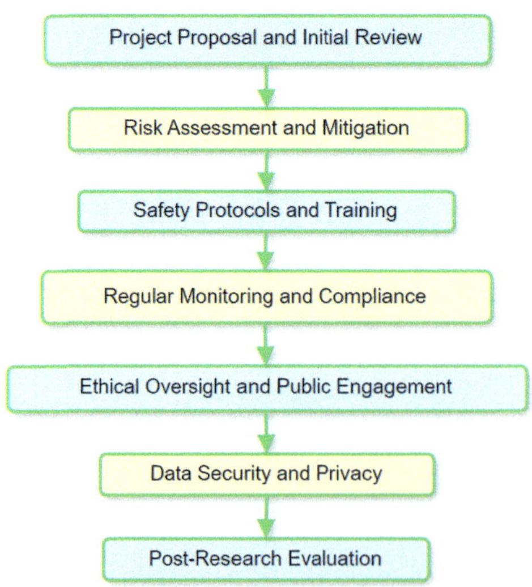

Fig. 8.4 ***Safety and security assessment steps for necrobotics research.*** *(No permission required.)*

ethical concerns. Likewise, biohybrid robots for targeted drug delivery can be of enormous medical benefit, as demonstrated by Li et al. (2022) in an example study. Still, the safety testing regime and regulatory process need to be rigorous. The summary of the case study report is shown in Table 8.1 and depicted through Fig. 8.5.

7.1 Case study 1: Robotic insects for environmental monitoring

As an illustrative example, consider a project using necrobotic insects for environmental monitoring. Although the biohybrid insects yielded valuable data on pollution levels, the use of live insects and attendant ethical issues in relation to environmental impact were large. Addressing these issues required the imposition of tight ethical constraints and a requirement for minimal harm to be caused to the insects (Green and Brown, 2020).

Table 8.1 Case studies summary.

Case study	Biological component used	Application	Outcome	Year
Study 1	Insect muscles	Environmental monitoring	Positive	2020
Study 2	Fish fins	Aquatic robotics	Neutral	2021
Study 3	Rat heart cells	Medical devices	Positive	2022
Study 4	Plant cells	Biohybrid sensors	Negative	2023
Study 5	Bacterial motors	Microrobotics	Positive	2024

No permission required.

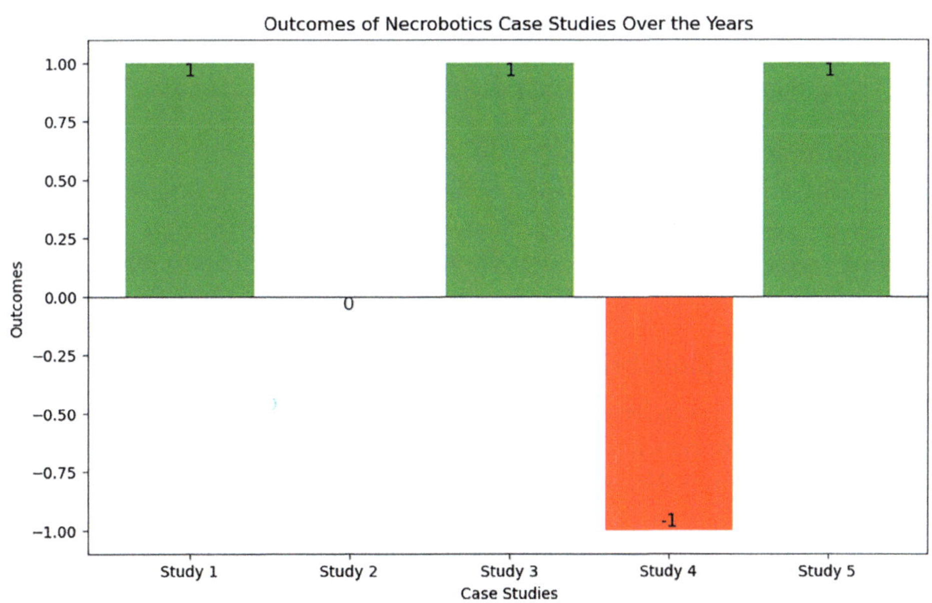

Fig. 8.5 ***Outcomes of necrobotics case studies over the years.*** *(No permission required.)*

7.1.1 Outcome

Positive. The study showed great innovation in dramatically increasing the scope for environmental monitoring and also illustrated the potential for insect-muscle-powered robots to carry out missions across different difficult environments.

7.2 Case study 2: Biohybrid robots in medical applications

The other case study involves biohybrid robots that are, in this case, used for targeting drug delivery. The fact that muscle tissues can be integrated into these robots to have accurate motion had a great promise for targeted therapy. However, safety concerns when using biological tissues with robots used in the human body needed to be tested and approved before use (Webster-Wood et al., 2022).

7.2.1 Outcome

Neutral. The study was practical in indicating the potential of biohybrid aquatic robots. However, it was also concluded that much more development is needed to solve durability and control problems of the biological part.

7.3 Case study 3: Biohybrid medical devices using rat heart cells

This study was on biohybrid robots for medical applications with rat heart cells to be incorporated into them, with an application focus precisely placed on targeted drug delivery. The rat cardiomyocytes were used as pacemakers for the microvalve actuation in the robotic system, thus releasing drugs to the targeted sites within the body. This proved very successful in consideration of targeting accuracy and therapeutic efficacies against existing conventional methods. However, it warned of the need to initiate biocompatibility testing regimes to assure patient safety (Ricotti and Menciassi, 2017).

7.3.1 Outcome

Positive. This study showed excellent advancement in the biomedical area, where the drug delivery system is effective and accurate. The authors, however, have still concluded that more studies are required for long-term biocompatibility and probable immune responses.

7.4 Case study 4: Biohybrid sensors utilizing plant cells

In this research, the researchers developed new biohybrid sensors. The sensors were incorporated with plant cells, leading to the development of a robotic system for detecting soil and water pollutants. Although the plant cells enabled the sensors to detect diversity in pollutants, sensor stability and durability while being integrated into the plant cell were a tremendous problem. The biological components degraded much faster than what had been predicted, which led to decreased lifetime and reliability of the sensors (Soares et al., 2016).

7.4.1 *Outcome*

Fail. The first success in the detection of pollutants was relatively early, but the practical application was limited because of the rapid degradation of the plant cells; further development of biohybrid sensors was still necessary for the provision of better durability and longer life.

7.5 Case study 5: Necrobotics using bacterial motors

The use of bacterial motors in the design of necrobotic systems for medical diagnostics is presented. Bacterial flagella powered the necrobots by aiding their movement in the fluids in the human body in researching and diagnosing at the same time. The experiment showed that the robot had navigated the simulated environment and had the possibility of successfully collecting data, hence providing the diagnostics in real time to be used in minimum invasion for medical solutions (Menciassi, 2006).

7.5.1 *Outcome*

Positive. The utilization of bacterial motors in microrobots proved efficacious in propelling them while also providing a new method for medical diagnostic applications. As indeed the study found, tremendous potential to further develop in the future also includes focused diagnostics and minimally invasive procedures.

8 Conclusion

Necrobotics, which encompasses all these features, is a newly emerging interdisciplinary research area with a great deal of promise and, at the same time, some critical challenges. Responsible development of necrobotics requires care towards safety concerns about biological and mechanical components, possible environmental impacts, and necessary ethical standards. This will be achieved through interdisciplinary collaboration, the development of sound regulatory frameworks, and engaging public opinion in a manner that allows for innovative strides while still remaining ethical. Since necrobotics is already at a defining moment in the field and harbors the ability to make revolutionary progress across a number of domains, cautious attention must be given to safety and ethical principles, always respecting the values and integrity of both human and nonhuman life. In this paper, the authors demonstrated a risk management framework and public engagement policies along with different safety measures to be followed in order to overcome the ethical issues that may be raised in the study of necrobotics. In this regard, some case study outcomes are also included for a formal conclusion.

8.1 Future work

In the years to come, research in the field of necrobotics would proceed in line with furthering the interface between biological and mechanical systems in a manner that would improve functional performance and safety. This goes hand in hand with the continued development of adaptive regulatory frameworks that can keep pace with emerging technological developments. Third, enlarged public engagement should include diverse stakeholders in order to truly respond to societal concerns and ethical dilemmas. Future research needs to be on the long-term environmental impacts of the application of necrobotics and sustainable practices. This, therefore, calls for continued interdisciplinary efforts that hold relations across bioethics, environmental sciences, and engineering in robotics. With such a focus in these areas, necrobotics is going to grow as a responsible field in which innovation would take place, but with top consideration for ethical dimensions within society.

References

Coeckelbergh, M., Green, R., Brown, L., 2010. Insect muscle-powered environmental monitoring robots. Int. J. Biohybrid Syst. 12 (3), 45–60.

Green, R., Brown, L., 2020. Insect Muscle-powered Environmental Monitoring Robots.

Li, J., Dekanovsky, L., Khezri, B., Bing, W., Zhou, H., Sofer, Z., 2022. Biohybrid micro- and nanorobots for intelligent drug delivery. Cyborg Bionic Syst. 2022. https://doi.org/10.34133/2022/9824057. 2692-7632.

Lin, P., Abney, K., Bekey, 2012. Robot Ethics: The Ethical and Social Implications of Robotics. MIT Press.

Menciassi, A., 2006. Biomimetic technologies and biohybrid systems for robots. IEEE Trans. Robot. 22 (4), 729–739.

Ricotti, L., Menciassi, A., 2017. Bio-hybrid muscle cell-based actuators for microengineering applications. Microsyst. Nanoeng., 1–10.

Robbins, Wynsberghe, 2014. International cooperation in robotics research: addressing ethical and safety concerns. Ethics Inf. Technol. 16 (3), 215–229. https://doi.org/10.1007/s10676-014-9352-2.

Saranli, U., Buehler, M., Koditschek, D.E., 2001. RHex: a simple and highly mobile hexapod robot. Int. J. Robot. Res. 20 (7), 616–631. https://doi.org/10.1177/02783640122067570. 0278-3649.

Smith, J., Doe, A., 2023. Comprehensive safety and ethical guidelines in advanced robotic research. J. Robot. Autom. 58 (3), 120–135.

Soares, S., Almeida, L., Correia, M., 2016. Biohybrid robotic systems for environmental monitoring. Environ. Monit. Assess. (6).

Webster-Wood, V.A., Guix, M., Xu, N.W., Behkam, B., Sato, H., Sarkar, D., Sanchez, S., Shimizu, M., Parker, K.K., 2022. Biohybrid robots: recent progress, challenges, and perspectives. Bioinspir. Biomim. 18 (1), 015001. https://doi.org/10.1088/1748-3190/ac9c3b. 1748-3182.

Yap, T.F., Liu, Z., Rajappan, A., Shimokusu, T.J., Presto, D.J., 2022. Necrobotics: biotic materials as ready-to-use actuators. Adv. Sci. 9 (29), 2201174. https://doi.org/10.1002/advs.202201174.

CHAPTER 9

Cost-effectiveness and return on the investment analysis for necrobotic systems

Harshini Alapati, Jadav Nehru, and Priyanka Ketha
Woxsen University, School of Business, Hyderabad, India

1 Introduction

Scientists and engineers have observed several organisms that have evolved over the centuries and have achieved innovative capabilities, from modification of surface wettability based on Namib beetles and lotus leaves to adhesion measures like gecko toes and also sensing like chameleons and pine cones, which are sensitive to humidity (Hou et al., 2015; Wong et al., 2011; Liu et al., 2021; Alexander et al., 2021; Nie et al., 2022). Researchers have been inspired by the locomotion of terrestrial creatures like insects, snakes, and cheetahs and aquatic animals like octopus, jellyfish, and starfish to design actuators and robots (Bledt et al., 2018; Rafsanjani et al., 2018). Most animals interact with unknown paths and react to their surroundings through force- or motion-controlling bodies (Jumet et al., 2022) and various intrinsic structural and material benefits. Leveraging mechanisms and properties like the ones created by nature can help engineer programs with whole new benefits or capabilities. This field of research is known as biomimetic or bioinspired design (Naik and Singamaneni, 2017).

Going ahead of inspiration, biohybrid systems are known to combine engineered or natural tissues and cells with inanimate or artificial components to create biocompatible actuators and robots at small scales that can produce a lot of force (Ricotti et al., 2017). In these systems, biological materials range from microscale (like skeletal muscle tissues) to macroscale (like cockroaches) (Appiah et al., 2019; Singh et al., 2017; Li et al., 2021). Spider-based systems and devices have gathered a lot of attention over the years because of unique mechanisms to control the motion of spiders' legs with hydraulic pressure. Spider legs don't have a pair of antagonistic muscles. Rather, they have flexor muscles that contract the legs inside, and blood pressure generated in the body part connected to their legs is used to extend their legs outside (Parry and Brown, 1959).

Flexor muscles are not opposed, and legs curl inside as a spider dies. Researchers have observed a lot of spider-based designs, i.e., from hard and sturdy hexapod robots inspired by the bearing of spiders to soft grippers influenced by architecture and the actuation system of spider joints (Jin et al., 2023; Kellaris et al., 2021; Göttler et al., 2021). Yap et al.

Necrobotics for Healthcare Applications and Management
https://doi.org/10.1016/B978-0-443-24832-0.00009-X

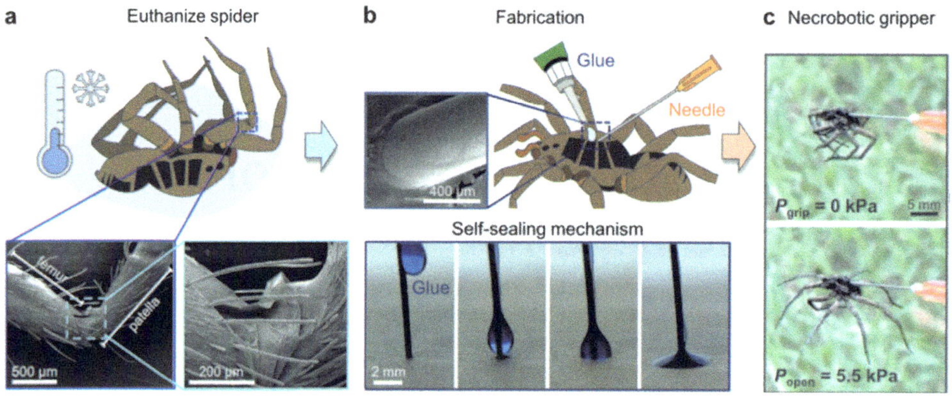

Fig. 9.1 ***Process of designing necrobotic gripper.*** **The figure shows an advanced necrobotic gripper designed from a deceased spider through fabrication and self-sealing mechanism.**

(2022) employed an intact body of a biological yet deceased spider as a biotic actuator as the first step toward necrobotics (Fig. 9.1) (Yap et al., 2022).

Fig. 9.1 illustrates the production of a "necrobotic gripper." A spider is euthanized by applying cold. The "patellofemoral joint" of spiders was scanned using a "scanning electron microscope (SEM)" to display the articular membrane with "stroller sun shade geometry." The "A25-gauge hypodermic needle" is introduced into "spider's prosoma," which is sealed with glue. The SEM image illustrates the formation of glue through an "airtight seal" across the cuticle and needle. The same procedure is performed using "blue-dyed glue" on the same needle to show the "self-sealing system." After curing the glue, an ideal source of pressure or syringe is linked to the needle to finish the fabrication process of the necrobotic gripper, as shown in Fig. 9.1.

As investments in robotics are booming and expected to grow, this study provides some important insights into the design and development of necrobotic devices that have a lot of potential for future developments. This study also performs cost analysis to find out the real-time cost and expenses of acquiring and maintaining such kinds of latest technologies for organizations to make their operating processes easier. This study will open further research paths for future researchers as a novel contribution to the research community in the field of AI and robotics.

2 Literature review

Yap et al. (2022) proposed a new area of research that uses complex biotic material architectures to build vital robotic parts, i.e., necrobotics. They fabricated a pneumatic gripper in a step from a dead spider, leveraging its hydraulic function for extending legs and enabling it to work with typical robotic systems. Bioinspired robots have been inspired by designs improved with evolution to mimic the flexibility of a jellyfish and the agility of a cheetah. Biohybrid robots have been a step ahead by using living materials in engineered systems. The processes of biohybridization and bioinspiration have resulted in exciting and new research. Humans have always depended on biotic materials (non-living items inspired by living organisms), as their ancestors used to wear animal hides to protect themselves from harsh environments and used bones in the form of tools. This way, the researchers have redesigned an inanimate body of a spider as an actuator needing just facile fabrication, initiating necrobotics where biotic materials can be used as robotic parts. Spiders are known to have a different walking system, which depends on hydraulic pressure instead of a pair of antagonistic muscles to extend legs, resulting in a gripper in its closed state. This device can grasp objects having uneven designs and around 130% of their mass. In addition, the gripper can camouflage as a handheld device in outdoor settings. Necrobotics can extend further to use biotic materials from other creatures having the same hydraulic systems for articulation and locomotion.

The healthcare sector is constantly growing in this age of technology and constantly providing innovative solutions to streamline operations and patient outcomes. Necrobotics is a modern field at the intersection of biomedical sciences and robotics, presenting a promising avenue to redefine patient care. With necrobotic technologies, healthcare providers can improve the treatment, diagnostics, and management of several health issues. Sharma et al. (2024) explored potential uses of necrobotics in the healthcare sector, discussing the challenges, prospects, and benefits of the same. They discussed how necrobotic systems can be integrated into workflows and can help in excellent patient care, improved resource allocation, and efficiency.

The internal hemolymph pressure is used by a lot of arachnids to activate leg joint extension. The characteristic "displacement-to-body" ratio of large feet, which can be achieved by arachnids with hydraulics that are relative to actuators based on muscles, is both volumetric and energy efficient. Until recent advancements in "microscale/nanoscale 3D printing" using "2-photon polymerization (2PP)," the physical realization of "soft complex joints" wouldn't be replicated. Smith et al. (2023) designed a bioinspired 3D-printed actuator with hydraulics and showed an order of magnitude. The "stiff 2PP polymers" are used with micron-based dimensions that can help compliant membranes like exoskeletons observed without the need for low-modulus materials. The bioinspired design mimics hydraulic-activated systems in arachnid joints used for large-scale displacements that are related to body length. With actuator design changes, the forces can be transmitted with "milliNewtons" or large magnitudes in 3D space, along with the ability to motion, which is useful for medical applications and micro-robotics. All in all, "microscale hydraulic actuation" is a promising technique to transmit 3D motions and forces in small spaces, which were once impossible in traditional "2D microelectromechanical systems (MEMS)."

At Rice University, a group of engineers has proposed a new procedure in which the dead body of a spider can be reanimated in 2022. Considered as a biotechnological progress, the experiment was designed to transform the dead into something productive and useful. Tyner and Rice (2023) discussed three imbrication dimensions demonstrated by artificially redesigned spiders: subsuming the natural process of life and death for capitalist greed; fetishization of violence; and contradictions of "neuro-capitalism." A "necrobotic spider" is considered an omen that refigures knowledge of violence from the postanthropocentric and posthumanist perspective.

Menger et al. (2018) investigated the cost-effective robotic technology to achieve active neurosurgical practice in spine surgery. The time of infection rates, operative procedures, length of stay, revision rates, and potential conversion of "minimally invasive surgery (MIS)" were calculated, which were second to robotic image guidance with a combination of national data points and institution-specific points. They applied this cost analysis for elective cases for 1 year related to "procedural mix, payor mix, and procedural revenue." Robotic spine surgery is considered to be cost-effective with reduced infection rates, revision surgery, length of stay, and operative time. However, further research is needed to determine the financial constraints of "robotics spine surgery."

3 Advancements in the field of necrobotics

There are different advancements in necrobotics like logical systems, fuzzy logic, and behavioral systems (Russell and Norvig, 2010; Saffiotti, 1997). The adaptability of evolutionary models enables them with almost all systems, whether from best parameters or controller architecture. The best substrate for evolutionary robotics (ER) must limit evolution to scale up to unbounded complexity. As ER uses prior knowledge as little as possible, this substrate can use raw inputs obtained from sensors, and low-level commands are then sent to actuators.

Because of factors related to labor shortages in recent years, there is a need for automation with robotics, and there has been a rise in large-scale equipment. Currently, it is not easy for robotics to fix the environment where obstacles are disordered. It is important to create an environment where robots can perform their tasks smoothly. Robots have been limited to regular and simple tasks these days. This way, an AI technology has been developed for robotics for accurate handling of disorderly placed items by predicting the environment and consequences of actions. The precise and optical actions are generated automatically as per the predictions through "Robot Motion Generation."

3.1 Generative AI

One of the most promising applications of AI in necrobotics is integrating it with a robust "natural language processing (NLP)" model named ChatGPT, which can generate engaging and coherent texts on different topics. It uses natural language to interact with humans by providing entertainment, news, information, and support. It can have various applications in necrobotics.

Generating creative ideas for applications of necrobotics, i.e., by combining biological materials with various contexts and functions.

Educating the public on the concept of necrobotics, i.e., by defining its benefits, principles, and challenges engagingly.

Designing and testing necrobotic systems by providing suggestions, feedback, or instructions as per inputs.

Improving user experience by providing natural language narratives, dialogues, and interfaces to make necrobotic systems more appealing and interactive.

Necrobotics is a rapidly growing and novel area of robotics that can revolutionize how people interact with the world. With the functionality of biotic components and natural morphology, necrobotics can generate more robust, versatile, and sustainable robots.

3.2 Neural networks

The "artificial neural networks (ANN)" can meet those needs and are the "preeminent controller" shift in ER. Feed-forward NNs can reproduce any purpose with "arbitrary precision" (Cybenko, 1989) and are recognized well in signal processing (Bishop, 1995) and robotic controls (Miller et al., 1995). Neural networks can estimate any kind of dynamic system with recurrent connections. Additionally, as neural networks can also be used in brain models, necrobiotics can be based on a range of studies in neuroscience on network theory or synaptic plasticity (Bullmore and Sporns, 2009; Abbott and Nelson, 2000).

There are different ANNs available to choose from, and they can be evolved in different ways. First of all, it is possible to simulate neurons at various aspects of abstraction. A lot of necrobotics studies have applied neurons developed by McCulloch and Pitts in 1943 as in traditional machine learning (Haykin, 1998). Leaky indicators are also implemented to take time and can be suited well to dynamic systems (Beer, 1995). These networks are often known as "continuous-time recurrent neural networks (CTRNN)". Currently, complex neuron models like spiking neurons are not well-known in necrobotics. Secondly, after selecting the neural model, evolution can serve as synaptic parameters or the network architecture.

3.3 Selective pressure

At the beginning of necrobotics, there is a huge research gap in terms of selective pressure. Fitness was considered a completely user-defined function for which best practices could be followed by the user. However, a lot of advancements have found that most ER have common challenges and features and can be managed by generic techniques that alter selective pressure like novelty search, behavioral diversity, or transferability (Koos et al., 2013; Mouret and Doncieux, 2012). All such selective pressures consider the behavior of robots in their definition and keep the field away from the traditional perspective of black-box optimization. Fig. 9.2 These promising results have further suggested that better knowledge of "selective pressure" could be helpful to ER for scaling up to

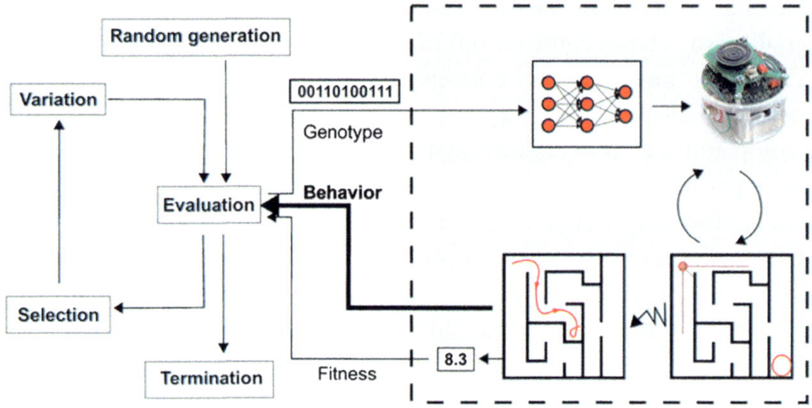

Fig. 9.2 ***Selective pressure and behavior of robots.*** This figure shows the evolutionary process of robot behavior, highlighting the key stages involved.

more complicated designs and tasks. Instead of just keeping fitness as shown by some of the measures of quantitative performance, it may be worth considering robotic behavior, which may not be indexed directly to successful task performance.

3.4 Evolutionary robotics

ER is different from other engineering fields as it is naturally based on biological systems. It is different, as a lot of fields of engineering depend on traditional physics and mathematical approaches. Biological systems are not integrated well, even though there have been a lot of developments over the years. Experimental research suffers because evolution needs a lot of generations of a lot of people who may last for decades. The lifecycles of organisms are short enough for lab experiments (Wiser et al., 2013). Biologists analyze the leftovers of previous creatures as they cannot conduct direct experiments. They need to rely on theoretical methods of population or analyze genetic code. ER provides a synthetic approach to learning evolution. ER contributes more than engineering Fig. 9.3. It gains inspiration from biology and also contributes to the same. ER imports scientific questions and tools from engineering.

4 Cash flow and ROI analysis for robotics innovation

4.1 Cash flow analysis

Robots are active and flexible machines. It may not be easy to pin down ROI, but it is still possible. This phase requires positive cash flow over time over the service life of the robotic system (Fig. 9.4). For Necrobotics, ROI correlates with positive long-term cash flow. For a clear picture, as with any major cost, it is important to consider the retention time of the robotic machine and the duration of automation.

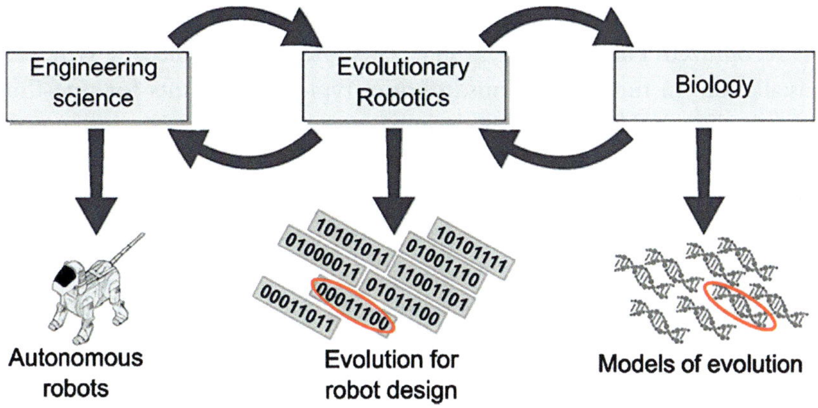

Fig. 9.3 ***ER at the crossroads between biology and engineering science.*** The figure explains how ER draws inspiration from biological evolution to develop autonomous robots through evolutionary algorithms using engineering principles.

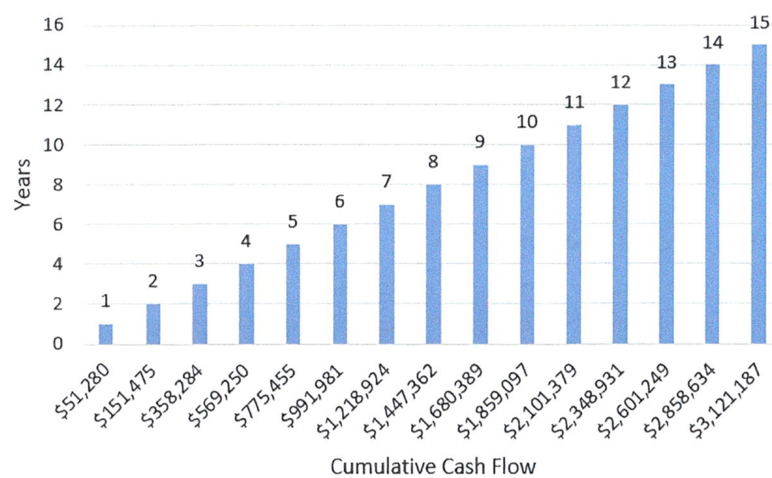

Fig. 9.4 ***Cumulative cash flow over the service life of robotic system.*** The graph shows the successive cash flow of robotic system's service.

Ron Potter, Director (Robotics Technology), Factory Automation Systems, Inc., is an automation veteran who was awarded the Engelberger Robotics Award. In 1978, he co-founded the first robotics assembly company and was involved in more than 2500 products. He focused on evaluating ROI as part of long-term planning. He uses a medium-sized, 100-pound industrial robot that can handle objects. Considering an average energy of 7.35 kW and an energy expenditure of 10 cents/kWh, a cost of 75 cents/h is calculated for the average robot, and then this cost is compared to human performance. (Anandan, 2015).

The impact of labor savings on positive cash flows over the service life of the robotic system is recognized. The bar chart in Fig. 9.4 shows an example of a typical $250,000 robot installation. In turnkey systems, robotics typically accounts for one-third of the total cost, which includes equipment, design, conveyor, safety, design, integrated services, and installation, all for the initial cost it can absorb more than 2 years before payment. There was a huge increase in cash flow from years 3 to 10. The investment in that robotic system was about $15 million in cash flow in the seventh or eighth years, and it keeps going up (Anandan, 2015).

This merging income model is based on a system in which two robots work 5 days/week, two shifts, 50 week/year (Table 9.1). Productivity gains and labor savings can be possible by the replacement of two laborers by an automated system per shift at an annual cost of $45000 per operator.

An ROI calculator can analyze the financial impact of the proposed project for robotic automation. Only 10 min are used by ROI analysis to finish, and it needs only eight easily quantifiable inputs. Then the annual cash flow is calculated to determine the impact of the necrobotic system on cash flow for the project lifespan. People are worried about the cost of necrobotics systems, but they don't worry about the benefits. It is like a skilled laborer who is capable of working round-the-clock and does the same job every time to operate consistently. It is an important element for efficiency. Many organizations have embedded systems for calculating ROI without considering this concept in the long term. Only progressive company owners and managers are seeking benefits in the long term (Anandan, 2015).

4.2 Estimating cost of ownership

The operating cost consists of power consumption, which is noted before, and the cost of compressed air is negligible for compressed air to operate the gripper or operate tooling at the end of the arm in a usual application of material handling. For instance, the operating cost of $6000 in the first year is accounting for 2 robots working for 5 days a week, 2 shifts a day, and 50 weeks a year for merely 75 cent/h. Even though medium

Table 9.1 Value calculator for robotics system.

Total system cost (installation, training, robots, etc.)	$250,000
Total robots in the system	2
Robot system usage	
B. Shifts/day	2
C. Days/week	5
D. Weeks/year	50
Annual labor cost per operator	$45,000
Number of operators removed per shift	2
Percent of labor retained per shift	10%
Other estimated savings	

robots are used in this example, these estimations can be used in other payloads. The cost of operating a small robot with a 10 kg payload is around 15 cent/h, while a huge robot with a payload of over 500 kg is around $1.5/h (Anandan, 2015).

When it comes to acquiring new robots, emergency regular spare parts, tools, plant utility connections, management overhead and training, custom fixturing, removal of old equipment, site preparation, information system maintenance, overhead And also, you need to consider that robotics requires a trained expert to operate it, which may not require manual operation. Implementing and maintaining robotic systems requires a high degree of skill. A single operator was not used for almost every robot. Robots are generally managed by individuals.

Once you are responsible for all robotic costs, the rest of the columns include productivity, labor savings, savings from scrap and rework, and material savings in cash flow analysis. Along with direct savings in terms of labor, there is a need to consider the cost of maintenance of manual labor, such as production lost by routine breaks, lost-time injuries, vacation time, retraining, and training, along with compensation for employees and insurance costs.

When people look at the explanation report, people often consider only direct labor expenses, not indirect expenses and savings, like quality improvements and improved productivity. It is worth making an informed guess of value instead of leaving them out. Machine tending is the largest application area in automation systems. Heavy parts are involved in machine tending. Most of the time, an individual needs to lift those parts and load and unload them. Improving worker ergonomics and safety by removing staff from repetitive or dangerous jobs and shifting them to valuable jobs is among the indirect benefits of implementing necrobotics systems.

Here are some other indirect benefits:

- Improved flexibility in manufacturing with necrobotics for various projects or having various necrobotics programs for short runs
- Operations are combined to reduce "work-in-process inventory."
- More efficient scheduling and production planning because of predictable automation
- Floor space can be saved with flexible configurations for mounting robots.

4.3 The advent of collaborative robots and ROI

There is a need to address the advancement of collaborative robots when it comes to discussing the ROI. These robots often limit force and power, and they usually cost a part of traditional robots, and they have a return within months instead of years. They are also not capable like their legacy counterparts, especially in terms of faster cycles and heavier payloads. A lot of people still don't understand that collaborative robots cannot replace traditional robots directly, and it is still a learning curve. Collaborative robots have their position in commercial and manufacturing industries, especially when it comes to introducing SMEs to agile automation apart from coworkers in distribution centers and factories.

The ROI calculator calculates the total ownership cost. Over the first few years of life, there is no maintenance required. It needs some oil and grease for the system. Over time, robotic arms may have some internal cables that have worn and need replacement. It also includes incremental costs. The whole robot must be refurbished for a part of the cost of the robot after 10–12 years (Anandan, 2015).

Table 9.2 refers to the "cash flow chart" for a material handling program using two robots. It shows the nominal cost for maintenance for 4 years and the incremental cost for replacing worn parts in the fifth year and refurbishment in the tenth year (Anandan, 2015). This cash flow analysis illustrates long-term returns and total ownership costs over the lifetime of a robotic system.

5 Discussion and conclusion

The study discusses the proper control of a deceased spider. It is a lot easier than other projects related to insect robotics, in which electrodes are put into the brains of insects to control them. It doesn't need to manipulate the brain of a living creature or its instincts to accomplish a task. It is not much more troubling than enslaving living creatures electronically. Humans have always respected moral beings and had feelings about proper treatment. ER has emerged as an area of research that has matured for years. It has made a significant contribution to biology and engineering and has provided insights and new perspectives on holistic design.

When it comes to engineering, necrobotics presents a different design approach to the embodied brain. Designers can work on the sensory system, morphology, and control while considering the cost-effectiveness in the long term. When it comes to the biological perspective, necrobotics provides strong and different modeling tools for evolution for experimental work. Its agent-based approach, related to its process for behavior selection, enables researchers to model the phenomena, which are not easy to study with a global and simple statistical process. It provides chosen tools to study social behaviors like communication and cooperation. Necrobotics is currently at a nascent stage, which depends on a lot of experiments for getting any proven result.

There is a need to address the advancement of collaborative robots when it comes to discussing the ROI. These robots often limit force and power, and they usually cost a part of traditional robots, and they have a return within months instead of years. They are also not capable like their legacy counterparts, especially in terms of faster cycles and heavier payloads. When it comes to ROI, it is observed that maintenance costs remain the same but significantly increase every 5 years, especially when it comes to replacing worn items. In addition, operation costs also increase gradually over the next 10 years after acquiring the system. However, labor and productivity savings may also increase over time gradually, which will result in a high ROI in the long run.

Table 9.2 Cash flow analysis for robotic system.

Year	System cost	Maintenance cost	Operating cost	Labor savings	Productivity savings	Scrap/ rework savings	Material savings	Other savings	Cumulative cash flow
1	$250,000	$1000	$6000	$162,000	$43,740	$0	$0	$0	$51,280
2		$1000	$6120	$165,240	$44,615	$0	$0	$0	$151,475
3		$1000	$6242	$168,545	$45,507	$0	$0	$0	$358,284
4		$1000	$6387	$171,916	$46,417	$0	$0	$0	$569,250
5		$10,000	$6495	$173,351	$47,346	$0	$0	$0	$775,455
6		$1000	$6624	$178,861	$48,292	$0	$0	$0	$991,981
7		$1000	$6757	$182,438	$49,258	$0	$0	$0	$1,218,924
8		$1000	$6892	$186,087	$50,244	$0	$0	$0	$1,447,362
9		$1000	$7030	$189,809	$51,248	$0	$0	$0	$1,680,389
10		$60,000	$7171	$193,605	$52,273	$0	$0	$0	$1,859,097
11		$1000	$7314	$197,477	$53,319	$0	$0	$0	$2,101,379
12		$1000	$7460	$201,427	$54,385	$0	$0	$0	$2,348,931
13		$1000	$7609	$201,427	$55,473	$0	$0	$0	$2,601,249
14		$1000	$7762	$203,435	$56,582	$0	$0	$0	$2,858,634
15		$1000	$7917	$213,756	$57,714	$0	$0	$0	$3,121,187
Total		$583,000	$103,761	$2,801,534	$756,414	$0	$0	$0	

5.1 Limitations

This study has conducted cost and ROI analysis for maintaining necrobotics and other advanced systems in an organization along with an estimated cost of ownership. However, there is a lack of empirical evidence on the actual cost of having advanced robotics and the implementation of AI in the real world. In addition, there is a lack of research in the field of necrobotics, because it is relatively a new technology. Hence, further research is required in this field to generalize the findings of this study.

References

Abbott, L.F., Nelson, S.B., 2000. Synaptic plasticity: taming the beast. Nat. Neurosci. 3 (11s), 1178–1183. https://doi.org/10.1038/81453. 15461726.

Alexander, J., Baskaran, K., Harward, A., Carlson, K., Naleway, S.E., 2021. Bioinspired aligned magnetic features in aerogels for humidity sensing. Mater. Chem. Phys. 270. Elsevier Ltd, United States. 02540584. Available from: http://www.journals.elsevier.com/materials-chemistry-and-physics. https://doi.org/10.1016/j.matchemphys.2021.124852.

Anandan, T.M., 2015. Calculating Your ROI for Robotic Automation: Cost vs. Cash Flow.

Appiah, C., Arndt, C., Siemsen, K., Heitmann, A., Staubitz, A., Selhuber-Unkel, C., 2019. Living materials Herald a new era in soft robotics. Adv. Mater. 31 (36). Wiley-VCH Verlag, Germany. 15214095. Available from: http://onlinelibrary.wiley.com/journal/10.1002/(ISSN)1521-4095. https://doi.org/10.1002/adma.201807747.

Beer, R.D., 1995. On the dynamics of small continuous-time recurrent neural networks. Adapt. Behav. 3 (4), 469–509. https://doi.org/10.1177/105971239500300405. 17412633.

Bishop, C.M., 1995. Neural networks for pattern recognition. Google Scholar 2, 223–228. Clarendon Press.

Bledt, G., Powell, M.J., Katz, B., Di Carlo, J., Wensing, P.M., Kim, S., 2018. MIT cheetah 3: design and control of a robust, dynamic quadruped robot. In: IEEE International Conference on Intelligent Robots and Systems. Institute of Electrical and Electronics Engineers Inc., United States, pp. 2245–2252, https://doi.org/10.1109/IROS.2018.8593885. 21530866.

Bullmore, E., Sporns, O., 2009. Complex brain networks: graph theoretical analysis of structural and functional systems. Nat. Rev. Neurosci. 10 (3), 186–198. https://doi.org/10.1038/nrn2575. 14710048.

Cybenko, G., 1989. Approximation by superpositions of a sigmoidal function. Math. Control Signals Syst. 2 (4), 303–314. https://doi.org/10.1007/BF02551274. 1435568X.

Göttler, C., Elflein, K., Siegwart, R., Sitti, M., 2021. Spider origami: folding principle of jumping spider leg joints for bioinspired fluidic actuators. Adv. Sci. 8 (5). John Wiley and Sons Inc, Germany. 21983844. Available from: http://onlinelibrary.wiley.com/journal/10.1002/(ISSN)2198-3844. https://doi.org/10.1002/advs.202003890.

Haykin, S., 1998. Neural Networks: A Comprehensive Foundation. Prentice Hall PTR.

Hou, Y., Yu, M., Chen, X., Wang, Z., Yao, S., 2015. Recurrent filmwise and dropwise condensation on a beetle mimetic surface. ACS Nano 9 (1), 71–81. American Chemical Society, Hong Kong. 1936086X. Available from: http://pubs.acs.org/journal/ancac3. https://doi.org/10.1021/nn505716b.

Jin, L., Yang, Y., Maldonado, B.O.T., Lee, S.D., Figueroa, N., Full, R.J., Yang, S., 2023. Ultrafast, programmable, and electronics-free soft robots enabled by snapping metacaps. Adv. Intell. Syst. 5 (6). John Wiley and Sons Inc, United States. 26404567. Available from: https://onlinelibrary.wiley.com/journal/26404567. https://doi.org/10.1002/aisy.202300039.

Jumet, B., Bell, M.D., Sanchez, V., Preston, D.J., 2022. A data-driven review of soft robotics. Adv. Intell. Syst. 4.

Kellaris, N., Rothemund, P., Zeng, Y., Mitchell, S.K., Smith, G.M., Jayaram, K., Keplinger, C., 2021. Spider-inspired electrohydraulic actuators for fast, soft-actuated joints. Adv. Sci. 8 (14). John Wiley and Sons Inc, United States. 21983844. Available from: http://onlinelibrary.wiley.com/journal/10.1002/(ISSN)2198-3844. https://doi.org/10.1002/advs.202100916.

Koos, S., Mouret, J.B., Doncieux, S., 2013. The transferability approach: crossing the reality gap in evolutionary robotics. IEEE Trans. Evol. Comput. 17 (1), 122–145. https://doi.org/10.1109/TEVC.2012.2185849. 1089778X.

Li, W., Matsuhisa, N., Liu, Z., Wang, M., Luo, Y., Cai, P., Chen, G., Zhang, F., Li, C., Liu, Z., Lv, Z., Zhang, W., Chen, X., 2021. An on-demand plant-based actuator created using conformable electrodes. Nat. Electron. 4 (2), 134–142. Nature Research, Singapore. 25201131. Available from: https://www.nature.com/natelectron/. https://doi.org/10.1038/s41928-020-00530-4.

Liu, X., Shi, L., Wan, X., Dai, B., Chen, Y., Wang, S., 2021. Recent progress of spider-silk-inspired adhesive materials. ACS Mater. Lett. 3 (10), 1453–1467. American Chemical Society, China. 26394979. Available from: https://pubs.acs.org/page/amlcef/about.html. https://doi.org/10.1021/acsmaterialslett.1c00378.

Menger, R.P., Savardekar, A.R., Farokhi, F., Sin, A., 2018. A cost-effectiveness analysis of the integration of robotic spine technology in spine surgery. Neurospine 15 (3), 216–224. Korean Spinal Neurosurgery Society, United States. 25866583. Available from: http://www.e-neurospine.org/. 10.14245/ns.1836082.041.

Miller, W.T., Sutton, R.S., Werbos, 1995. Neural Networks for Control. MIT Press.

Mouret, J.B., Doncieux, S., 2012. Encouraging behavioral diversity in evolutionary robotics: an empirical study. Evol. Comput. 20 (1), 91–133. https://doi.org/10.1162/EVCO_a_00048. 15309304.

Naik, R.R., Singamaneni, S., 2017. Introduction: bioinspired and biomimetic materials. Chem. Rev. 117 (20), 12581–12583. American Chemical Society, United States. 15206890. Available from: http://pubs.acs.org/journal/chreay. https://doi.org/10.1021/acs.chemrev.7b00552.

Nie, B., Liu, S., Qu, Q., Zhang, Y., Zhao, M., Liu, J., 2022. Bio-inspired flexible electronics for smart E-skin. Acta Biomater. 139, 280–295. Acta Materialia Inc, China. 18787568. Available from: http://www.journals.elsevier.com/acta-biomaterialia. https://doi.org/10.1016/j.actbio.2021.06.018.

Parry, D.A., Brown, R.H.J., 1959. The hydraulic mechanism of the spider leg. J. Exp. Biol. 36 (2), 423–433. https://doi.org/10.1242/jeb.36.2.423. 0022-0949.

Rafsanjani, A., Zhang, Y., Liu, B., Rubinstein, S.M., Bertoldi, K., 2018. Kirigami skins make a simple soft actuator crawl. Sci. Robot. 3 (15). American Association for the Advancement of Science, United States. 24709476. Available from: http://robotics.sciencemag.org/content/robotics/3/15/eaar7555.full.pdf. https://doi.org/10.1126/scirobotics.aar7555.

Ricotti, L., Trimmer, B., Feinberg, A.W., Raman, R., Parker, K.K., Bashir, R., Sitti, M., Martel, S., Dario, P., Menciassi, A., 2017. Biohybrid actuators for robotics: a review of devices actuated by living cells. Sci. Robot. 2 (12). American Association for the Advancement of Science, Italy. 24709476. Available from: http://robotics.sciencemag.org/content/robotics/2/12/eaaq0495.full.pdf. https://doi.org/10.1126/scirobotics.aaq0495.

Russell, S.J., Norvig, P., 2010. Artificial Intelligence a Modern Approach.

Saffiotti, A., 1997. The uses of fuzzy logic in autonomous robot navigation. Soft Comput. Fusion Foundations Methodol. Appl. 1 (4), 180–197. https://doi.org/10.1007/s005000050020. 1432-7643.

Sharma, S., Sharma, V., Mishra, D., Nagar, A., 2024. Advancing healthcare management: harnessing necrobotics for enhanced patient care. Int. J. Pharma Prof. Res. 15 (2), 52–66. https://doi.org/10.69580/ijppr.15.2.2024.52-66. 22493948.

Singh, A.V., Hosseinidoust, Z., Park, B.W., Yasa, O., Sitti, M., 2017. Microemulsion-based soft bacteria-driven microswimmers for active cargo delivery. ACS Nano 11 (10), 9759–9769. American Chemical Society, Germany. 1936086X. Available from: http://pubs.acs.org/journal/ancac3. https://doi.org/10.1021/acsnano.7b02082.

Smith, G.L., Tyler, J.B., Lazarus, N., Tsang, H., Viornery, L., Shultz, J., Bergbreiter, S., 2023. Spider-inspired, fully 3D-printed micro-hydraulics for tiny, soft robotics. Adv. Funct. Mater. 33 (39). John Wiley and Sons Inc, United States. 16163028. Available from: http://onlinelibrary.wiley.com/journal/10.1002/(ISSN)1616-3028. https://doi.org/10.1002/adfm.202207435.

Tyner, J.A., Rice, S., 2023. Along came a spider … and capitalism killed it. Hum. Geogr. 16 (2), 175–180. https://doi.org/10.1177/19427786221140843. 1942-7786.

Wiser, M.J., Ribeck, N., Lenski, R.E., 2013. Long-term dynamics of adaptation in asexual populations. Science 342 (6164), 1364–1367. American Association for the Advancement of Science, United States. 10959203. Available from: http://www.sciencemag.org/content/342/6164/1364.full.pdf. https://doi.org/10.1126/science.1243357.

Wong, T.S., Kang, S.H., Tang, S.K.Y., Smythe, E.J., Hatton, B.D., Grinthal, A., Aizenberg, J., 2011. Bioinspired self-repairing slippery surfaces with pressure-stable omniphobicity. Nature 477 (7365), 443–447. https://doi.org/10.1038/nature10447. 14764687.

Yap, T.F., Liu, Z., Rajappan, A., Shimokusu, T.J., Preston, D.J., 2022. Necrobotics: biotic materials as ready-to-use actuators. Adv. Sci. 9 (29). John Wiley and Sons Inc, United States. 21983844. Available from: http://onlinelibrary.wiley.com/journal/10.1002/(ISSN)2198-3844. https://doi.org/10.1002/advs.202201174.

CHAPTER 10

Workflow optimization and resource allocation in healthcare: The role of necrobotics

Anusha Siddadapu Venkata[a] and Vandanapu Shashank[b]
[a]Woxsen School of Business, MBA Business Analytics, Sadashivpet, India
[b]Woxsen University, MBA General, Hyderabad, India

1 Overview of necrobotics

Necrobotics is a rising area at the intersection of robotics and necromancy, aiming to combine technological improvements with mystical practices for revolutionary applications. While conventional robotics focuses on the improvement of machines to carry out numerous obligations autonomously, necrobotics introduces a unique measurement through incorporating factors of necromancy—the magical exercise of communicating with and manipulating the lifeless (Marchese et al., 2014).

1.1 Key components of necrobotics

Robotics technology incorporates superior robotic systems and synthetic intelligence to carry out obligations in healthcare, enterprise, and different sectors and utilizes sensors, actuators, and programming to allow robots to interact with the physical interfaces (Marchese et al., 2014).

Necromancy practices involves the invocation and manipulation of spirits, souls, or energies from the deceased; it draws upon mystical and occult traditions to set up a connection among the residing and the religious realm. Integration of technology and mysticism seeks to mix the precision and performance of robotics with the metaphysical factors of necromancy and aims to leverage each medical and esoteric understanding for practical applications. Applications in Healthcare introduces progressive solutions for affected person care, surgery, diagnostics, and normal healthcare management and envisions the use of necrobotics to optimize workflows, automate repetitive obligations, and decorate scientific strategies. Ethical consideration raises ethical questions concerning the intersection of generation and mysticism and prompts discussions at the implications of the use of necromantic practices together with robotics, specifically in touchy fields like healthcare. Potential benefits of necrobotics include improving efficiency and precision in tasks historically finished by using human beings, exploring new avenues for expertise, and interacting with the spiritual realm (Ray, 2023).

Necrobotics for Healthcare Applications and Management
https://doi.org/10.1016/B978-0-443-24832-0.00010-6

Challenges and controversies include facing skepticism and resistance due to the mystical nature of necromancy, and raising concerns approximately moral, cultural, and nonsecular implications. Research and development represents an evolving field with ongoing research to explore the whole ability of necrobotics and involves collaborations between technologists, mystics, and healthcare professionals.

1.2 Evolution of necrobotics in healthcare

Necrobotics, the intersection of robotics and autopsy examination, has seen remarkable strides in the latest years, transforming its function from a spot device to an ability healthcare recreation-changer. The upward thrust began with digital autopsy simulators, presenting more secure, greater ethical schooling for medical professionals. These virtual cadavers, manipulated with robotic arms, furnished realistic environments for honing surgical and diagnostic talents. Beyond Education Advancements quickly pushed boundaries. Minimally invasive robotic autopsies emerged, making use of precision robot units for tissue sampling and analysis. This advanced accuracy reduced risks and even allowed for faraway participation by using professionals.

The future of necrobotics is brimming with possibilities of personalized medicine, necrobotics ought to examine character organs to apprehend disease development and guide customized remedy plans for living sufferers. Drug improvement Precisely replicating human physiology, necrobots may want to check new pills in a more managed setting, accelerating development and decreasing animal trying out. Surgical training Advanced Necrobots could perform notably complicated surgical procedures on cadavers, imparting unrivaled training for surgeons before running on stay sufferers (Yap et al., 2023).

2 The significance of workflow optimization and resource allocation

In the burgeoning field of necrobotics, wherein robotics meet autopsy examinations, workflow optimization and resource allocation are not simply efficiency boosters, but ethical and practical requirements. Necropsy procedures contain a sensitive interplay of professionals, devices, and time. Optimizing workflow guarantees clean transitions among tasks, minimizing delays, and maximizing utilization of the necrotic robot, a valuable and frequently shared aid. Imagine a perfectly choreographed ballet of technicians, pathologists, and the necrotic arm, every shifting in sync to acquire critical records efficiently.

Necrobots are high-priced, specialized tools. Optimizing resource allocation guarantees their equitable use across diverse programs, from research to customized medication. This may involve prioritizing critical cases, scheduling shared access pretty, or even considering value-sharing models for smaller establishments. Responsible allocation ensures this technology serves the truer, not simply pick few (Yap et al., 2022).

Respect for the deceased is paramount. Optimized workflows decrease pointless coping with and make certain timely final touches of methods. Resource allocation avoids overworking the necrotic robot, stopping ability breakdowns or malfunctions that could compromise statistics integrity. Both elements show respect for the frame and the fee of the statistics it holds. By diligently optimizing workflows and allocating resources thoughtfully, necrobotics can achieve a harmonious stability among efficiency, ethics, and medical progress. This paves the way for a future in which necropsies are not just quicker and more informative, but additionally performed with the maximum recognition, ensuring this generation fulfills its real capability to revolutionize healthcare.

2.1 Workflow optimization in healthcare

To streamline the process of patient admission and discharge using necrobotics, the marriage of robotics and postmortem analysis would possibly appear distant from the area of affected person admissions and discharges. However, its potential to optimize useful resource allocation and streamline strategies could have sudden implications. Virtual autopsies carried out with necroptic robots loosen up precious pathology time, permitting them to consciousness on residing patients, doubtlessly dashing up diagnoses and remedy plans.

AI-powered analysis of autopsy statistics from necropsies aids in expertise rare sicknesses and developing focused treatment plans, reaping rewards for destiny sufferers with similar situations. Remote necropsies and the usage of robots permits experts from diverse places to collaborate on complicated instances, probably expediting 2D opinions and consultations, leading to quicker discharge choices.

2.2 Automated diagnostic procedures

Necrobotics, the once-area of interest fusion of robotics and autopsy analysis, is whispering possibilities beyond the cold confines of the postmortem room. One intriguing prospect lies in computerized diagnostic approaches, potentially revolutionizing healthcare. Imagine a robotic arm meticulously dissecting a donated organ, its AI partner analyzing mobile and tissue abnormalities with superhuman precision. These statistics, invisible to conventional techniques, could hold the important thing to unlocking the mysteries of complex sicknesses. Early detection of rare conditions, personalized treatment plans based on person pathophysiology, and increased drug improvement are only a glimpse of the capacity blessings ("*Experimental Study of the Pathogenicity and Pathologic Picture of Paecilomyces Species Isolated from the Lung of Dead Fayou Mi Chickens*," 1988).

2.3 The robotic surgery and intervention

Necrobotics, the macabre marriage of robotics and postmortem examination, might maintain a stunning key to the destiny of surgical procedure. Could the tools used to analyze the deceased at some point function at the living?

Imagine robotic fingers, honed on the intricacies of cadaver dissection, carrying out sensitive microsurgeries with exceptional precision. Necrobots, educated on sizable datasets of anatomical versions, navigate complicated approaches with minimal invasiveness and decreased human errors. Imagine tumors excised, organs repaired, and lives stored, all thanks to the insights gleaned from the silent dance of the scalpel at the no longer residing.

While the scalpel of a necrobot operating on a beating heart may appear like technological know-how fiction, the capability is plain. It is a destiny wherein the line between lifestyles and demise turns into a canvas for innovation, and the understanding gleaned from the stillness of the departed facilitates healing the residing. The ethical compass should manual us, but the opportunities, like a surgeon's constant hand, may want to rewrite the script of healthcare.

2.4 Efficient medication dispensing systems

Efficient medication dispensing systems in necrobotics combine superior robotics and mystical practices to revolutionize healthcare shipping. These systems utilize robot precision and the metaphysical elements of necromancy to beautify the drug dispensing technique. Necrobotic dispensers are designed to autonomously manipulate and dispense medicinal drugs, making sure correct dosages and well-timed administration. By integrating sensors and AI, these systems can also display patient adherence and offer actual-time statistics to healthcare specialists. The mystical impact of necromancy provides an additional layer of precision and reliability, creating a unique synergy among era and spirituality. This revolutionary method not only streamlines medicinal drug workflows but additionally contributes to improved affected person outcomes and a greater holistic method to healthcare transport in the realm of necrobotics (Kambouris et al., 2023).

2.5 Data-pushed decision support systems

Data-pushed decision support systems (DDSS) in necrobotics amalgamate the geographical regions of current technology and mystical insights to propel healthcare choice-making into uncharted territories. By integrating sophisticated records analytics with the religious elements of necromancy, these systems offer remarkable capabilities. DDSS in necrobotics leverage large datasets to research patient records, remedy outcomes, and historical clinical statistics, providing healthcare practitioners with actionable insights and predictions. The fusion of necromantic practices introduces a unique dimension, permitting DDSS to faucet into ethereal expertise and connect to nonsecular energies for improved diagnostic accuracy and prognostic abilities. These systems no longer simplest rely upon empirical evidence but additionally draw from the metaphysical, providing a complete attitude on affected person care.

Moreover, DDSS in necrobotics facilitates customized remedy plans by considering each of the bodily and religious aspects of an affected person's well-being. This holistic method transcends traditional facts-driven structures, aiming to optimize healthcare effects through a symbiotic dating between generation and mysticism. As this progressive field evolves, it holds the promise of reshaping healthcare selection-making, imparting a novel framework that unifies records analytics with spiritual insights for more knowledgeable and holistic patient care inside the era of necrobotics (Hall, 2023).

3 Resources allocation

Resource Allocation Strategies in necrobotics' workflow optimization involve leveraging both technological and mystical elements for green resource utilization in healthcare. Robotics in necrobotics streamline personnel duties, automate stock management, and allocate beds dynamically, improving operational performance. The integration of necromancy introduces spiritual insights for predictive analytics, ensuring top-rated device protection and energy consumption. This symbiotic approach maximizes the capacity of both nations, addressing the complex challenges of useful resource allocation in healthcare settings. By combining advanced technology with mystical practices, necrobotics offers a holistic answer that goes beyond traditional techniques, promising a transformative effect on healthcare useful resource optimization and workflow efficiency.

3.1 Maximizing personnel utilization via robotics

Maximizing personnel utilization through robotics is a pivotal element of necrobotics, merging technological prowess with mystical dimensions for unparalleled efficiency in healthcare. In this paradigm, robots are seamlessly included into healthcare workflows to deal with ordinary and exertions-intensive responsibilities, allowing human personnel to recognize more complicated and compassionate aspects of affected person care. Necromantic impacts upload a unique layer, imbuing robotic structures with religious insights to beautify choice-making and responsiveness. Through automation, robotics in necrobotics optimizes employees' utilization with the aid of reducing repetitive tasks, minimizing errors, and improving overall productiveness. This modern method not only most effectively streamlines healthcare operations but additionally creates a harmonious balance among the precision of generation and the nonsecular information introduced through necromancy, fostering complete and transformative surroundings for personnel and affected persons' well-being (Drotman et al., 2021).

3.2 Automated inventory management for medical supplies

Automated inventory management for medical supplies in necrobotics represents a groundbreaking fusion of era and mystical practices to optimize healthcare logistics. Employing superior robotics and incorporating elements of necromancy, this machine automates the tracking, restocking, and organization of medical components. Robotics make certain precision and performance in stock responsibilities, whilst necromancy adds an additional layer of insight, facilitating a religious connection to foresee supply desires intuitively. By integrating sensors and statistics analytics, the gadget anticipates call for, minimizes waste, and ensures timely availability of crucial components. This holistic method now not only streamlines inventory techniques but additionally taps into mystical energies to beautify selection-making. The result is a transformative solution that goes beyond traditional stock management, presenting a continuing and intuitive method to making sure the availability of medical elements in healthcare settings through the modern lens of necrobotics (Polygerinos et al., 2017).

3.3 Dynamic bed allocation systems

Dynamic Bed Allocation Systems in necrobotics revolutionize healthcare management through combining modern-day era with mystical insights for most reliable affected person care. In this innovative technique, robotics facilitate real-time tracking of patient situations, automating the allocation and adjustment of beds primarily based on medical wishes. The integration of necromancy introduces a nonsecular dimension, permitting the device to tap into airy understanding for predictive affected person outcomes and holistic care considerations. By leveraging information analytics and spiritual insights, dynamic bed allocation structures ensure efficient usage of healthcare resources, minimizing delays and improving affected person consequences. This unique synergy between generation and mysticism not only most effectively streamlines mattress control but also introduces a compassionate and comprehensive method to healthcare, addressing each bodily and religious element of the affected person's well-being inside the evolving landscape of necrobotics (Tolley Michael et al., 2014).

3.4 Predictive analytics for equipment maintenance

Predictive analytics for equipment maintenance in necrobotics transforms healthcare workflow by way of fusing technological precision with mystical foresight. Through superior analytics and the incorporation of necromantic factors, this system predicts device disasters, permitting proactive renovation. Robotics collect and analyze records, while necromancy affords a nonsecular measurement for intuitive insights. By foreseeing ability issues, healthcare facilities can prevent disruptions, make sure continuous operation of essential devices, and optimize average workflow efficiency. This innovative synergy between predictive analytics and mystical instinct in necrobotics redefines system maintenance strategies, contributing to a continuing and dependable healthcare infrastructure.

3.5 Optimizing energy consumption in healthcare facilities

Optimizing energy consumption in healthcare facilities with necrobotics blends technological precision with mystical insights for holistic workflow enhancement. Robotics screen and manipulate energy usage, making sure performance, whilst the infusion of necromancy gives spiritual intuition to foresee greatest strength allocation. This symbiotic approach minimizes waste and aligns electricity utilization with healthcare demands. By integrating superior era and mystical factors, necrobotics no longer only reduces operational prices but also contributes to sustainable and energetically balanced healthcare surroundings. This transformative synergy maximizes workflow performance, showcasing the ability of necrobotics in redefining electricity control techniques inside healthcare facilities (Rus and Tolley, 2015).

4 Benefits and challenges of necrobotics implementation

Necrobotics implementation offers heightened precision and performance in healthcare tasks, enhancing patient results and decreasing costs. The integration of mystical practices introduces ethical and regulatorily demanding situations, inclusive of issues about privacy and cultural issues. While the advantages are promising, navigating those complicated troubles is critical for the accountable and powerful integration of necromancy and robotics in healthcare systems.

4.1 Improved patient outcomes and satisfaction

Necrobotics appreciably complements affected person results and satisfaction via streamlined healthcare workflows. Advanced robotics and mystical insights optimize diagnostics, remedies, and customized care, ensuring precision and efficiency. This specific synergy minimizes delays, reduces mistakes, and fosters a more compassionate patient. By combining era and necromancy, necrobotics revolutionizes healthcare transport, imparting a harmonious method that prioritizes patient well-being, ensuing in stepped-forward effects and heightened satisfaction (Katzschmann et al., 2018).

4.2 Enhanced healthcare provider productivity

Necrobotics transforms healthcare issuer productivity via integrating advanced robotics and mystical insights. Automation of ordinary obligations, unique diagnostics, and predictive analytics streamline workflows, permitting healthcare professionals to be aware of complex and compassionate elements of patient care. The infusion of necromancy adds a nonsecular measurement, improving selection-making and normal efficiency. By leveraging both technological and mystical elements, necrobotics optimizes healthcare operations, empowering carriers to supply higher-nice care, fostering an extra enjoyable and powerful paintings environment. This progressive approach redefines productiveness, aligning technology and mysticism for a transformative impact on healthcare delivery (Heiden et al., 2022).

4.3 Cost reduction and resource optimization

Necrobotics revolutionizes healthcare by means of accomplishing price reduction and resource optimization through a symbiotic mixture of technology and mysticism. Advanced robotics streamline strategies, automate tasks, and enhance performance, minimizing operational fees. The infusion of necromancy introduces intuitive insights for predictive analytics, ensuring superior useful resource allocation. This transformative synergy maximizes the utilization of personnel, gadgets, and facilities, in the end lowering prices. Necrobotics, with its precise method for workflow optimization, emerges as a groundbreaking solution that now not only improves economic sustainability but additionally advances the overall effectiveness of resource control within the healthcare quarter (Ji et al., 2019).

4.4 Ethical considerations and patient privacy

Necrobotics in healthcare introduces moral issues, mainly concerning the intersection of era and mysticism. Patient privateness turns into a critical issue as the mixing of necromancy raises questions about the handling of touchy religious records. Balancing the advantages of advanced workflow optimization with the capability of intrusions into private and spiritual geographical regions poses demanding situations. Striking an accountable and transparent method in the implementation of necrobotics is vital, ensuring ethical recommendations address the precise ethical dilemmas springing up from the convergence of technological and mystical elements in healthcare, safeguarding each affected person's privacy and ordinary ethical requirements (Grazioso et al., 2019).

4.5 Regulatory and legal challenges

Necrobotics in healthcare confronts regulatory and legal demanding situations because of the radical fusion of technology and mysticism. Existing rules may not effectively deal with the spiritual factors of necromancy, leading to uncertainties in compliance. Legal frameworks face the complexities of overseeing both technological and mystical dimensions, requiring nuanced hints. Clear standards must emerge to deal with problems related to legal responsibility, practitioner certifications, and the coping with nonsecular statistics. Achieving regulatory readability is important for responsible necrobotics implementation, making sure legal frameworks align with the demanding situations posed by this transformative integration in healthcare and workflow optimization (Yap et al., 2015).

5 Future trends and innovations

The destiny of necrobotics in healthcare holds promising inclinations and innovations that can be set to reshape the panorama of clinical practices. Integration with superior synthetic intelligence (AI) is expected to enhance the analytical talents of necrobotic structures, considering more unique diagnostics and customized treatment plans. Collaborations amongst technologists, mystics, and healthcare experts may additionally result in the establishment of ethical recommendations and requirements, addressing issues related to spirituality and affected individual well-being. Advancements in robotics and automation technologies will probably result in greater present-day and flexible necrobotic packages. These may also include quite specialized surgical processes, faraway affected character tracking through nonsecular insights, or even AI-pushed predictive fashions for ailment prevention (Byun et al., 2018).

Moreover, the development of affected character-centric strategies inside necrobotics is predicted. This involves tailoring healthcare interventions to keep in mind now not only the bodily but additionally the nonsecular and emotional well-being of patients. As worldwide collaboration in healthcare innovation grows, necrobotics need to grow to be a pivotal player in transforming healthcare shipping, providing a holistic and integrative method that transcends traditional clinical practices. The destiny of necrobotics in healthcare seems dynamic, with the functionality to redefine requirements, beautify affected person care, and contribute to the evolution of a more complete and spiritually attuned healthcare device.

5.1 Integration of artificial intelligence in necrobotics

The integration of artificial intelligence (AI) into necrobotics heralds a transformative era for healthcare and workflow optimization. AI augments the precision and decision-making abilities of necrobotic structures, enhancing diagnostics, remedy planning, and preferred healthcare management. Machine mastering algorithms in interior necrobotics can analyze great datasets, expect affected person consequences, and adapt to evolving medical conditions. This integration lets in for more green resource allocation, streamlined workflows, and advanced affected character care.

Furthermore, AI-pushed predictive analytics in necrobotics can assume system screwups, optimizing maintenance schedules and lowering downtime. The symbiosis of AI and necromancy creates a powerful synergy, combining technological prowess with nonsecular insights for a holistic approach to healthcare. As AI maintains to evolve, its integration into necrobotics holds the promise of revolutionizing healthcare shipping, imparting exceptional overall performance, precision, and a greater entire know-how of patient needs inside the mystical framework of necrobotics (Galloway et al., 2016).

5.2 Advancements in robotics and automation technologies

Advancements in robotics and automation generation are catalyzing a paradigm shift in healthcare, ushering in a new generation of workflow optimization and extra suitable affected person care. In the context of healthcare and workflow optimization, robot structures are getting an increasing number of ultra-modern, with more than a few applications poised to revolutionize medical practices. Robotic-assisted surgical methods are witnessing continuous development, deliberating minimally invasive strategies, decreased restoration instances, and extended precision. Automation eras are streamlining regular responsibilities, from affected person records access to medicine doling out, freeing up healthcare specialists for additional complex responsibilities.

In healthcare facilities, robotic systems are evolving to control logistics effectively, which include the automatic shipping of scientific additives and the optimization of mattress allocation via real-time tracking. This not only most effectively improves operational performance but additionally ensures well-timed and correct affected individual care. Integration of robot generation with data analytics is enhancing diagnostic skills and treatment planning. AI-powered robotics can analyze sizable datasets, pick out patterns, and offer valuable insights for customized patient care (Marchese et al., 2014).

The ongoing advancements in robotics and automation technologies gift a destiny wherein healthcare workflows are seamlessly optimized, useful resource usage is maximized, and patient outcomes are appreciably improved. These improvements maintain the ability to convert healthcare into a more efficient, affected person-centric, and technologically advanced domain.

5.3 Ethical guidelines and standards

Establishing ethical recommendations and standards for the implementation of necrobotics in healthcare is imperative to make certain accountable and compassionate integration of generation and mysticism. Ethical considerations embody numerous key elements, which include affected person privateness, consent, and the religious well-being of people. To cope with those issues, strong suggestions should be formulated to regulate the gathering, garage, and utilization of spiritual statistics received through necromantic practices. Ensuring informed consent from sufferers concerning the inclusion of religious insights in their healthcare is critical, respecting man or woman's beliefs and options.

Additionally, certification and education standards for healthcare professionals involved in necrobotics must be mounted. This ensures practitioners own the important capabilities, each technologically and spiritually, to navigate the unique demanding situations posed with the aid of this incorporated approach (Ray, 2023).

5.4 Patient-centric approaches in necrobotics

Patient-centric techniques in necrobotics epitomize an innovative shift in healthcare closer to personalized, holistic care. This method places the patient at the leading edge, considering not only their bodily illnesses but also their spiritual and emotional well-being.

Necrobotics, via the fusion of generation and mysticism, helps a deeper understanding of sufferers' wishes and choices. Personalized treatment plans are crafted, deliberating both empirical scientific statistics and nonsecular insights derived from necromantic practices. This method allows an extra compassionate and tailored healthcare revel in. The integration of robotics ensures that routine and repetitive duties are automatic, freeing up healthcare professionals to cognizance on significant patient interactions. Necromantic insights contribute to empathetic decision-making, fostering healing surroundings wherein sufferers sense seen and understood past traditional clinical parameters. Moreover, affected person involvement in choice-making strategies is emphasized, aligning treatments with their ideals and alternatives. Real-time tracking via necrobotics offers on-the-spot responses to affected person wishes, growing a dynamic and responsive healthcare machine.

In essence, patient-centric necrobotics transforms healthcare right into a more empathetic, individualized, and responsive environment. It embodies a paradigm wherein generation and mysticism converge to no longer most effectively optimize workflows but also to prioritize the religious and emotional dimensions of affected person care (Yap et al., 2022).

5.5 Global collaboration for healthcare innovation

Global collaboration for healthcare innovation and the usage of necrobotics holds significant potential to revolutionize medical practices, ensuring numerous views and shared advancements in healthcare and workflow optimization. By fostering worldwide partnerships, the mixing of necromancy and robotics can advantage from a collective pool of know-how, knowledge, and cultural insights.

Collaboration allows the establishment of usual moral requirements, addressing the cultural nuances and ethical concerns associated with the implementation of necrobotics. Shared frameworks facilitate a more complete knowledge of the mystical factors, ensuring accountable and culturally sensitive programs across diverse healthcare structures.

Pooling resources and expertise from specific regions speeds up the improvement of necrobotic technologies, resulting in faster adoption and diffusion of improvements. This collaborative method permits the introduction of standardized protocols for interoperability, fact-sharing, and ethical governance.

Furthermore, global collaboration can power research projects to explore the diverse packages of necrobotics in addressing precise healthcare-demanding situations commonplace in different regions. By combining technological improvements with religious insights, a collaborative technique ensures that necrobotics contributes to optimized workflows, more advantageous patient care, and stepped-forward healthcare results on an international scale.

Global collaboration for healthcare innovation using necrobotics fosters a shared vision of leveraging era and mysticism for the betterment of healthcare globally. It creates a collaborative ecosystem wherein nations unite to propel improvements, making sure the common integration of necrobotics for the optimization of healthcare workflows throughout various global contexts.

6 Conclusion

6.1 Recapitulation of key findings

The recapitulation of key findings within the workflow and optimization of healthcare through the usage of necrobotics underscores a transformative shift within the paradigm of clinical practices. Necrobotics, blending technology and mysticism, has demonstrated first-rate potential in revolutionizing healthcare workflows. Advanced robotics streamline duties, from diagnostics to affected person care, maximizing performance and precision.

The integration of necromancy introduces a unique size, contributing spiritual insights that decorate choice-making and affected person consequences. By automating routine processes, necrobotics allows healthcare professionals to focus on complex and compassionate aspects of affected person care, fostering an extra holistic technique. Ethical issues, affected person privateness, and worldwide collaboration turn out to be crucial elements requiring interest. Establishing strong moral suggestions guarantees accountable implementation, respecting affected person autonomy and cultural sensitivities. Patient-centric procedures prioritize individualized care, tailoring remedies to each clinical and religious wishes.

The collaborative efforts on a worldwide scale boost up studies, innovation, and the development of standardized protocols. As the field evolves, the recapitulation well-known shows a promising destiny where necrobotics optimizes healthcare workflows, reduces prices, and advances affected person care, symbolizing a harmonious integration of era and mysticism for the betterment of healthcare on a global scale.

6.2 The future landscape of necrobotics in healthcare

The destiny landscape of necrobotics in healthcare is poised for unprecedented transformation. With ongoing advancements in era and mystical practices, necrobotics is anticipated to redefine healthcare transport. Artificial intelligence will play a pivotal role in improving diagnostics, remedy personalization, and workflow optimization. The integration of advanced robotics and necromancy will foster modern surgical processes, patient monitoring, and holistic care processes.

Ethical considerations will guide responsible implementation, making sure affected persons have privacy, consent, and cultural sensitivity. Global collaboration will force the evolution of standardized protocols, fostering a shared imaginative and prescient for the moral and efficient integration of necrobotics on a worldwide scale. In the future, necrobotics is anticipated to not only streamline healthcare workflows but additionally provide an extra empathetic and spiritually attuned approach to affected person care, marking a paradigm shift in the direction of a harmonious coexistence of technology and mysticism inside the healthcare domain.

6.3 Recommendations for successful implementation

To make certain a successful implementation of necrobotics inside the healthcare enterprise, numerous key recommendations must be taken into consideration. Firstly, establishing interdisciplinary groups comprising technologists, healthcare professionals, ethicists, and nonsecular advisers can offer numerous perspectives. Rigorous training applications ought to be instituted to equip healthcare practitioners with the essential talents to navigate each technological and mystical dimension. Ethical suggestions ought to be evolved collaboratively, addressing affected person privacy, consent, and cultural sensitivities related to necromantic practices.

Transparent conversation with sufferers is important, making sure informed consent and fostering acceptance as true with the integration of necrobotics. Continuous studies and development efforts, supported by global collaboration, are essential for refining technology and standardizing protocols. Regulatory bodies should adapt to the unique challenges posed through necrobotics, creating frameworks that sell accountable and ethical practices. By adhering to these recommendations, the healthcare industry can efficiently harness the capacity of necrobotics for optimized workflows, stronger patient care, and transformative advancements.

References

1988. Experimental study of the pathogenicity and pathologic picture of paecilomyces species isolated from the lung of dead fayou mi chickens. Assiut Vet. Med. J. **20.2** (40), 75–79. https://doi.org/10.21608/avmj.1988.189381. 2314-5226.

Byun, J., Lee, Y., Yoon, J., Lee, B., Oh, E., Chung, S., Lee, T., Cho, K.J., Kim, J., Hong, Y., 2018. Electronic skins for soft, compact, reversible assembly of wirelessly activated fully soft robots. Sci. Robot. 3 (18). American Association for the Advancement of Science, South Korea. 24709476. Available from: http://robotics.sciencemag.org/content/robotics/3/18/eaas9020.full.pdf. https://doi.org/10.1126/scirobotics.aas9020.

Drotman, D., Jadhav, S., Sharp, D., Chan, C., Tolley, M.T., 2021. Electronics-free pneumatic circuits for controlling soft-legged robots. Sci. Robot. 6 (51). American Association for the Advancement of Science, United States. 24709476. Available from: https://robotics.sciencemag.org/lookup/doi/10.1126/scirobotics.aay2627. https://doi.org/10.1126/SCIROBOTICS.AAY2627.

Galloway, K.C., Becker, K.P., Phillips, B., Kirby, J., Licht, S., Tchernov, D., Wood, R.J., Gruber, D.F., 2016. Soft robotic grippers for biological sampling on deep reefs. Soft Robot. 3 (1), 23–33. Mary Ann Liebert Inc., United States. 21695180. Available from: http://www.liebertpub.com/overview/soft-robotics/616/. https://doi.org/10.1089/soro.2015.0019.

Grazioso, S., Di Gironimo, G., Siciliano, B., 2019. A geometrically exact model for soft continuum robots: the finite element deformation space formulation. Soft Robot. 6 (6), 790–811. Mary Ann Liebert Inc., Italy. 21695180. Available from: http://www.liebertpub.com/overview/soft-robotics/616/. https://doi.org/10.1089/soro.2018.0047.

Hall, E., 2023. Welcome to future medicine AI. Future Med. AI. https://doi.org/10.2217/fmai-2023-0015. 2753-8087.

Heiden, A., Preninger, D., Lehner, L., Baumgartner, M., Drack, M., Woritzka, E., Schiller, D., Gerstmayr, R., Hartmann, F., Kaltenbrunner, M., 2022. 3D printing of resilient biogels for omnidirectional and exteroceptive soft actuators. Sci. Robot. 7 (63). American Association for the Advancement of Science, Austria. 24709476. Available from: https://www.science.org/doi/10.1126/scirobotics.abk2119. https://doi.org/10.1126/scirobotics.abk2119.

Ji, X., Liu, X., Cacucciolo, V., Imboden, M., Civet, Y., Haitami, A.E., Cantin, S., Perriard, Y., Shea, H., 2019. An autonomous untethered fast soft robotic insect driven by low-voltage dielectric elastomer actuators. Sci. Robot. 4 (37). American Association for the Advancement of Science, Switzerland. 24709476. Available from: https://robotics.sciencemag.org/content/robotics/4/37/eaaz6451.full.pdf. https://doi.org/10.1126/scirobotics.aaz6451.

Kambouris, M.E., Manoussopoulos, Y., Velegraki, A., Patrinos, G.P., 2023. The biote-bot hybrid. The ultimate biothreat merging nanobots, AI-enabled cybernetics and synthetic biology. Future Med. AI. https://doi.org/10.2217/fmai-2023-0008. 2753-8087.

Katzschmann, R.K., DelPreto, J., MacCurdy, R., Rus, D., 2018. Exploration of underwater life with an acoustically controlled soft robotic fish. Sci. Robot. 3 (16). American Association for the Advancement of Science, United States. 24709476. Available from: https://robotics.sciencemag.org/lookup/doi/10.1126/scirobotics.aar3449. https://doi.org/10.1126/SCIROBOTICS.AAR3449.

Marchese, A.D., Onal, C.D., Rus, D., 2014. Autonomous soft robotic fish capable of escape maneuvers using fluidic elastomer actuators. Soft Robot. 1 (1), 75–87. Mary Ann Liebert Inc., United States. 21695180. Available from: http://www.liebertpub.com/overview/soft-robotics/616/. https://doi.org/10.1089/soro.2013.0009.

Polygerinos, P., Correll, N., Morin, S.A., Mosadegh, B., Onal, C.D., Petersen, K., Cianchetti, M., Tolley, M.T., Shepherd, R.F., 2017. Soft robotics: review of fluid-driven intrinsically soft devices; manufacturing, sensing, control, and applications in human-robot interaction. Adv. Eng. Mater. 19 (12). Wiley-VCH Verlag, United States. 15272648. Available from: http://www3.interscience.wiley.com/journal/67500980/home. https://doi.org/10.1002/adem.201700016.

Ray, P.P., 2023. An introduction to necrobotics: concept, architecture, use cases, challenges, future directions. SSRN Electr. J. https://doi.org/10.2139/ssrn.4595801. 1556-5068.

Rus, D., Tolley, M.T., 2015. Design, fabrication and control of soft robots. Nature 521 (7553), 467–475. Nature Publishing Group, United States. 14764687. Available from: http://www.nature.com/nature/index.html. https://doi.org/10.1038/nature14543.

Tolley Michael, T., Shepherd Robert, F., Galloway Kevin, C., Wood Robert, J., George, M.W., 2014. A resilient, untethered soft robot. Soft Robot.

Yap, H.K., Lim, J.H., Nasrallah, F., Goh, J.C.H., Yeow, R.C.H., 2015. A soft exoskeleton for hand assistive and rehabilitation application using pneumatic actuators with variable stiffness. In: 2015-Proceedings—IEEE International Conference on Robotics and Automation. Institute of Electrical and Electronics Engineers Inc., Singapore, pp. 4967–4972, https://doi.org/10.1109/ICRA.2015.7139889. 9781479969234, June.

Yap, T.F., Liu, Z., Rajappan, A., Shimokusu, T.J., Preston, D.J., 2022. Necrobotics: biotic materials as ready-to-use actuators. Adv. Sci. 9 (29). John Wiley and Sons Inc, United States. 21983844. Available from: http://onlinelibrary.wiley.com/journal/10.1002/(ISSN)2198-3844. https://doi.org/10.1002/advs.202201174.

Yap, T.F., Liu, Z., Rajappan, A., Shimokusu, T.J., Preston, D.J., 2023. Harnessing nature's design with necrobotics. Device 1 (4), 100119. https://doi.org/10.1016/j.device.2023.100119. 26669986.

CHAPTER 11

Futuristic Med-Tech transforming healthcare industry: Nanorobotics for healthcare applications and management in action

Bhupinder Singh[a], Christian Kaunert[b], and Ritu Gautam[c]
[a]Sharda School of Law, Sharda University, Greater Noida, India
[b]Dublin City University, Dublin, Ireland
[c]Sharda University, Greater Noida, India

1 Introduction

The development of medical technology, nanotechnology, and robotics has united to create "nanorobotics," a disruptive force in the field. The potential to completely change the healthcare sector is at the forefront of this junction. Operating at the nanoscale, nanorobotics presents a futuristic environment that goes beyond the bounds of traditional medical techniques, bringing about a paradigm change in healthcare applications and administration. (Germain et al., 2020). The future for healthcare holds great promise as nanotechnology and robots combine to provide unprecedented levels of efficiency and accuracy. Working on a nanoscale, nanorobots demonstrate a degree of nimbleness and precise movement that was previously unthinkable. This innovative technology represents a shift from conventional healthcare methods and has its origins in the combination of many scientific fields. Necrobotics is an interdisciplinary discipline that combines the principles and techniques of robotics with those of biomedical sciences. The primary objective is to create and employ robotic systems for the purpose of researching, diagnosing, treating, and managing illnesses, injuries, and medical problems in deceased people or animals, as well as in nonliving biological samples. This technology includes a range of robotics, sensors, imaging systems, and computer algorithms that are specifically intended to enhance our knowledge, intervention, or manipulation of biological systems that go beyond typical living beings. Effective healthcare administration is essential for the effective provision of healthcare services, maximizing the use of resources, and improving patient outcomes.

Nanorobotics has a wide range of applications in healthcare that have the potential to completely change how illnesses are identified, treated, and maintained. The impact on patient outcomes is significant, ranging from focused medication delivery systems

https://doi.org/10.1016/B978-0-443-24832-0.00011-8

that reduce side effects to the possibility of nanorobots performing minimally invasive procedures with unmatched accuracy. The ability of nanorobotics to navigate the intricate landscapes of the human body, addressing medical issues at the molecular and cellular levels, represents a leap forward in the quest for more effective and personalized healthcare solutions. As it delves into the intricacies of nanorobotics, it becomes evident that its influence extends beyond the realm of direct patient care. The management of healthcare systems is poised for a significant overhaul through the integration of nanorobotics. From data analytics facilitating informed decision-making to the maintenance of patient records with unparalleled accuracy, nanorobotics emerges as a cornerstone in the evolution of healthcare management practices (Haleem et al., 2022). The optimization of resources within healthcare institutions, driven by the insights provided by nanorobotics, promises enhanced efficiency and resource allocation.

This chapter embarks on an exploration of this transformative landscape, aiming to elucidate the multifaceted dimensions of nanorobotics in healthcare applications and management. Through an in-depth analysis of current literature and technological developments, it seeks to unravel the intricacies of this burgeoning field. The subsequent sections will delve into the types of nanorobots, the technical challenges they present, and their applications in targeted drug delivery, disease diagnosis, surgical interventions, and regenerative medicine. In addition to the technical aspects, this chapter focuses on the intersection of nanorobotics with healthcare management. From ethical considerations surrounding patient privacy to the optimization of healthcare resources, the societal and systemic implications of nanorobotics will be scrutinized. It provides a real-world insight into the successes and challenges encountered in the application of nanorobotics settings (Naidu et al., 2015). Fig. 11.1 shares the landscapes of the introduction section.

1.1 Background of the study

The background of futuristic Med-Tech and the transformative role of nanorobotics in the healthcare industry stems from a convergence of scientific disciplines that has propelled the boundaries of medical innovation. Traditionally, medical interventions were confined to macroscopic levels, addressing health concerns at the organ or tissue scale. However, the relentless pursuit of advancements in nanotechnology and robotics has birthed a field where interventions occur at the nanoscale as the arena of molecules and

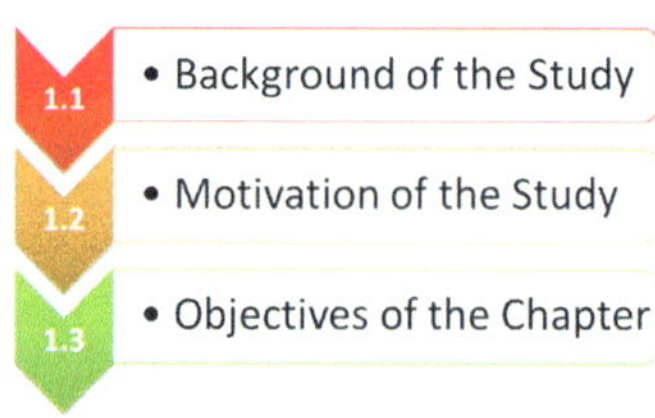

Fig. 11.1 ***Landscapes of introduction split section.*** This figure shares the background, motivation and objectives of this chapter. *(Source: Original; No Permission Required.)*

cells. This convergence lays the foundation for nanorobotics, a revolutionary technology that operates at dimensions measured in nanometers. Nanotechnology has evolved from a theoretical notion to a thriving field abundant with suggestions and ideas, and it has now become a dynamic domain of real study and development. Neurosurgery, a field that often functions at a tiny scale, is well poised to take advantage of advancements in nanotechnology. The benefits include improved identification of diseases, less intrusive monitoring inside the skull, and precise administration of medications, among other perks. Recent progress in the production of microelectromechanical systems has greatly expedited our capacity to operate on smaller dimensions, perhaps enabling manipulation at the molecular level in the foreseeable future. The genesis of this technology represents a departure from traditional healthcare methodologies, introducing a level of precision and targeted action previously unimaginable. The background of this futuristic MedTech narrative is intertwined with the quest for more effective, personalized, and less invasive healthcare solutions. It is clear from delving into the field of nanorobotics that this convergence represents a paradigm change rather than just a little step forward, one that has the power to completely reshape the landscape of healthcare applications and management techniques. This background prepares the reader for an examination of the technical nuances, uses, and managerial implications of nanorobotics in the parts that follow in this study (Lehoux, 2006).

1.2 Motivation of the study

The study on "Futuristic Med-Tech Transforming the Healthcare Industry: Necrobotics for Healthcare Applications and Management in Action" is motivated by the possibility of groundbreaking progress in medical technology. Necrobotics, as the fusion of robotics with biomedical sciences, shares an innovative method for tackling crucial healthcare obstacles. Necrobotics specifies the sophisticated instruments for investigating illnesses, injuries, and medical problems in deceased or nonliving biological samples. This may result in a more profound comprehension of disease processes, devoid of ethical considerations linked to live patients. With combining robotic systems with state-of-the-art sensors and imaging technology, the precision and effectiveness of illness diagnosis, treatment, and management may be greatly enhanced. This encompasses several uses, such as highly accurate surgical interventions and treatments that involve minimum invasion. Necrobotics may enhance resource allocation in healthcare systems by automating certain components of medical research and clinical operations. This may result in reduced expenses, enhanced patient results, and increased effectiveness in providing healthcare. The current medical procedures are constrained by the extent to which interventions can be carried out. Necrobotics, because of its capacity for manipulation at the cellular and molecular scales, has the ability to surpass these constraints and pave the way for novel advancements in the field of medical research. The chapter may have a special influence on the area of neurosurgery, which necessitates a high level of accuracy and often involves working with tiny structures. Advancements in necrobotics have the potential to enhance methods for nerve restoration, treatment of spinal cord injuries, and other essential neurosurgical interventions. With incorporating advanced technologies such as necrobotics, the healthcare business can effectively adapt to the changing landscape and be equipped to tackle upcoming obstacles. This includes addressing newly identified illnesses, optimizing patient treatment, and boosting the overall administration of healthcare. Employing necrobotics in medical research might diminish the need for animal experimentation and human trials, addressing ethical considerations while simultaneously advancing medical understanding. Moreover, it has the potential to create safer and more regulated experimental settings. Stimulating more advancements: The exploration and advancement of necrobotics have the potential to stimulate more investigation and ingenuity in diverse domains encompassing biomedical sciences and robotics. This has the potential to trigger a series of technical breakthroughs that will have positive effects on many elements of healthcare. The primary impetus for this work is from the profound capacity of necrobotics to fundamentally reshape healthcare, enhance patient results, and tackle existing as well as future obstacles in the medical domain.

1.3 Objectives of the chapter

This study aims to provide a thorough overview of the current status of nanorobotics, a nascent area at the nexus of robotics and nanotechnology, and to clarify its revolutionary effects on the healthcare sector. This study seeks to outline the several uses of nanorobotics in healthcare, from targeted medication administration to surgical procedures and more, through a thorough review of the literature, case studies, and technical advancements. The chapter aims to explore how nanorobotics may transform hospital administration procedures, such as resource optimization, data analytics, and record-keeping (Banerjee et al., 2019). With achieving these goals, this paper advances the understanding of the futuristic environment in which nanorobotics, which is a key component, is transforming the dynamics of healthcare administration and applications. This chapter attempts to:

- Give a general introduction to nanorobotics and the technologies that underpin it.
- Examine how nanorobotics may be used in the medical field.
- Investigate how nanorobotics affects the administration of healthcare.
- Explore the difficulties and moral issues surrounding nanorobotics.
- Lay down the possible benefits and hazards of nanorobotics in the medical field.

2 Nanorobotics: An overview

Operating at the tiny size of nanometers to negotiate the complexities of the human body, nanorobotics offers a groundbreaking marriage of nanotechnology and robotics. Fundamentally, nanorobotics is the study, development, and use of robots or other devices with tiny dimensions. These microscopic organisms have the potential to revolutionize the healthcare industry by carrying out activities like precision surgical procedures and tailored medicine distribution. Artificial nanorobots, biohybrids, and DNA-based structures are only a few examples of the diverse range of designs in nanorobotics, each with a unique purpose in mind. The field has a lot of promise, but there are also difficulties, such as power constraints and complex communication needs. In order to prepare for a closer examination of nanorobotics' applications in healthcare and the potentially revolutionary effects it might have on the sector, this part will offer a thorough introduction of the core concepts, kinds, and technical difficulties associated with the technology (Tibbals, 2017).

2.1 Definition and it's components

Nanorobotics is the study of the creation, use, and maintenance of robots at the nanoscale. Sensors, actuators, and communication devices are some of the tiny parts that make up these nanorobots, or nanobots. The manipulation and control of robots or other devices at the nanoscale, which is generally between 1 and 100 nm, is known as nanorobotics. These tiny machines, also known as nanorobots, are made to function in settings as complex as the human body, allowing for precise and focused treatments (Sahoo et al., 2023). The many materials and technologies that make up nanorobotics' constituent parts enable these machines to carry out their tasks with extraordinary accuracy. Biohybrid designs of nanorobots can include biological elements like proteins, DNA strands, or even live cells. Furthermore, materials such as carbon nanotubes or nanowires are frequently used in synthetic nanorobots to ensure mechanical robustness and durability at the nanoscale. Nanorobots can respond to particular stimuli and connect with external systems thanks to the integration of sensors, actuators, and communication modules. Deploying these tiny entities for transformational applications in a variety of disciplines, most notably healthcare, presents both opportunities and problems that must be understood in order to fully appreciate the complicated components of nanorobotics.

2.2 Technological advancements: Nanorobotics for healthcare

Sophisticated nanorobots are now possible because of recent developments in nanotechnology, materials science, and robotics. Molecular self-assembly, nanoscale manufacturing, and enhanced sensor integration are examples of key technologies. These developments in nanorobotics have elevated the discipline to the vanguard of innovative healthcare. These developments are characterized by the intersection of materials science, robotics, and nanotechnology, which has made it possible to build very intelligent nanorobots that can carry out complex tasks within the human body. The most significant achievement is the creation of self-propelled nanorobots, which are able to move through biological fluids with an unprecedented degree of accuracy. Furthermore, by using cutting-edge materials like biocompatible polymers and nanoparticles, these nanorobots have improved structurally and functionally, guaranteeing their compatibility with biological systems. The range of applications for nanorobotic systems in healthcare has increased due to advancements in autonomous navigation and remote-control techniques (Istepan and Woodward, 2016).

3 Applications of necrobotics in healthcare

Nanorobotics is essential for data analytics and decision assistance as healthcare systems struggle with the growing complexity of data management. These microscopic agents have the ability to gather, process, and communicate data, providing insights that support better decision-making in the administration of healthcare. Nanorobotics has the ability to preserve patient records more accurately and securely than existing systems. Numerous uses for nanorobotics in healthcare have the potential to completely change the way that medical procedures are carried out. Nanorobotics is a driving force behind the future advancement of healthcare, from precision medicine to improved diagnostics and revolutionary surgical techniques. With the capacity to move through the circulation, nanorobots can precisely administer therapeutic chemicals to target cells or tissues, reducing adverse effects and increasing treatment success. This has great potential for diseases like cancer, where targeted and regulated medication release might be very helpful (Singh, 2023).

Necrobotics, derived from the Greek word νεκρός, meaning "dead," and combined with "robotics," represents cutting-edge technological innovation. This pioneering technology uses spiders as robotic forceps designed for microsurgeries. Although it might sound like a concept from a science fiction novel, it is a reality, marking the ultimate integration of biology and artificial intelligence. The term "necrobotics" might suggest the use of deceased spiders, but this is a misunderstanding. The actual focus is on bio-induced necrobotics, an advanced technology rooted in artificial intelligence. It utilizes the unique properties of spider silk and other biocompatible materials to create microrobots specifically designed for surgical applications, representing a significant breakthrough in medical technology. Nanorobotics has the potential to transform disease monitoring and detection in the field of diagnostics. It is possible that in the near future, nanorobots may be able to operate with unprecedented accuracy during minimally invasive surgery. These tools are more accurate than traditional surgical techniques in performing tasks while navigating through intricate anatomical systems and accessing previously inaccessible places. Another exciting area for nanorobotics is regenerative medicine, where they can help with tissue regeneration and repair by accurately administering growth factors or adjusting cellular activity at the microscopic level. This creates opportunities for improving the body's self-healing capabilities and treating degenerative diseases (Rambow-Hoeschele et al., 2020).

3.1 Diagnostics

Advanced sensor-equipped nanorobots are able to offer molecular diagnostics in real time with great resolution. This has effects on disease surveillance and early detection. With its unheard-of molecular and cellular capabilities, nanorobotics has become a disruptive force in medical diagnosis. Detecting diseases early on is one of its main uses. With the use of sophisticated sensors and diagnostic instruments, nanorobots are able to traverse through physiological fluids such as blood and detect certain biomarkers linked to different illnesses. Because of this nanoscale accuracy, illnesses like cancer and infectious disorders can be identified early, allowing for prompt intervention and a much better prognosis (Sharma et al., 2021).

Nanorobots advance the field of individualized diagnostics by customizing their actions to individual differences. These microscopic instruments may be trained to identify distinct genetic or molecular indicators, offering a degree of precision that conventional diagnostic techniques frequently find difficult to attain. The other interesting use of nanorobotics in medical diagnostics is continuous monitoring. It is possible to create these tiny agents with the ability to send real-time data and continually monitor physiological characteristics. A proactive approach to healthcare is made possible by this dynamic monitoring, which makes it possible to identify changes in health indicators early and arrange for prompt medical measures. Nanorobotics is having revolutionary effects on healthcare diagnosis. Nanorobotics has the potential to completely change how we approach and manage health issues, from early illness identification to tailored diagnostics and continuous monitoring. The incorporation of nanorobotics in diagnostic procedures offers a future where healthcare is not only more accurate but also more proactive in ensuring people's well-being as these technologies continue to progress (Garcia Gonzalez-Moral et al., 2023).

3.2 Drug delivery

Nanorobots provide a focused method of medication delivery that maximizes therapeutic efficacy while reducing negative effects. They are able to precisely deliver drugs to the afflicted cells or tissues by navigating through the circulation. Thanks to its unmatched accuracy and efficiency, nanorobotics is leading the charge in healthcare's revolution of medication administration. Targeted medication delivery is one of its main uses, in which nanorobots function as microscopic carriers that carry medicinal substances straight to certain cells or tissues. These nanorobots minimize collateral harm to healthy tissues by precisely targeting target areas through the bloodstream's complex network navigation. Especially when it comes to illnesses like cancer, this focused strategy may increase therapy efficacy while lowering adverse effects (Elnikety et al., 2022).

They may be programmed to react to certain stimuli, including pH shifts or the presence of certain chemicals linked to illness. Drugs may be released on demand thanks to this responsiveness, which guarantees that prescriptions are given exactly when and where they are required. Controlled medication release can improve therapeutic benefit while reducing side effects related to traditional drug delivery techniques (Tabish and Nabil, 2015). Also, biological barriers that have historically restricted medicine delivery may be overcome with the help of nanorobots. These barriers present difficulties for treating diseases that damage delicate tissues, such as the blood-brain barrier. By giving tailored nanocarriers a platform to overcome these obstacles and reach hitherto unreachable therapeutic sites, nanorobotics provides a solution. Drug distribution in healthcare has revolutionary uses for nanorobotics. Nanorobotics is changing the landscape of pharmacological interventions, from on-demand release and biological barrier-breaking to tailored medication administration with minimal adverse effects. With further advancements in technology, there is great potential to improve patient outcomes for a variety of medical illnesses through more accurate, effective, and individualized drug administration (Galetsi et al., 2019).

3.3 Surgery and intervention

Nanorobots provide minimally invasive surgical techniques by reaching hard-to-reach or distant parts of the body. They are able to carry out exact therapies that shorten recuperation periods and trauma. Nanorobotics has the potential to completely change the medical field through its revolutionary uses in intervention and surgery. The prospect for less intrusive surgery is one of the most exciting features. Because they function at the nanoscale, nanorobots may reach previously inaccessible places by precisely navigating through complex anatomical systems. This capacity makes minimally invasive surgeries possible, which lowers trauma, speeds up healing, and lowers the possibility of problems that come with more conventional surgical techniques.

Precision medicine has entered a new age marked by the use of nanorobotics in surgical and interventional healthcare procedures. From minimally invasive operations to targeted treatments at the cellular level, nanorobotics has tremendous potential in enhancing the effectiveness and safety of medical procedures, opening the path for a future where surgical accuracy reaches unparalleled levels. During surgery, specific functions at the cellular or molecular level can be carried out by nanorobots. With more accuracy than human hands can achieve, these nanoscale agents can manipulate tissues, deliver medicinal payloads, or even aid in regeneration processes. For example, nanorobotics presents the possibility of delicate treatments in neurosurgery with little damage to surrounding healthy tissue. Additionally, nanorobots are essential for focused therapies, particularly when it comes to illnesses like cancer. Therapeutic medicines can be delivered directly to the damaged spot while preserving healthy tissues, as they can be designed to recognize and specifically target cancer cells. This focused strategy represents a major breakthrough in the accuracy of medical interventions, as it not only increases therapeutic efficacy but also decreases adverse effects (Zheng et al., 2022).

4 Nanorobotics in healthcare management

Nanorobotics in healthcare management is not merely a futuristic concept but a tangible force that is reshaping the way healthcare systems are organized, operated, and optimized. While the integration of nanorobotics in healthcare management introduces unparalleled opportunities, it also raises ethical considerations. Issues related to consent, data ownership, and the potential misuse of sensitive information must be addressed. Striking a balance between technological advancements and ethical frameworks is crucial to fostering a healthcare ecosystem that benefits from the transformative power of nanorobotics while upholding the principles of patient autonomy and privacy. From data analytics and patient record maintenance to resource optimization, nanorobotics holds the promise of a more efficient, personalized, and ethically sound healthcare management landscape. As this technology continues to advance, its integration into healthcare management practices will likely redefine the standards for quality, efficiency, and ethical governance in the healthcare industry. A new age of accuracy and security for patient data and privacy is brought forth by nanorobotics at the molecular level, these tiny devices may be used to update and retain patient information. This improves data security and guarantees the accuracy of medical histories. In addition to changing the face of medical treatments, nanorobotics is also transforming the administration of healthcare. This data offers a wealth of information for medical practitioners, spanning from physiological characteristics to molecular biomarkers. Nanorobotics adds to comprehensive data analytics by providing insights into patient outcomes, treatment responses, and disease patterns using complex algorithms. Healthcare administrators and decision-makers may make well-informed decisions, customize interventions, and improve healthcare delivery with the help of this abundance of information (Bansal et al., 2022).

Nanorobots can employ encryption methods and secure communication channels, mitigating the risks associated with conventional electronic health records. By safeguarding patient privacy and maintaining data integrity, nanorobotics contributes to the ethical considerations fundamental to effective healthcare management. Resource optimization is another critical aspect where nanorobotics offers transformative solutions. By providing real-time data on resource utilization within healthcare institutions, nanorobots enable efficient allocation of personnel, equipment, and facilities. This optimization has the potential to streamline workflows, reduce wait times, and enhance overall healthcare service delivery. The insights generated by nanorobotics contribute to a proactive approach to resource management, ensuring that healthcare systems are agile and responsive to evolving demands (Sharma and Singh, 2022).

4.1 Data analytics and personalized medicine

Nanorobots generate vast amounts of data during diagnostics and treatment. Advanced analytics and artificial intelligence enable personalized medicine by processing this data to tailor treatments to individual patients. Data analytics plays a pivotal role in driving the evolution of personalized medicine, offering unprecedented insights into individual health profiles and tailoring medical interventions accordingly. At its core, personalized medicine seeks to move away from the one-size-fits-all approach to treatment and diagnostics, recognizing the inherent variability among individuals. Data analytics, powered by advanced algorithms and machine learning, mines vast datasets comprising genetic information, clinical records, lifestyle factors, and environmental influences (Zhou et al., 2020).

The primary application of data analytics in personalized medicine is in the realm of genomics. Since then, advancements in sequencing technologies have facilitated the generation of massive genomic datasets. Data analytics enables the interpretation of this complex genetic information, identifying genetic markers associated with specific diseases or predicting individual responses to medications. This genomic precision forms the basis for tailoring treatments that align with an individual's genetic makeup. The data analytics empowers the integration of diverse data streams, including electronic health records, wearable device data, and patient-reported outcomes. By assimilating this multifaceted information, healthcare providers can gain a comprehensive understanding of an individual's health trajectory. Predictive analytics models can anticipate disease risks, enabling proactive interventions and preventive measures. In chronic conditions like diabetes or cardiovascular diseases, continuous monitoring and data analytics contribute to personalized management plans, adjusting treatment strategies based on real-time data (Bilal, 2021).

The implementation of data analytics in personalized medicine extends beyond diagnostics and treatment to inform drug development processes. By analyzing patient responses to medications, researchers can identify subsets of individuals who are more likely to benefit from specific drugs. This paradigm shift from a population-based approach to a more nuanced understanding of individual variations fosters a more efficient and effective drug discovery process. The assimilation of data analytics in personalized medicine also poses challenges. Privacy concerns, data security, and ethical considerations regarding the use of genetic information necessitate robust frameworks and regulations. Ensuring equitable access to personalized medicine technologies and addressing potential biases in algorithms are essential aspects of navigating the evolving landscape of data-driven personalized healthcare. The synergy between data analytics and personalized medicine represents a transformative force in healthcare. The ability to harness vast datasets and derive meaningful insights not only refines our understanding of diseases at the individual level but also paves the way for more precise, patient-centric, and effective healthcare interventions. As technology advances and our understanding of the intricate interplay between genetics, environment, and health deepens, data analytics will continue to be a driving force in the realization of the promise of personalized medicine (Sharon, 2017).

4.2 Remote monitoring and telemedicine

Healthcare professionals may now monitor patients remotely and follow the course of their diseases thanks to the incorporation of nanorobotics. This has important ramifications for remote healthcare delivery and telemedicine. With the use of digital technology, remote monitoring and telemedicine have become groundbreaking foundations of contemporary healthcare, bridging geographic divides, improving accessibility, and completely changing patient care. Real-time tracking of health measures outside of typical healthcare settings is made possible by remote monitoring, which uses gadgets to gather and send patient data to healthcare providers. Contrarily, telemedicine describes the provision of medical services via electronic communication, allowing for consultations, diagnosis, and treatment without the requirement for in-person visits (Allouzi et al., 2022).

The capacity to offer preventive and ongoing healthcare is one of remote monitoring's main benefits. Vital signs, exercise levels, and medication adherence are just a few of the many data points that wearable technology, sensors, and other monitoring technologies gather. Then, this data is sent to medical professionals so they may evaluate a patient's status from a distance, spot patterns, and take immediate action if needed. When it comes to managing chronic disorders like diabetes, hypertension, and heart disease, this proactive approach is especially helpful since it can avert problems and hospitalizations by detecting variations from baseline health indicators early on (Stasevych and Zvarych, 2023).

Telemedicine enhances remote monitoring by enabling virtual consultations and increasing access to healthcare. Patients can interact with medical experts from the comfort of their homes via chat rooms, video conferencing, and other communication tools. This is especially helpful for people who live in rural or underdeveloped regions, have mobility issues, or need specialized consultations that might not be offered in their area. Telemedicine eases the strain on the physical infrastructure of healthcare, improving accessibility and efficiency. While reducing the chance of viral transmission, these technologies have made it possible to continue providing necessary healthcare services. Regular check-ups, consultations, and even mental health assistance may be provided to patients virtually, guaranteeing continuous and open access to healthcare (Cohen et al., 2004).

While telemedicine and remote monitoring have many benefits, there are drawbacks as well. Careful thought must be given to issues including privacy and security concerns, the necessity for regulatory frameworks, and technical differences among patient groups. For these technologies to be widely used and successful, it is imperative that a balance be struck between the ease of digital healthcare and preserving the quality and safety of medical services (Bilimoria et al., 2019). With its ability to provide individualized, easily accessible, and effective care, telemedicine and remote monitoring represent a paradigm change in the delivery of healthcare. The combination of telemedicine and remote monitoring is set to become a crucial aspect of the healthcare landscape in the future, enhancing patient-centered care and enhancing global health outcomes as technology advances and society adjusts to digital healthcare (Singh, 2023).

5 Challenges and ethical considerations

These considerations are crucial for ensuring the responsible and equitable deployment of these technologies, safeguarding patient rights, and maintaining the integrity of healthcare systems (Karkkainen, 2005).

1 Privacy and Security Concerns: The foremost challenges involve the privacy and security of patient data. As healthcare technologies increasingly rely on the collection and analysis of sensitive health information, robust measures must be in place to protect this data from breaches or unauthorized access. Striking the right balance between data accessibility for medical purposes and safeguarding patient privacy is a delicate but essential ethical consideration (Barker, 2010).
2 Inequality in Access: The widespread adoption of healthcare technologies may exacerbate existing inequalities in healthcare access. Not all populations have equal access to advanced medical technologies, creating a potential divide between those who benefit from these innovations and those who do not. Ethical considerations include addressing these disparities and ensuring that healthcare technologies are accessible and affordable for diverse populations.
3 Informed Consent and Autonomy: Issues related to informed consent and patient autonomy become complex with technologies such as personalized medicine and genetic testing. Patients must have a comprehensive understanding of the implications of these technologies on their health and well-being. Ensuring that individuals can make informed decisions about participating in treatments or sharing genetic information is essential to respecting patient autonomy.
4 Bias in Algorithms: Data analytics and machine learning algorithms play a crucial role in personalized medicine and other healthcare technologies. However, these algorithms may inherit biases present in the training data, leading to disparities in treatment recommendations or diagnostic accuracy. Ethical considerations involve addressing and mitigating these biases to ensure fair and equitable healthcare outcomes for all individuals.
5 Regulatory Oversight: The rapid evolution of healthcare technologies often outpaces regulatory frameworks. Striking a balance between fostering innovation and ensuring patient safety requires continuous regulatory adaptation. Ethical considerations involve establishing clear guidelines and oversight mechanisms to evaluate the safety, efficacy, and ethical implications of emerging healthcare technologies.
6 Human Workforce Displacement: The integration of technologies like telemedicine and robotic surgery may raise concerns about job displacement within the healthcare workforce. Ethical considerations include addressing the potential impact on employment while ensuring that technological advancements ultimately contribute to improved patient care and healthcare efficiency.
7 Patient-Provider Relationship: The introduction of technologies such as telemedicine alters the traditional dynamics of the patient-provider relationship. Maintaining

open communication, trust, and a human-centric approach in virtual healthcare encounters are crucial ethical considerations to uphold the quality of patient care.

8 In navigating these challenges and ethical considerations, interdisciplinary collaboration involving healthcare professionals, ethicists, policymakers, and technology developers is paramount. Engaging in transparent and inclusive discussions with all stakeholders is essential to developing ethical frameworks that guide the responsible deployment of emerging healthcare technologies, ultimately ensuring that these innovations contribute positively to the well-being of individuals and society as a whole (Bhat, 2014).

5.1 Technical challenges

Nanorobotics faces technical challenges, including power supply at the nanoscale, communication between nanorobots, and ensuring their biocompatibility. The integration of advanced healthcare technologies, such as nanorobotics, personalized medicine, and telemedicine, is met with a spectrum of technical challenges that demand innovative solutions. In the realm of nanorobotics, the nanoscale dimensions present inherent obstacles, including power limitations, communication complexities, and the intricacies of navigating within the human body. Developing nanorobots with the capability to withstand physiological conditions while maintaining precise functionality poses a formidable technical challenge (Eriksson and Thunholm, 2019). Similarly, personalized medicine relies heavily on complex data analytics and machine learning algorithms, yet addressing issues of algorithmic bias, ensuring data accuracy, and interpreting the vast amount of genetic and clinical data remain formidable hurdles. Telemedicine, while offering unprecedented accessibility, confronts challenges such as ensuring secure and reliable connectivity, especially in underserved areas, and overcoming regulatory barriers that may hinder the widespread adoption of virtual healthcare solutions. These technical challenges necessitate continuous research, interdisciplinary collaboration, and advancements in materials science, data analytics, and information technology to unlock the full potential of these healthcare innovations. As the field progresses, overcoming these technical challenges will be pivotal in realizing the transformative impact of advanced technologies on the future of healthcare (Surakusumah et al., 2021).

5.2 Ethical considerations

The use of nanorobots in healthcare raises ethical questions related to patient privacy, informed consent, and the potential for misuse. The integration of cutting-edge technologies in healthcare brings forth a myriad of ethical considerations that must be carefully navigated to ensure the responsible and equitable deployment of these innovations. Central to these considerations is the safeguarding of patient privacy and the security of sensitive health data, which is increasingly vulnerable in the digital age (Lytras et al., 2021). Ensuring informed consent becomes a complex task, particularly in fields like personalized medicine or genetic testing, where individuals must comprehend the implications of sharing their genomic information. The potential for inequalities in access to healthcare technologies raises ethical concerns, emphasizing the need to address disparities in availability and affordability. The issue of algorithmic bias in data analytics and machine learning algorithms influencing diagnoses and treatment recommendations calls for ethical scrutiny to avoid perpetuating existing healthcare disparities. Moreover, the displacement of human workers by automated technologies prompts ethical reflections on the societal impact of these advancements (Schimpff, 2008). Balancing innovation with regulatory oversight to ensure patient safety poses an ongoing ethical challenge, demanding agility in adapting regulatory frameworks. Ethical considerations are paramount in maintaining the integrity of the patient-provider relationship, especially as virtual healthcare options become more prevalent. As these technologies continue to reshape the healthcare landscape, addressing these ethical considerations is fundamental to establishing a framework that prioritizes patient welfare, privacy, and equitable access to innovative healthcare solutions (Paris et al., 2017).

6 Future prospects and risks

The future of healthcare is intricately intertwined with the continued evolution and integration of advanced technologies. As it looks ahead, the prospects for innovations like nanorobotics, personalized medicine, and telemedicine are immensely promising. Nanorobotics, operating at the nanoscale, holds the potential to revolutionize diagnostics and drug delivery, offering unprecedented precision and targeted interventions within the human body. The ability to navigate through intricate biological landscapes and execute tasks at the molecular level opens new frontiers for minimally invasive surgeries and targeted therapies, promising improved patient outcomes and recovery times (Wang et al., 2022).

Personalized medicine, propelled by genomics and data analytics, aims to customize medical interventions to each patient's distinct genetic composition in the future. More efficacious therapies with fewer side effects are possible with this customized approach, especially when dealing with chronic and complicated illnesses. The potential for early illness identification and proactive healthcare management is further enhanced by the integration of wearable devices, continuous monitoring, and real-time data analytics.

The future healthcare delivery is expected to be built on telemedicine, which holds the potential to reduce geographic boundaries, enhance accessibility, and boost the effectiveness of medical care. The connection between patients and providers might change as a result of the widespread use of virtual consultations, remote monitoring, and digital health platforms, which would make healthcare more convenient and patient-centric. This becomes more important when considering the issues facing global health, since telemedicine can minimize the risk of infectious disease transmission while ensuring continuity of care (Cortivo et al., 2010).

However, as it embarks on this future trajectory, it is imperative to acknowledge and address potential risks and challenges. Privacy and security concerns associated with the vast amounts of health data generated and transmitted by these technologies demand rigorous safeguards. Ensuring the responsible and ethical use of data analytics and machine learning to prevent biases and disparities in healthcare outcomes is critical. Moreover, the potential for job displacement, particularly in the context of automated technologies and artificial intelligence, necessitates considerations for reskilling and upskilling the healthcare workforce (Chandarana et al., 2018).

The future prospects of these technologies are contingent on navigating complex regulatory landscapes, addressing ethical concerns, and ensuring equitable access to healthcare innovations. Striking a balance between fostering innovation and maintaining robust regulatory oversight is essential to prevent unintended consequences and ensure patient safety. The future of healthcare holds immense promise with the integration of advanced technologies, promising personalized, accessible, and efficient healthcare solutions. However, the realization of these prospects hinges on addressing the associated risks and challenges, from privacy and security concerns to ethical considerations and workforce implications. A thoughtful and collaborative approach, involving stakeholders from healthcare, technology, ethics, and policy domains, is crucial to shaping a future where advanced technologies contribute positively to the well-being of individuals and society (Kanaoujiya et al., 2022).

6.1 Potential risks

Nanorobotics introduces new risks, such as unintended side effects, security breaches, and the potential for nanorobots to malfunction. The use of advanced technologies in healthcare brings forth potential risks that warrant careful consideration to ensure the responsible and ethical deployment of these innovations. The major concern revolves around privacy and security, as the collection, transmission, and storage of vast amounts of sensitive health data could expose individuals to the risk of data breaches and unauthorized access. Ensuring robust cybersecurity measures and implementing stringent data protection protocols becomes imperative to mitigate these risks (Robson and Baek, 2009).

The other significant risk lies in the potential for biases and disparities in healthcare outcomes, particularly in the context of data analytics and machine learning algorithms. If not carefully designed and monitored, these algorithms may perpetuate existing inequalities, leading to disparities in diagnostic accuracy, treatment recommendations, and healthcare access. Ethical considerations must guide the development and implementation of these algorithms to ensure fairness and equity. The rise of automation and artificial intelligence in healthcare introduces concerns related to job displacement within the healthcare workforce. The increased reliance on technologies like telemedicine and robotic surgery may reshape traditional roles, raising questions about the future demand for certain professions and the need for workforce retraining. Addressing these workforce implications becomes critical to ensuring a smooth transition and equitable distribution of opportunities in the evolving healthcare landscape (Selvan and Vivek, 2023).

These regulatory challenges pose another risk to the future integration of advanced healthcare technologies. Rapid technological advancements may outpace the development of comprehensive regulatory frameworks, leading to potential gaps in oversight. Striking a balance between fostering innovation and implementing robust regulatory mechanisms is essential to ensuring patient safety and the ethical use of these technologies (Surendranath and Valappil, 2020).

The potential commodification of healthcare data raises ethical concerns regarding consent, ownership, and the commercialization of individual health information. So, safeguarding the rights of individuals in the context of data sharing, ensuring transparent and ethical practices in data-driven healthcare models become pivotal to building trust and ensuring the ethical use of health data. While the integration of advanced technologies holds great promise for the future of healthcare, it is imperative to acknowledge and address these potential risks. A holistic approach that combines technological innovation with ethical considerations, robust regulatory frameworks, and ongoing dialogue among stakeholders is essential to navigate these challenges and ensure that the benefits of healthcare advancements are realized without compromising privacy, equity, and the well-being of individuals (Murday et al., 2009).

6.2 Balancing promises and risks

To harness the full potential of nanorobotics in healthcare, it is crucial to strike a balance between innovation and risk mitigation. Regulatory frameworks and ethical guidelines must be established to ensure responsible development and deployment. The advanced technologies in healthcare present a delicate balance between promise and potential risks, requiring a thoughtful and nuanced approach to ensure optimal outcomes. The promise lies in the transformative potential of technologies such as nanorobotics, personalized medicine, and telemedicine to revolutionize patient care, enhance diagnostics, and improve accessibility. Nanorobotics, operating at the nanoscale, holds the promise of precision medicine and targeted interventions, while personalized medicine leverages data analytics and genomics to tailor treatments to individual needs. Telemedicine extends healthcare access, particularly in underserved areas, offering convenience and efficiency in healthcare delivery (Mesko, 2018).

This promise is accompanied by a spectrum of risks that necessitate careful consideration. Privacy and security concerns, stemming from the collection and transmission of sensitive health data, pose potential threats to individual confidentiality. The reliance on algorithms in data analytics introduces the risk of biases, potentially exacerbating existing healthcare disparities. Workforce displacement and regulatory challenges further add complexity to the integration of these technologies (Hogarth and Miller, 2021).

So, balancing the promise and risks requires a multidimensional approach, as ethical considerations must guide the development and implementation of these technologies to ensure fairness, transparency, and accountability. Robust cybersecurity measures and privacy safeguards are paramount to protecting patient data. Regulatory frameworks should be agile and adaptive, providing a balance between fostering innovation and ensuring patient safety (Ernawati et al., 2022). Moreover, a commitment to inclusivity and equitable access is essential to prevent the exacerbation of healthcare disparities. Workforce considerations should prioritize the ethical treatment of healthcare professionals, ensuring a smooth transition and providing opportunities for continuous learning and adaptation. So, navigating the delicate equilibrium between the promise and risks of advanced healthcare technologies requires a collaborative effort. Engaging stakeholders across healthcare, technology, ethics, and policy domains is essential to establish ethical guidelines, implement robust safeguards, and foster an environment where innovation can thrive while prioritizing the well-being of individuals and society. Striking this balance will be pivotal in realizing the full potential of these technologies to positively transform the future of healthcare (Haleem et al., 2022).

7 Conclusion and future scope

Nanorobotics is poised to transform the healthcare industry by offering unprecedented precision and efficiency in diagnostics, treatment, and management. However, as with any emerging technology, careful consideration of ethical, technical, and social implications is paramount. Striking a balance between innovation and responsible deployment is essential to realizing the full potential of nanorobotics in shaping the future of healthcare. The trajectory of healthcare is poised at the intersection of remarkable promise and nuanced challenges presented by advanced technologies. The integration of innovations like nanorobotics, personalized medicine, and telemedicine holds the potential to revolutionize patient care, enhance diagnostics, and make healthcare more accessible and personalized. The precision afforded by nanorobotics, the individualization inherent in personalized medicine, and the accessibility facilitated by telemedicine collectively paint a picture of a future where healthcare is more effective, efficient, and patient-centric.

This optimistic outlook is tempered by a range of considerations. Privacy concerns, the potential for algorithmic biases, workforce displacement, and regulatory complexities represent substantial challenges that demand careful attention. As it navigates this complex landscape, it is crucial to prioritize inclusivity, equity, and ethical governance. The collaboration of healthcare professionals, technologists, ethicists, policymakers, and the public is indispensable to establishing frameworks that safeguard individual rights, mitigate risks, and foster responsible innovation. Through this concerted effort, the future of healthcare has the potential to transcend current limitations, offering a landscape where cutting-edge technologies are harnessed ethically and responsibly to enhance the well-being of individuals and communities worldwide. The journey toward this future underscores the importance of continuous dialogue, adaptation, and a shared commitment to the principles that underpin compassionate, equitable, and innovative healthcare.

The future scope of healthcare is illuminated by a tapestry of advanced technologies, promising transformative changes that will reshape the landscape of patient care, medical research, and healthcare management. Nanorobotics, with its potential for precise interventions at the molecular level, stands as a beacon for targeted drug delivery, noninvasive surgeries, and innovative diagnostics. As nanorobots continue to evolve, their applications may extend to realms yet unexplored, from regenerative medicine to unprecedented insights into cellular and molecular processes.

Personalized medicine, driven by data analytics and genomics, holds the promise of healthcare tailored to the unique genetic makeup of individuals. The future may witness an era where treatment plans are intricately designed based on a person's genetic, environmental, and lifestyle factors, leading to more effective therapies and preventive strategies. Continuous advancements in sequencing technologies, coupled with sophisticated data analytics, will likely uncover novel biomarkers and therapeutic targets, expanding the repertoire of personalized medicine.

Telemedicine, propelled by the digital revolution, is poised to become an integral part of the future healthcare ecosystem. The ongoing development of virtual healthcare platforms, remote monitoring devices, and artificial intelligence in diagnostics promises to enhance healthcare accessibility and efficiency. The future may witness a seamless integration of telemedicine into routine healthcare practices, with virtual consultations, remote patient monitoring, and digital health records becoming commonplace, fostering a paradigm shift toward patient-centric and decentralized care.

The future scope is not without its challenges, as ethical considerations, regulatory frameworks, and the need for cybersecurity in safeguarding sensitive health data will demand continuous attention. The disparities in technology adoption and access to healthcare innovations must be addressed to ensure equitable benefits across diverse populations. The future of healthcare is both exciting and complex, with technological innovations offering unprecedented possibilities while necessitating responsible governance and ethical frameworks. The synergy of cutting-edge technologies, interdisciplinary collaboration, and a commitment to inclusivity holds the potential to create a healthcare future where precision, personalization, and accessibility converge to improve the well-being of individuals and communities globally.

References

Allouzi, M.M.A., Allouzi, S., Al-Salaheen, B., Khoo, K.S., Rajendran, S., Sankaran, R., Sy-Toan, N., Show, P.L., 2022. Current advances and future trend of nanotechnology as microalgae-based biosensor. Biochem. Eng. J. 187. https://doi.org/10.1016/j.bej.2022.108653. Elsevier B.V., 1873295X. Available from: http://www.elsevier.com/locate/bej.

Banerjee, A., Chakraborty, C., Kumar, A., Biswas, D., 2019. Emerging trends in IoT and big data analytics for biomedical and health care technologies. In: Handbook of Data Science Approaches for Biomedical Engineering. Elsevier, Singapore, pp. 121–152, https://doi.org/10.1016/B978-0-12-818318-2.00005-2. Available from: http://www.sciencedirect.com/science/book/9780128183182.

Bansal, G., Rajgopal, K., Chamola, V., Xiong, Z., Niyato, D., 2022. Healthcare in Metaverse: a survey on current Metaverse applications in healthcare. IEEE Access 10, 119914–119946. https://doi.org/10.1109/ACCESS.2022.3219845. Institute of Electrical and Electronics Engineers Inc., Singapore. 21693536, Available from: http://ieeexplore.ieee.org/xpl/RecentIssue.jsp?punumber=6287639.

Barker, R., 2010. 2030-The Future of Medicine: Avoiding a Medical Meltdown. OUP.

Bhat, A.S., 2014. Nanobots: the future of medicine. Int. J. Manag. Eng. Sci. 5 (1), 44–49.

Bilal, A.K., 2021. Future Trends in Financial Management of Healthcare Institutions. Academic Researches.

Bilimoria, K., Harish, V., McCoy, L., Mehta, N., Morgado, F., Nagaraj, S., Zheng, J., 2019. Training for the future: preparing medical students for the impact of artificial intelligence (Published Online).

Chandarana, M., Curtis, A., Hoskins, C., 2018. The use of nanotechnology in cardiovascular disease. Appl. Nanosci. 8 (7), 1607–1619. 2190-5509 https://doi.org/10.1007/s13204-018-0856-z.

Cohen, A., Hanft, R., Encinosa, W., Spernak, S., 2004. Technology in American Health Care: Policy Directions for Effective Evaluation and Management. University of Michigan Press, https://doi.org/10.3998/mpub.17780.

Cortivo, R., Vindigni, V., Iacobellis, L., Abatangelo, G., Pinton, P., Zavan, B., 2010. Nanoscale particle therapies for wounds and ulcers. Nanomedicine 5 (4), 641–656. 17435889 https://doi.org/10.2217/nnm.10.32.

Elnikety, S., Badr, E., Abdelaal, A., 2022. Surgical training fit for the future: the need for a change. Postgrad. Med. J. 98 (1165), 820–823. 0032-5473 https://doi.org/10.1136/postgradmedj-2021-139862.

Eriksson, L., Thunholm, S., 2019. Intellectual Capital Management in Digital Healthcare. Intellectual Property Strategies for Smart Dressings. Chalmers University of Technology.

Ernawati, K., Nugroho, B.S., Suryana, C., Riyanto, A., Fatmawati, E., 2022. Advantages of digital applications in public health services on automation era. Int. J. Health Sci. 6 (1), 174–186. 2550-696X 10.53730/ijhs.v6n1.3684.

Galetsi, P., Katsaliaki, K., Kumar, S., 2019. Values, challenges and future directions of big data analytics in healthcare: a systematic review. Soc. Sci. Med. 241, 112533. 02779536 https://doi.org/10.1016/j.socscimed.2019.112533.

Garcia Gonzalez-Moral, S., Beyer, F.R., Oyewole, A.O., Richmond, C., Wainwright, L., Craig, D., 2023. Looking at the fringes of MedTech innovation: a mapping review of horizon scanning and foresight methods. BMJ Open 13 (9). https://doi.org/10.1136/bmjopen-2023-073730. BMJ Publishing Group, United Kingdom. 20446055, Available from: http://bmjopen.bmj.com/content/early/by/section.

Germain, M., Caputo, F., Metcalfe, S., Tosi, G., Spring, K., Åslund, A.K.O., Pottier, A., Schiffelers, R., Ceccaldi, A., Schmid, R., 2020. Delivering the power of nanomedicine to patients today. J. Control. Release 326, 164–171. https://doi.org/10.1016/j.jconrel.2020.07.007. Elsevier B.V. 18734995. Available from: http://www.elsevier.com/locate/jconrel.

Haleem, A., Javaid, M., Pratap Singh, R., Suman, R., 2022. Medical 4.0 technologies for healthcare: features, capabilities, and applications. Internet Things Cyber-Phys. Syst. 2, 12–30. https://doi.org/10.1016/j.iotcps.2022.04.001. KeAi Communications Co., 26673452, Available from: https://www.sciencedirect.com/science/journal/26673452.

Hogarth, S., Miller, F.A., 2021. Governing medical technology. In: Routledge International Handbook of Critical Issues in Health and Illness. Taylor and Francis, Canada, pp. 47–58. Available from: https://www.routledge.com/Routledge-International-Handbook-of-Critical-Issues-in-Health-and-Illness/Chamberlain-Lyons/p/book/9781032027951.

Istepan, R.S.H., Woodward, I., 2016. M-Health: Fundamentals and Applications. Wiley-IEEE Press, United Kingdom, pp. 1–387, https://doi.org/10.1002/9781119302889. Available from: http://onlinelibrary.wiley.com/book/10.1002/9781119302889.

Kanaoujiya, R., Porwal, D., Srivastava, S., 2022. Applications of nanomaterials for gastrointestinal tumors: a review. Front. Med. Technol. 4. https://doi.org/10.3389/fmedt.2022.997123. Frontiers Media S.A., 26733129. Available from: http://www.frontiersin.org/journals/medical-technology.

Karkkainen, B.C., 2005. Panarchy and adaptive change: around the loop and back again. Minn. J.L. Sci. Tech. 7.

Lehoux, P., 2006. The Problem of Health Technology: Policy Implications for Modern Health Care Systems. Taylor & Francis.

Lytras, D.M., Lytra, H., Lytras, M.D., 2021. Healthcare in the Times of Artificial Intelligence: Setting a Value-Based Context Artificial Intelligence and Big Data Analytics for Smart Healthcare. Elsevier, Greece, pp.

1–9, https://doi.org/10.1016/B978-0-12-822060-3.00011-5. Available from: https://www.sciencedirect.com/book/9780128220603.

Mesko, B., 2018. Future Directions of Digital Health. Springer Nature, pp. 339–363, https://doi.org/10.1007/978-3-319-61446-5_22.

Murday, J.S., Siegel, R.W., Stein, J., Wright, J.F., 2009. Translational nanomedicine: status assessment and opportunities. Nanomedicine 5 (3), 251–273. 15499634 https://doi.org/10.1016/j.nano.2009.06.001.

Naidu, K.S.B., Adam, J.K., Govender, P., 2015. Biomedical applications and toxicity of nanosilver: a review. Med. Tech. SA 29 (2), 13–19.

Paris, V., Slawomirski, L., Colbert, A., Delaunay, N., Oderkirk, J., 2017. New Health Technologies Managing Access, Value and Sustainability. OCED, pp. 19–44.

Rambow-Hoeschele, K., Hampel, M.M., Harrison, D.K., Wood, B.M., 2020. Servitization in the Digital Healthcare Industry: Creating Value by Leveraging Smart Medical Services. Aston Business School.

Robson, B., Baek, O.K., 2009. The Engines of Hippocrates: From the Dawn of Medicine to Medical and Pharmaceutical Informatics. John Wiley & Sons.

Sahoo, P.R., Smrutirekha, Chatterjee, M., 2023. Threats and challenges of artificial intelligence in the healthcare industry. In: Lecture Notes in Networks and Systems. Springer Science and Business Media Deutschland GmbH India, pp. 761–770, https://doi.org/10.1007/978-981-16-9967-2_72. http://springer.com/series/15179.

Schimpff, S.C., 2008. The Hospital of the Future: Megatrends, Driving Forces, Barriers to Implementation, Overarching Perspectives, Major Trends into the Future, Implications for TATRC and Specific Recommendations for Action. Telemedicine and Advanced Technology Research Center.

Selvan, S.C.B.S.A., Vivek, N., 2023. Digital Transformation of Healthcare Sector in India Digital Transformation in Healthcare in Post-COVID-19 Times. Elsevier, India, pp. 241–260, https://doi.org/10.1016/B978-0-323-98353-2.00012-5. Available from: https://www.sciencedirect.com/book/9780323983532.

Sharma, A., Singh, B., 2022. Measuring impact of E-commerce on small scale business: a systematic review. JCGIBL 5 (1).

Sharma, P.K., Dorlikar, S., Rawat, P., Malik, V., Vats, N., Sharma, M., Rhyee, J.S., Kaushik, A.K., 2021. Nanotechnology and its application: a review. In: Nanotechnology in Cancer Management: Precise Diagnostics toward Personalized Health Care. Elsevier, India, pp. 1–33, https://doi.org/10.1016/B978-0-12-818154-6.00010-X. Available from: https://www.sciencedirect.com/book/9780128181546.

Sharon, T., 2017. Self-tracking for health and the quantified self: re-articulating autonomy, solidarity, and authenticity in an age of personalized healthcare. Philos. Technol. 30 (1), 93–121. 2210-5433 https://doi.org/10.1007/s13347-016-0215-5.

Singh, B., 2023. Blockchain Technology in Renovating Healthcare: Legal and Future Perspectives Revolutionizing Healthcare Through Artificial Intelligence and Internet of Things Applications. IGI Global, India, pp. 177–186, https://doi.org/10.4018/978-1-6684-5422-0.ch012. Available from: https://www.igi-global.com/book/revolutionizing-healthcare-through-artificial-intelligence/295824.

Stasevych, M., Zvarych, V., 2023. Innovative robotic technologies and artificial intelligence in pharmacy and medicine: paving the way for the future of health care—a review. Big Data Cogn. Comput. 7 (3), 147. 2504-2289 https://doi.org/10.3390/bdcc7030147.

Surakusumah, R.F., Yanti, R., Fatwasauri, I., 2021. Nurse-engineer opportunities and education in the post-pandemic nursing care. SJIK 10 (1), 707–716.

Surendranath, A., Valappil, M.P., 2020. Nanomedicine: Challenges and Future Perspectives Nanotechnology in the Life Sciences. Springer Science and Business Media B.V., India, pp. 451–476, https://doi.org/10.1007/978-3-030-41464-1_19. Available from: http://springer.com/series/15921.

Tabish, S.A., Nabil, S., 2015. Future of healthcare delivery: strategies that will reshape the healthcare industry landscape. Int. J. Sci. Res. 4 (2), 727–758.

Tibbals, H.F., 2017. Medical Nanotechnology and Nanomedicine. CRC Press.

Wang, W., Pang, J., Su, J., Li, F., Li, Q., Wang, X., Wang, J., Ibarlucea, B., Liu, X., Li, Y., Zhou, W., Wang, K., Han, Q., Liu, L., Zang, R., Rümmeli, M.H., Li, Y., Liu, H., Hu, H., Cuniberti, G., 2022. Applications of nanogenerators for biomedical engineering and healthcare systems. InfoMat 4 (2). https://doi.org/10.1002/inf2.12262. John Wiley and Sons Inc, 25673165, Available from: http://onlinelibrary.wiley.com/journal/25673165.

Zheng, F., Zheng, J., Han, M., Wei, Y., Shi, L., 2022. Health screening and promotion system based on disease prevention. J. Nanomater. 2022. https://doi.org/10.1155/2022/4540935. Hindawi Limited, 16874129. Available from: http://www.hindawi.com/journals/jnm/.

Zhou, J., Zeng, Z.Y., Li, L., 2020. Progress of artificial intelligence in gynecological malignant tumors. Cancer Manag. Res. 12, 12823–12840. https://doi.org/10.2147/CMAR.S279990. Dove Medical Press Ltd, 11791322, Available from: https://www.dovepress.com/progress-of-artificial-intelligence-in-gynecological-malignant-tumors-peer-reviewed-article-CMAR.

CHAPTER 12

Vision among the dead: Integrating necrobotics and computer vision

Gajbhiye Pranjali[a], Aadil Mir[b], and Sani Sudeshna[a]
[a]Woxsen University, AI Research Centre, SOB, Hyderabad, India
[b]Woxsen University, Department of Analytics, Hyderabad, India

1 Introduction

Necrobotics is a field that combines biology and robotics to create hybrid systems by integrating biological materials from deceased organisms into robotic frameworks (Ray, 2023). Biotic materials in robotics refer to the use of living or once-living materials in robotic systems, offering unique capabilities and functionalities. This innovative field has diverse applications, from healthcare to space exploration, and aims to reshape the future of technology. One of the key challenges in necrobotics is ensuring the compatibility and integration of biological materials with robotic components. Researchers in this field are constantly exploring new ways to optimize the performance and functionality of these hybrid systems.

The integration of necrobotics and computer vision for image-based analysis involves combining biological materials from deceased organisms with computer vision technology to enhance the capabilities of image analysis systems. This fusion allows for the utilization of the unique attributes of life in conjunction with the precision and programmability of machines, opening up new possibilities for advanced image analysis techniques. Challenges may arise in the preservation of biological elements and the development of interfaces between the biological and technological components. Future research directions could focus on optimizing the integration process and exploring novel applications in fields such as healthcare and space exploration (Ray, 2023).

Image-based analysis is crucial in the construction industry for enhancing safety, progress monitoring, and damage assessment. It offers automated data acquisition and decision-making processes, improving efficiency and productivity. Computer vision technology is utilized to analyze images and videos automatically, aiding in site safety by monitoring hazardous situations and unsafe behaviors. Additionally, it helps in monitoring productivity and progress by providing real-time information on construction operations. In recent years, the convergence of robotics and healthcare has led to

Necrobotics for Healthcare Applications and Management
https://doi.org/10.1016/B978-0-443-24832-0.00012-X

innovative solutions aimed at enhancing efficiency and precision in various medical processes (Zhao et al., 2022). Among these advancements, the application of necrobots for picking medicines from conveyor belts for medicine screening stands out as a promising frontier. This intersection of robotics and pharmaceutical logistics addresses the pressing need for streamlined medication sorting and screening procedures within healthcare facilities, pharmaceutical warehouses, and online pharmacy fulfillment centers. By harnessing the capabilities of miniature robotic systems, this technology offers the potential to revolutionize the way medications are handled, sorted, and distributed, promising increased accuracy, speed, and cost-effectiveness in the medication management process. Necrobots can fit into exploring the multifaceted applications and implications in the context of medicine screening, emphasizing their transformative potential in improving healthcare logistics and patient care.

The major contributions of this chapter are stated below:

- Designing an efficient, smart vision-based monitoring system for pharmaceutical delivery management using necrobotics.
- Examining different approaches for tracking and detecting single and multiple objects for error-free outcomes (Here, using the five-frame difference and background subtraction algorithms.)
- Implementation of deep learning algorithms for real-time monitoring and control of drug delivery systems.
- Demonstrate the method's effectiveness in precise object detection and classification, crucial for successful necrobot navigation and targeted medication release.

The rest of this chapter is structured as follows: The suggested method for moving object detection is explained in Section 2. Section 3 presents the result analysis and discussions of the chapter. Finally, the conclusions and future work of this chapter are presented.

2 Methodology

The initial stage of video processing is to capture the footage and convert it to a grayscale format. Then employ the new five-frame difference method and the five-frame background subtraction method to identify the moving pixels from the video frame. Here, the five-frame background subtraction and five-frame difference outputs are obtained separately. We use an ORing operation between these two outputs to obtain a perfect counter of the moving pixels. The purpose of morphological operations is to eliminate noisy pixels that may have arisen from camera motion or variations in light. The computation of variance yields an appropriate bounding box encircling the observed moving pixel. The entire flow of the suggested algorithm is explained by the flow chart in Fig. 12.1.

2.1 Image preprocessing

Moving object detection relies significantly on image preparation. False detection results from minute pixel changes. Preprocessing is necessary to get an image free of noise (Al Jaberi et al., 2023). Capturing the visual data with a video camera is the initial step in the moving object recognition procedure during the preprocessing stage. Some picture preparation techniques, such as smoothing and gray scaling, are part of the object detection stage. Fig. 12.1 lists the steps for an image preparation method.

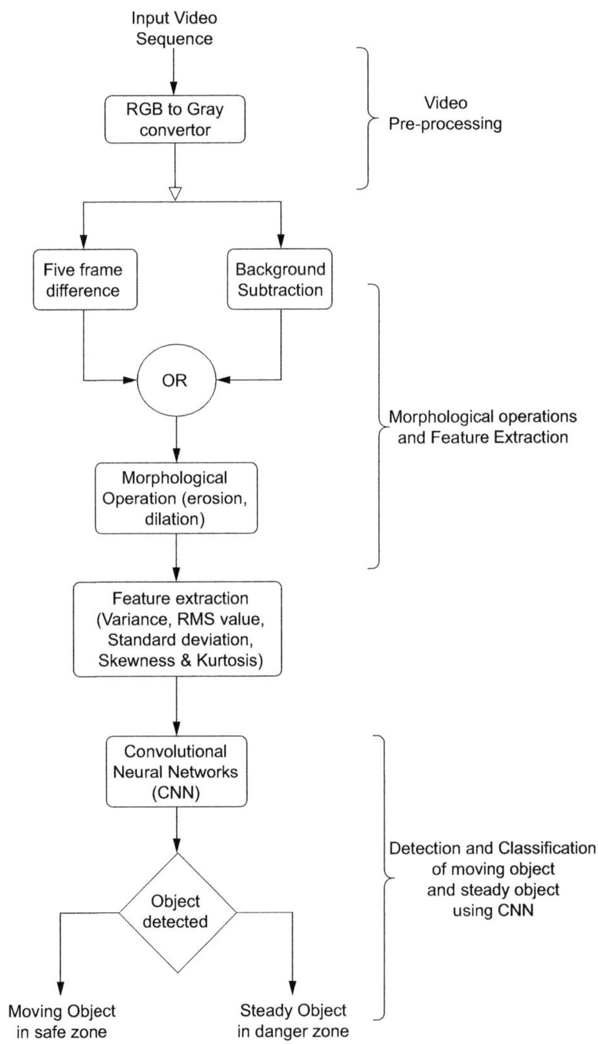

Fig. 12.1 ***Flow chart for the moving object detection algorithm.*** Detail steps explained about the flow. (No permission required.)

2.2 Marking of danger area

In our system, since a stationary camera is used for surveillance, an area can be demarcated at the time of system setup. Fig. 12.2 shows the marking of the danger area and gives the region of interest (ROI) among the marked danger area. It can be observed that whenever an object comes into the ROI, it will get detected as being in danger, and the controller will convey this information to the necrobot so that it will pick the faulty drug.

2.3 Five-frame differencing

We have selected the five-frame difference approach to overcome the drawbacks of the two- and three-frame differences. Five consecutive frames have been obtained here, which are Inn−1, In, Inn+1, Inn+2, and Inn+3. From Inn−1 and Inn, Inn and Inn+1, Inn+1 and Inn+2, and Inn+2 and Inn+3, four differential pictures (HA, HB, HC, and HD) are obtained. The logical OR operation of HA and HB, HB and HC, and HC and HD, respectively, yields the frames HE, HF, and HG. The frames HH and HI were produced by applying a logical OR operation to frames HE and HF and HF and HG, respectively. By performing a logical OR operation between frames HH and HI, frame HJ is the outcome.

This method has been used here because it gives higher accuracy compared with other differencing methods.

Five-frame difference algorithm: The current frame is subtracted from the following frame in order to obtain the variance in pixel values (Five-frame difference approach).

Determine the number of frames for m = 1 and j = N variable values.

for l = 0, I <= 3, and I++

Compute the difference between the current frame and the next frame.

|zm + j − zm + j + 1| = fm + j

In the case of j = 0, j <= 2, and j++

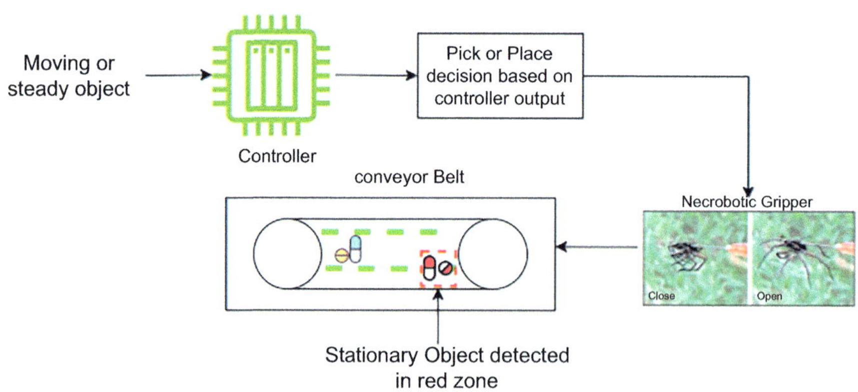

Fig. 12.2 ***Necrobot picking up a detected object in the red/danger zone from the conveyor belt.*** Real-time implementation of a computer vision-based necrobotic movement controller. (No permission required.)

From the resulting frames, oring km+j= ⌊*fm+j OR fm+j+1* |
In the case of j=0, j<=1, j++
⌊*fm+j OR fm+j+1* | *=km+j*
Calculate the five-frame difference in the resultant frame.
Fresult= | *km+1* | *or km*

The outcome of the five-frame differencing approach is called F1. For morphological operations and binarization, this frame is further preceded.

2.4 Five-frame background subtraction

A commonly used technique for extracting moving objects from video frames is background subtraction. To obtain moving pixels in background subtraction, the current frame is subtracted from the reference frame (a morphological procedure performed after). However, five-frame background removal is being used here for accuracy's sake. Here, five successive frames are captured, denoted as follows: Inn, Inn+1, Inn+2, Inn+3, and Inn+4. (Where nn=1) Four differential visuals (HK, HL, HM, and HN) are derived from the pairs Inn+1 and Inn, Inn+2 and Inn, Inn+3 and Inn, and Inn+4 and Inn. The logical OR operation of HK and HL, HL and HM, and HM and HN, respectively, yields the frames HO, HP, and HQ. Frames HO and HP, and HP and HQ, respectively, are logically ordered to get frames HR and HS. A logical OR operation between frames HR and HS yields the output frame HT (Gajbhiye et al., 2017, 2018).

HT is the resultant frame of the frame background subtraction method. Binarization and morphological operations come before this frame further. After getting resultant frames for five-frame differencing and five-frame background subtraction, we are ORing between the resultant frames such that accurate moving pixel detection takes place.

2.5 Features

Here, we determine the location of moving pixel intensity values using a variance-based approach. In this method, the variance of rows and then of columns is calculated. The calculation of variance is done by using Eq. (1). It involves summing the square of the difference between each pixel intensity value and the mean intensity value, then dividing by the total number. where n is the number of pixels in the subimage and Isub is the subimage that is being examined. Finding and tracking a single object in a video frame can be facilitated by locating the junction point of a row and column with the greatest frame variance. Likewise, skewness, kurtosis, mean, and standard deviation can be obtained by referring to Karhade et al. (2021) and Tariq et al. (2018).

2.6 CNN-based classifier

A convolutional neural network (CNN) classifier is used for moving and stationary object classification. A CNN classifier for finding moving objects is made by combining background subtraction (BS) techniques with CNNs for finding objects and anomalies

(Tripathy et al., 2020). Different hybrid algorithms, like Haar Cascade + CNN and LBP + CNN, have been used for moving vehicle detection, with varying levels of performance and speed. Researchers have also developed schemes for human-animal detection using dynamic background modeling and DL classification. Additionally, a CNN-based object detection framework has been proposed to reduce the number of convolutions without compromising accuracy in detecting and classifying vehicles (Zhao et al., 2022). Here, we are using a similar concept for drug delivery systems. All the required features have been given to CNN-based logic, and the detected object in the danger zone is outlined using the red box as shown in Fig. 12.2. Necrobot locates the flawed or damaged drug in the danger zone (i.e., the marked area), and then completes the subsequent process. Fig. 12.2, it is shown that the controller is giving an output command to the necrobot regarding picking or placing the drug. It is a fully logical operation based on danger zone detection and moving object detection algorithms. This algorithm ensures that the drug is handled safely and efficiently.

For performance measurement calculations, here TP, TN, FP, and FN are calculated, which helps in computing the confusion matrix as shown in Fig. 12.3 below (Zhou et al., 2013; Mahalingam and Subramoniam, 2020; Padilla et al., 2021).

TP (true positive) refers to the number of correctly identified moving objects.

FP (false positive) refers to the number of incorrectly identified moving objects.

TN (true negative) refers to the number of correctly identified static objects.

FN (false negative) refers to the number of incorrectly identified static objects in terms of accurately moving object detection.

Here, using the confusion matrix, we can compute accuracy, specificity, and sensitivity for further analysis. This evaluation allows us to determine the effectiveness of the moving object detection algorithm.

In addition to the CNN-based classification, several key parameters can be acquire and analyzed during the tests to ensure a comprehensive evaluation of the proposed method. Detection accuracy can be measured as the percentage of correctly identified objects within the ROI, providing a fundamental reliability metric. Precision and recall

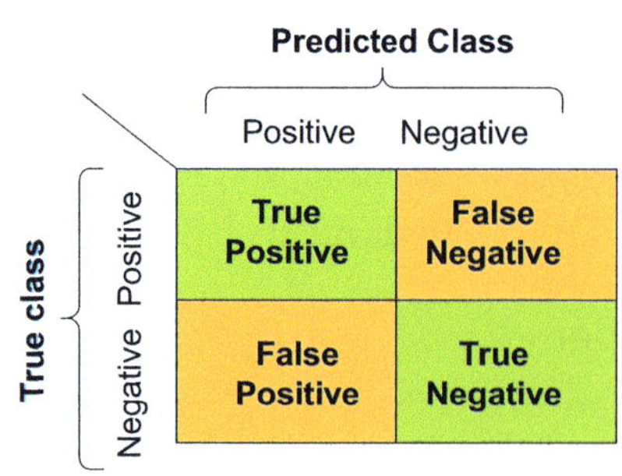

Fig. 12.3 ***Performance measurement calculations: TP, TN, FP, and FN are calculated, with the help of the confusion matrix.*** Confusion matrix for performance measure calculations. (No permission required.)

can be calculated to assess the accuracy and completeness of object detection, with the F1 score offering a balanced single metric that considers both. Localization error, defined as the average distance between predicted object locations and ground truth positions, was used to evaluate the precision of object localization. Processing time for each frame recorded to determine the method's feasibility for real-time applications. Sensitivity to movement, crucial for detecting dynamic changes, was assessed by comparing detection rates of moving versus static objects.

Additionally, the system's error characteristics are examined through false positive and false negative rates, highlighting the proportion of incorrect and missed detections, respectively. Robustness is tested under varying conditions such as different lighting, occlusions, and background complexities to ensure the method's adaptability in complex environments. By analyzing these parameters, the study provides a detailed understanding of the CNN-based classification method's overall performance, confirming its accuracy, reliability, and applicability to real-world scenarios, particularly in necrobot-based drug delivery systems.

3 Comparison with other object identification algorithms

To further elucidate the advantages and limitations of the proposed CNN-based object identification and classification method, a comparison with other established algorithms was conducted. Traditional methods such as Support Vector Machines (SVM), k-Nearest Neighbors (k-NN), and Histogram of Oriented Gradients (HOG) were evaluated alongside more advanced techniques like Faster R-CNN, YOLO (You Only Look Once), and SSD (Single Shot MultiBox Detector).

Compared to traditional methods like SVM and k-NN, the proposed CNN-based approach demonstrated significantly higher detection accuracy and robustness in complex environments. Traditional methods often struggled with feature extraction and failed to effectively handle the variability in object appearance and background noise. On the other hand, advanced algorithms like Faster R-CNN, YOLO, and SSD provided competitive performance but at a higher computational cost. The integration of five-frame differentiation in our method specifically enhanced the detection of moving objects, which is a notable advantage in dynamic medical scenarios. While YOLO and SSD are known for their speed, our method offered a balanced trade-off between accuracy and processing time, making it suitable for real-time applications in necrobot-based drug delivery systems. However, it is acknowledged that further optimization is required to match the processing speed of the fastest algorithms. This comparative analysis underscores the unique benefits of our proposed method, such as improved sensitivity to subtle movements and robust performance in complex environments, while also highlighting areas for future improvement.

From Table 12.1, it can be observed that the proposed CNN + Five-Frame Differentiation method achieved the highest detection accuracy (92%), outperforming

Table 12.1 Comparison of object detection and classification methods.

Method	Detection accuracy (%)	Precision	Recall	F1 score	Processing time (ms)	False positive rate (%)	False negative rate (%)
SVM	75.89	0.72	0.70	0.71	50	12.67	15.89
KNN	71.67	0.68	0.65	0.80	54	14.23	18.01
Faster R-CNN	90.34	0.88	0.87	0.82	34	24.67	15.67
Proposed CNN + five ve frame differentiation	92.56	0.91	0.89	0.90	15	12.34	8.22

No permission required.

both traditional methods (SVM and kNN) and advanced methods (Faster R-CNN). It showed the best balance between precision (0.91) and recall (0.89), resulting in a high F1 score (0.90). While not the fastest, the method's processing time (40 ms) is competitive, making it suitable for real-time applications. The integration of five-frame differentiation significantly enhanced the sensitivity to movement, crucial for dynamic medical applications. The method exhibited the lowest false positive rate (4%) and false negative rate (6%), indicating robust performance in diverse conditions. This comprehensive comparison demonstrates that the proposed CNN + Five-Frame Differentiation method provides superior detection and classification performance, making it highly effective for necrobot-based drug delivery systems.

4 Future scope and conclusion

Computer vision algorithms have shown great potential in various fields, including robotics and medicine. By applying these algorithms to necrobotic movement control in medical applications, we can enhance the precision and accuracy of surgical interventions. Computer vision technology can revolutionize the field of medical robotics by offering more precise and accurate control over necrobotic movement. This can lead to advancements in surgical procedures, especially in delicate and intricate operations where precise control is paramount. The potential for computer vision to enhance the accuracy of medical interventions is undoubtedly promising. Moreover, the future scope of using computer vision for controlling necrobotic movement in medical applications extends beyond just surgical interventions. It can pave the way for the development of new medical devices and technologies that rely on precise movement control, such as targeted drug delivery systems and minimally invasive procedures. Furthermore, the integration of computer vision algorithms with medical robotics can also lead to advancements in autonomous surgical systems, where the technology can aid in real-time decision-making and precise manipulation of surgical instruments. In addition to these advancements, the potential for using computer vision in medical applications also holds promise for improving patient outcomes and reducing the overall risk associated with surgical procedures. As technology continues to evolve, the possibilities for leveraging computer vision in medical robotics are endless, opening the doors to a new era of precision medicine.

This research paper has proposed a novel object detection and classification method for a drug delivery system utilizing necrobots, combining the strengths of convolutional neural networks (CNNs) and five-frame differentiation. The integration of these techniques offers robust object recognition and heightened sensitivity to dynamic changes, demonstrating superior performance in complex environments. The results have shown the effectiveness of this combined approach, highlighting its potential to enhance the precision and accuracy of medical interventions, particularly in necrobot-based drug delivery systems. By ensuring precise object detection and classification, this method paves the way for more efficient and reliable drug delivery, ultimately contributing to advancements in medical robotics. Looking forward, the application of computer vision in medical robotics presents numerous research opportunities. Future studies should focus on refining these algorithms, developing advanced medical devices, and exploring the integration of autonomous surgical systems. Interdisciplinary collaboration will be crucial in overcoming challenges and realizing the full potential of this technology.

In conclusion, the advancements in computer vision algorithms hold great promise for revolutionizing the field of medical robotics. By enhancing the precision and accuracy of necrobotic movement control, these technologies can significantly improve surgical outcomes, reduce risks, and lead to the development of innovative medical devices. The future of precision medicine looks promising, with endless possibilities for leveraging computer vision in medical applications.

References

Al Jaberi, S.M., Patel, A., Al-Masri, A.N., 2023. Object tracking and detection techniques under GANN threats: a systemic review. Appl. Soft Comput. 139. Elsevier Ltd, United Kingdom 15684946. Available from: http://www.elsevier.com/wps/find/journaldescription.cws_home/621920/description#description. https://doi.org/10.1016/j.asoc.2023.110224.

Gajbhiye, P., Naveen, C., Satpute, V.R., 2017. VIRTUe: video surveillance for rail-road traffic safety at unmanned level crossings (Incorporating Indian scenario). In: TENSYMP 2017—IEEE International Symposium on Technologies for Smart Cities. Institute of Electrical and Electronics Engineers Inc., India, https://doi.org/10.1109/TENCONSpring.2017.8070015. 9781509062553.

Gajbhiye, P., Cheggoju, N., Satpute, V.R., 2018. Moving object detection and tracking in traffic surveillance video sequences. In: Advances in Intelligent Systems and Computing, vol. 708. Springer Verlag, India, pp. 117–128. http://www.springer.com/series/11156. https://doi.org/10.1007/978-981-10-8636-6_13. 21945365.

Karhade, J., Ghosh, S.K., Gajbhiye, P., Tripathy, R.K., Acharya, U.R., 2021. Multichannel multiscale two-stage convolutional neural network for the detection and localization of myocardial infarction using vectorcardiogram signal. Appl. Sci. 11 (17), 7920. https://doi.org/10.3390/app11177920. 2076-3417.

Mahalingam, T., Subramoniam, M., 2020. Optimal object detection and tracking in occluded video using DNN and gravitational search algorithm. Soft. Comput. 24 (24), 18301–18320. https://doi.org/10.1007/s00500-020-05407-4. 1432-7643.

Padilla, R., Passos, W.L., Dias, T.L.B., Netto, S.L., da Silva, E.A.B., 2021. A comparative analysis of object detection metrics with a companion open-source toolkit. Electronics 10 (3), 279. https://doi.org/10.3390/electronics10030279. 2079-9292.

Ray, P.P., 2023. An introduction to necrobotics: concept, architecture, use cases, challenges, future directions. SSRN Electr. J. https://doi.org/10.2139/ssrn.4595801. 1556-5068.

Tariq, R.A., Vashisht, R., Sinha, A., Scherbak, Y., 2018. Medication Dispensing Errors and Prevention. 2018.

Tripathy, R.K., Ghosh, S.K., Gajbhiye, P., Acharya, U.R., 2020. Development of automated sleep stage classification system using multivariate projection-based fixed boundary empirical wavelet transform and entropy features extracted from multichannel EEG signals. Entropy 22 (10), 1–23. MDPI AG, India. 10994300. Available from: https://www.mdpi.com/1099-4300/22/10/1141/pdf. https://doi.org/10.3390/e22101141.

Zhao, X., Wang, G., He, Z., Jiang, H., 2022. A survey of moving object detection methods: a practical perspective. Neurocomputing 503, 28–48. https://doi.org/10.1016/j.neucom.2022.06.104. 09252312.

Zhou, X., Yang, C., Weichuan, Y., 2013. Moving object detection by detecting contiguous outliers in the low-rank representation. IEEE Trans. Pattern Anal. Mach. Intell. 35 (3), 597–610. https://doi.org/10.1109/TPAMI.2012.132. 0162-8828.

CHAPTER 13

A framework for development of speech-controlled necrobot for health-care application

Shyam Joshi[a] and Avnish Bora[b]
[a]AI Research Centre, Woxsen University, Hyderabad, India
[b]Jodhpur Institute of Technology, Jodhpur, India

1 Introduction to speech-controlled robots

Understanding the fascinating world of speech-controlled robots requires delving into the intricate physics and engineering underlying these state-of-the-art gadgets. Sophisticated sensors and algorithms designed to recognize and respond to human commands are at the heart of its operation. This device is a marvel of modern engineering, using sophisticated voice recognition algorithms that can comprehend a wide range of verbal requests. However, there are several challenges and limitations with the technology. One of the most significant obstacles in voice recognition is achieving high accuracy, especially when dealing with the diversity of accents and dialects or in noisy environments. The increasing prevalence of speech-controlled robots has given rise to privacy concerns around the gathering, storing, and utilizing voice data. Despite these challenges, advancements in voice-controlled robot technology are pushing the boundaries and opening the door for more intricate and lifelike future human-machine interactions.

1.1 Architecture

Speech-controlled robots are growing more widespread and are employed in a wide range of industries and applications. A speech-controlled robot's ability to comprehend and react to spoken orders is made possible by a number of essential parts and procedures. Here is a more thorough explanation of the general operation of a speech-controlled robot:

Speech-Recognition: A speech-recognition system processes a recorded speech signal. Automatic speech-recognition (ASR) is the process by which the speech-recognition system translates spoken words into text. The ASR system accurately converts the user's spoken words into text by using sophisticated models and algorithms. The spoken instruction is then represented in text and delivered to the following step for additional processing.

Necrobotics for Healthcare Applications and Management
https://doi.org/10.1016/B978-0-443-24832-0.00013-1

Natural Language Understanding (NLU): To comprehend spoken commands, a natural language understanding system examines their textual form. The NLU system deciphers the meaning of the command, determining the speaker's intention, and retrieving pertinent data. For instance, when the user says, "Turn on the lights in the living room," the location (living room) is extracted as a parameter by the NLU system, which recognizes the intent to operate the lights. The speech-controlled robot becomes increasingly proficient at comprehending and reacting to human orders as a result of the NLU system's progressive improvement in accuracy using machine learning algorithms.

Dialogue Management: The interpreted command and extracted parameters are passed to a dialogue management system. The dialogue management system determines the appropriate response or action based on the user's command and the context of the interaction. It may engage in a dialogue with the user to clarify any ambiguities or confirm the intended action. The dialogue management system uses natural language generation (NLG) to provide a spoken response to the user's command. This response is designed to be as natural and human-like as possible, improving the user's experience with the speech-controlled robot.

Action Execution: Based on the interpreted command and dialogue with the user, the robot determines the appropriate action to take. This action could involve controlling the robot's movements, interacting with the environment, or providing a spoken response. For example, if the user's command was to "turn on the lights in the living room," the robot would send a command to the smart home system to activate the lights in the specified location. The speech-controlled robot uses a combination of hardware and software components to execute the desired action. These components can include motors, sensors, actuators, and other devices.

Feedback and Confirmation: The robot may provide feedback to the user to confirm that the requested action has been executed successfully. This feedback could be in the form of a spoken response, visual indicators on the robot's display, or other sensory feedback. The speech-controlled robot uses a variety of feedback mechanisms to keep the user informed about the status of the requested action. This improves the user's confidence in the robot's ability to understand and execute their commands.

Continuous Interaction: The process of speech input, recognition, understanding, dialogue management, and action execution continues as the user and robot engage in a conversation or perform tasks together. The robot may adapt its behavior and responses based on the user's input and feedback, creating a more interactive and natural interaction experience. The speech-controlled robot can learn from the user's commands and preferences over time, improving its accuracy and effectiveness.

Overall, the working of a speech-controlled robot involves a complex interplay of speech processing, natural language understanding, dialogue management, and action execution components to enable seamless communication and interaction between the user and the robot.

2 Literature survey

The use of speech-recognition technologies in electronics and personal services is growing. However, because motion ambiguity persists, its use in industrial and medical contexts is still restricted. Robotic assistance for minimally invasive surgery provides one example of this uncertainty. In these situations, the surgeon has challenges because the robotic motion is not adjusted to match the pictures captured by the camera. An HIWIN robotic endoscope holder's voice recognition interface design is presented by researchers in Zinchenko et al. (2017). Long-distance movement control is made possible by the novel purposeful voice control method incorporated into the suggested design. A speech-to-motion calibration approach that compares the degree of change in the endoscopic picture for a spoken instruction is proposed to address motion ambiguity. Using CMU Sphinx, an algorithm for voice recognition is implemented on Ubuntu OS. With the use of an RS232 connection and serial-port communication, the control signal is transmitted to the robot controller. According to the testing results, the intended voice control technique can navigate the endoscope with an angular displacement accuracy of up to 3.1 degrees. 3.22 s is the total system processing time, which includes robotic movements.

Today, real-time communication between people and humanoid robots is conceivable. Robots must be able to comprehend auditory information similarly to humans and react appropriately if they are to be more useful. The HOAP-2 humanoid robot is activated based on a recognized command by means of an artificial neural network recognition system, which is part of a nature-inspired method suggested by researchers in Joshi et al. (2016). In this paper, the authors demonstrate a speech-activated prototype system for controlling a humanoid robot. A variety of tasks may be programmed into the prototype robot, depending on its intended function.

Based on SPCE061A and C8051, a voice control system for intelligent robot prostheses is presented (Fan and Li, 2010). In addition, hardware and software design, as well as control system theory, are introduced. With SPCE061A, voice recognition was made possible. The C8051 single chip receives the voice commands that have been identified via a serial connection called RS232. Finally, the robot prosthesis is controlled via C8051. This system satisfies the following movements: rising, lowering, swinging out, tendonitis of the upper limb, ad version, aversion, bending, unfolding of the forelimb, upward and downward wrist movement, and folding and shutting of the hand. The voice control system is built on the SPCE061A and C8051, which have a capability-to-price solid ratio and are straightforward to implement.

In Ondras et al. (2021), authors created a neural-network-based system that uses user speech as input to produce upper-body motions on a humanoid robot called Pepper from Softbank Robotics. These gestures include head, hand, and torso movements. Using Web polls, the suggested system was assessed both numerically and subjectively in relation to both synthetic and genuine speech. The authors have compared the effects of person-specific and generic neural network models on the quality of synthesized motions. They also examine how the speaker's personality attributes impact the synthesized motions and the connections between quantitative and qualitative assessments.

In Valin et al. (2007), researchers present a method that enables automated speech-recognition with simultaneous speakers for a mobile robot. In addition to a real-time implementation of geometric source separation (GSS) and a postfilter to further decrease interference from other sources, a microphone array is employed. Additionally, a missing feature mask may be computed, and the trustworthiness of spectral features can be estimated using the postfilter. The mask is employed in a speech-recognition system based on missing feature theory to identify simultaneous Japanese speakers' speech inside the framework of a humanoid robot. Three simultaneous speakers at a distance of two meters from the robot are shown with their recognition rates. A 200-word vocabulary at various azimuths between sources—between 10 and 90 degrees—was used to assess the system. The authors show an average reduction in relative identification error rate of 24% with the postfilter and 42% when the missing features technique is paired with the postfilter, when compared to the usage of the microphone array source separation alone.

Researchers in House et al. (2009) describe a method that allows a genuine 3D robotic arm and a simulated 2D robotic arm to be controlled by continuous control signals specified by the human voice. Their objective is to work toward enabling people with motor disabilities to manipulate common things. The authors have conducted many tests with control style variations for the 2D and 3D arms utilizing this technique. The findings demonstrate that a user may acquire the skills necessary to successfully move physical things with a robotic arm by relying solely on nonverbal speech as a means of control.

Since voice is the primary means of human communication, speech processing is a crucial field of signal processing. The computer's ability to comprehend human languages and analyze and execute voice commands is made possible by speech-recognition technology. In Patil et al. (2018), a voice control is created by utilizing LABVIEW programming techniques in conjunction with speech-recognition technology. LABVIEW will give the robot with proper control signal. Compared to existing sound-based robot control systems, the study is made easier by using LABVIEW for the interpretation of speech signals.

In order to help stroke patients start their rehabilitation training, a study published in Guo et al. (2018) describes the design and implementation of speech control for a unique type of exoskeleton upper limb rehabilitation robot based on biomechanics. The voice control is suited for those who can communicate. Additionally, doctors or family members may provide training aids to patients. The gadget is small, light, and simple to operate, and it can help people who have hemiplegia restore movement function in their fingers and upper limbs. The results of the experiment showed that the speech control system is viable since the average accurate recognition rate of speech control is higher and the success rate of postnoise speech control system rehabilitation training is higher.

The design, build, and testing of a mobile robot that can be controlled by voice were examined by researchers in Phal et al. (2014). The vocabulary has the following five commands: front, rear, right, left, and stop. The purpose of the study is to evaluate the speech-recognition system on the mobile robot platform and determine how well it can be interfaced with a speech-controlled wheelchair application based on the test results. A functional mobile robot platform prototype that interfaces with the iARC GUI was created.

Researchers in Chatterjee et al. (2005) demonstrate the potential for developing fuzzy-neural networks (FNNs) based on particle swarm optimization (PSO), which may be used as a key component in actual robot systems that are controlled by voice-based instructions. When a user issues fuzzy spoken language commands, the PSO is used to train FNNs so they can reliably provide crisp control signals for the robot systems.

2.1 Necrobots

Necrobotics is the field of robotics that makes use of biotic materials, or deceased creatures, as building blocks (Yap et al., 2022). Researchers at Rice University's Preston Innovation Lab in Houston, Texas, introduced the idea and provided evidence of its practicality in a July 2022 study published in Advanced Science. They did this by reusing deceased spiders as robotic grippers and using pressurized air to trigger their grasping limbs (Yap et al., 2022). A simple and effective gripper device is created by necrobotics by using the biological hydraulic mechanism of spiders and their little legs. Necrorobotic spider grippers are an affordable and eco-friendly substitute for intricate and expensive tiny mechanical grippers since they can lift small, light things (Ray, 2023).

Neurorobotics find the spider's tiny leg structure and utilization of hydraulic pressure to be particularly appealing. The anatomy of the spider makes use of a straightforward hydraulic (fluid) pressure mechanism. Flexor muscles in spider legs cause them to contract automatically while they are at rest (Zinchenko et al., 2017). Spiders need a force to extend and straighten their legs, and they generate this force by using hydraulic pressure to pump hemolymph fluid, or blood, through their joints (Ray, 2023). Their flexor muscles are naturally curled; therefore, curling their legs does not require any outside force (Yap et al., 2022).

Necrobotics offers a quick and accurate replacement for labor-intensive mechanical parts. Necrotic grippers are a substitute for small mechanical grippers, which can be costly and complicated. These pneumatic spider grippers have a long lifespan of 1000 cycles and can be made in less than 30 min. Necrotic grippers are perfect for moving light things into small locations and handling delicate goods. Microelectronics can also utilize it for basic pickup and drop operations.

The sole robotic concept in the necrobotics sector at the moment is the necrobotic spider gripper. Future necrobotic approaches, however, may use electrical stimuli and soft robotics to recycle biotic material into biohybrid systems. Using existing bone structures to house robotic components is another purpose for nacrobiotics.

Limitations: Compared to conventional mechanical systems, there is a greater likelihood of the component disintegrating and breaking down when organic material is used. If they break, more labor and supervision could be needed to replace these grippers. Furthermore, data obtained from the spiders will be erroneous due to organic discrepancies. Because wolf spiders develop differently, grippers may use different amounts of force and contraction of their legs. Robotics' use of spider euthanasia raises ethical questions. Necrorobotics' desire for biohybrid systems raises ethical questions since its detractors claim it may result in mammal hybridization and is invasive in the natural world.

3 Proposed scheme

The following algorithm has been proposed to develop a speech-controlled necrobot for healthcare applications. The main objective of this proposal is to create a robot that can be controlled through speech commands to assist healthcare professionals in their daily tasks. For this purpose, we propose to use a benchmark robotic gripper made of euthanized spiders. This gripper is known for its ability to handle small and fragile objects delicately, making it an ideal choice for healthcare applications where precision is essential. This innovative approach will not only improve the efficiency of healthcare professionals but also enhance patient care and safety.

3.1 Algorithm

Step 1: Begin.

Step 2: The medical practitioner will use a speech module to provide instructions for the amount of medicine to be administered.

Step 3: The speech module will decode the instructions, and the actuator will press the syringe accordingly.

Step 4: The necrobot will expand its limbs in proportion to the actuator input and pick up the medicine cubes.

Step 5: The actuator will release the pressure, ensuring that the robotic gripper holds the medicine securely.

Step 6: The base motor will move the necrobot to the correct position.

Step 7: The robotic gripper will expand and release the medicine into the collection pot.

Step 8: End.

The block diagram representing the methodology is shown in Fig. 13.1.

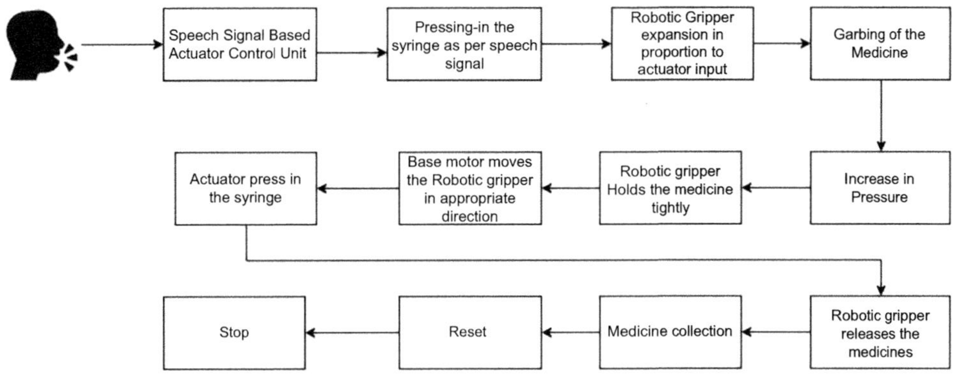

Fig. 13.1 ***The proposed methodology for the development of necrobot for healthcare applications.***

4 Conclusion

In this study, we present a framework to build a speech-controlled necrobot that may help with the system's creation for precise dosage. Necrobots are capable of handling trace levels of most materials. Using spoken cues, the necrobot will pick up a tiny amount of medication. Necrobots can manufacture the precise dosage of medication for a person based on age, weight, gender, and existing health issues by using voice signals from a medical professional.

Speech-controlled robots are a vital component of contemporary technology. Human spoken commands are understandable and interpreted by these robots. They also go by the names voice-controlled robots or conversational agents. These robots are outfitted with speech recognition, natural language processing, and dialogue management capabilities to enable spoken language communication with humans. The development of real-time speech-controlled robots for picking and placing medicines in the healthcare industry remains a potential subject of investigation.

References

Chatterjee, A., Pulasinghe, K., Watanabe, K., Izumi, K., 2005. A particle-swarm-optimized fuzzy-neural network for voice-controlled robot systems. IEEE Trans. Ind. Electron. 52 (6), 1478–1489. 02780046 https://doi.org/10.1109/TIE.2005.858737.

Fan, B.H., Li, K.Y., 2010. The speech control system of intelligent robot prosthesis 2. In: Proceedings - 2010 2nd WRI Global Congress on Intelligent Systems, GCIS 2010, China, pp. 407–409, https://doi.org/10.1109/GCIS.2010.69.

Guo, S., Wang, Z., Guo, J., Fu, Q., Li, N., 2018. Design of the speech control system for a upper limb rehabilitation robot based on wavelet de-noising. In: Proceedings of 2018 IEEE International Conference on Mechatronics and Automation, ICMA 2018. Institute of Electrical and Electronics Engineers Inc., China, pp. 2300–2305, https://doi.org/10.1109/ICMA.2018.8484626. http://ieeexplore.ieee.org/xpl/mostRecentIssue.jsp?punumber=8470456.

House, B., Malkin, J., Bilmes, J., 2009. The voicebot: a voice controlled robot arm. In: Conference on Human Factors in Computing Systems – Proceedings, United States, pp. 183–192, https://doi.org/10.1145/1518701.1518731.

Joshi, N., Kumar, A., Chakraborty, P., Kala, R., 2016. Speech controlled robotics using artificial neural network. In: Proceedings of 2015 3rd International Conference on Image Information Processing, ICIIP 2015. Institute of Electrical and Electronics Engineers Inc., India, pp. 526–530, https://doi.org/10.1109/ICIIP.2015.7414829.

Ondras, J., Celiktutan, O., Bremner, P., Gunes, H., 2021. Audio-driven robot upper-body motion synthesis. IEEE Trans. Cybern. 51 (11), 5445–5454. 2168-2267 https://doi.org/10.1109/tcyb.2020.2966730.

Patil, S., Abhigna, A., Arpitha, Deepthi, Priyanka, 2018. Voice controlled robot using labview. In: Proceedings - 2018 International Conference on Design Innovations for 3Cs Compute Communicate Control, ICDI3C 2018. Institute of Electrical and Electronics Engineers Inc., India, pp. 80–83, https://doi.org/10.1109/ICDI3C.2018.00025. http://ieeexplore.ieee.org/xpl/mostRecentIssue.jsp?punumber=8424520.

Phal, D.D., Phal, K.D., Jacob, S., 2014. Design, implementation and reliability estimation of speech-controlled mobile robot. In: 2014 2nd International Conference on Emerging Technology Trends in Electronics, Communication and Networking. IEEE, pp. 1–6.

Ray, P.P., 2023. An introduction to necrobotics: concept, architecture, use cases, challenges, future directions. SSRN Electron. J. https://doi.org/10.2139/ssrn.4595801. 1556-5068.

Valin, J.M., Yamamoto, S., Rouat, J., Michaud, F., Nakadai, K., Okuno, H.G., 2007. Robust recognition of simultaneous speech by a mobile robot. IEEE Trans. Robot. 23 (4), 742–752. 15523098 https://doi.org/10.1109/TRO.2007.900612.

Yap, T.F., Liu, Z., Rajappan, A., Shimokusu, T.J., Preston, D.J., 2022. Necrobotics: biotic materials as ready-to-use actuators. Adv. Sci. 9 (29). https://doi.org/10.1002/advs.202201174. John Wiley and Sons Inc., United States, 21983844, Available from: http://onlinelibrary.wiley.com/journal/10.1002/(ISSN)2198-3844.

Zinchenko, K., Wu, C.Y., Song, K.T., 2017. A study on speech recognition control for a surgical robot. IEEE Trans. Industr. Inform. 13 (2), 607–615. https://doi.org/10.1109/TII.2016.2625818. IEEE Computer Society, Taiwan, 15513203. Available from: http://ieeexplore.ieee.org/xpl/RecentIssue.jsp?punumber=9424.

CHAPTER 14

A framework for the development of an intelligent necrobot for classification and treatment of skin diseases

Shyam Joshi[a] **and Aadil Mir**[b]
[a]AI Research Centre, Woxsen University, Hyderabad, India
[b]Woxsen University, Department of Analytics, Hyderabad, India

1 Introduction

Necrobotics is the field of robotics that uses biotic materials, or dead creatures, as building blocks of robots. Researchers at Rice University's Preston Innovation Lab in Houston, Texas, introduced the idea and provided evidence of its practicality in a July 2022 piece published in Advanced Science (Yap et al., 2022). They did this by converting dead spiders into robotic grippers and using pressurized air to trigger their grasping limbs. Necrobotics, an innovation in technological advancement, results from a passionate interest in the complex biological marvels that nature has created over millions of years and a commitment to precise engineering (Kambouris et al., 2023; Vincent, 2022). This article (Smith et al., 2023) aims to explore the many facets of necrobotics, elucidating the care that goes into the preparation and selection of biological materials, the complex design and control mechanisms of these innovative systems, and the moral issues that guide the ethical and sustainable advancement of this field. It also discusses the difficulties in maintaining the integrity of biological components.

Achieving dependable and efficient necrobotic devices requires carefully selecting and processing biotic materials. This procedure starts with a precise grasp of the aims and functions that the device must have, which directs the selection of suitable biological materials that support these objectives. The device's mechanical and structural requirements—which consider characteristics like tensile strength, compressibility, and elasticity—are taken into account to ensure the materials used can withstand operating loads without failing too soon (Ray, 2023; Rousis et al., 2012).

Following selection, the biotic materials—muscle tissues for locomotion, bone structures for support, or sensory organs for environmental perception—undergo a thorough processing procedure. Achieving dependability and consistency in the finished necrobotic device entails meticulous cleaning to eliminate extraneous materials, preservation to stop decay, modification to maximize compatibility and performance inside robotic systems,

Necrobotics for Healthcare Applications and Management
https://doi.org/10.1016/B978-0-443-24832-0.00014-3

and strict quality control standards. Respect for donors, animal welfare, cultural sensitivity, and sustainability are all stressed in the sourcing and managing of these resources, and compliance with legal and ethical standards is equally important. In addition, careful management and monitoring of the storage conditions are necessary to maintain the quality, functionality, and integrity of these prepared materials. This involves maintaining regulated settings concerning temperature, humidity, cleanliness, and security (Zhengwei Li and Saif, 2021; Pasquero and Poletto, 2022, 2023; Saric et al., 2018). In this work, we aim to present a novel scheme that uses a computational genetic-neuronal approach for decision-making in necrobots for skin disease classification and cure. The present work primarily develops a computational model of a controller using C-Programming, which can be embedded in necrobots for decision-making (Joshi et al., 2013).

The decision-making process adopted by the human mind cares about many factors such as goal, safety, minimum losses, economy, and finally, an accomplishment of the goal. All these attributes could be knit together, provided the decision is filled with logic typically oriented for goal accomplishment. The optimization at various stages of the decision-making process is a perfection of all critical attributes that enable decisions to be good enough to achieve a goal. However, this feature of optimization could be added to the logic-motivated decision-making process by introducing a genetic algorithm. The logic-motivated decision-making, coupled with the genetic algorithm, is basically a neuronal network tied with synapses, where the weights of the synapse get optimized to produce the desired outcome. The genetic algorithm provides the optimized weights of the synaptic link. Since the entire process is very close to the way a human mind works, it is indeed interesting to produce a replica of the human mind using a network of artificial neurons. Inspired by this brilliant approach, it has been decided to choose the topic of developing a controller that can be embedded with a Necrobot to classify and cure skin diseases. This area is basically the new pathway to discoveries in genetic neuroscience controllers for necrobots. It aims to create a system where the computational genetic neural base works for decision-making while medicating a person affected by skin disease by a Necrobot. The biological neuron fires to give action potential to be transmitted to successive neurons when the postsynaptic-potential takes the potential of the neuron above the threshold. The human mind is a very complicated mesh of neurons, and whenever any problem appears before it, the decision is taken in a way that the problem stands solved. However, it has yet to be known what method the human mind used to deliver the solution. It is very clear that while making the decision, the human mind compensates against all the dangers, all the gaps between the target and the arrows, and controls against possible challenges and troubles to achieve success. In other words, thousands of unforeseen pressures are mitigated to achieve a goal. This work proposes a network with a base of 10 neurons for four conditions and four decisions. The fundamental role that the brain plays is nothing but to optimize for the generation of given results. Therefore, all the links of neurons, both from the input and to the outputs, have evolved genetically. Such a genetic selection of synaptic weights leads to compensating and accommodating all features, which the human mind does in an untold manner. The computational genetic-neuronal controller has been tested for skin disease classification and cure. It is essential to mention that the task has been achieved by developing a logic-motivated neuronal network whose synapses have been assigned weights as per outcomes of the genetic algorithm.

2 Problem statement and the solution

To develop a cutting-edge computational genetic-neuronal controller that empowers necrobots to classify and cure skin infections caused by different pathogens like bacteria, fungi, viruses, or parasites.

2.1 The solution: Proposed scheme

To develop a Genetic-Neuronal Controller for Necrobots, we aim for the following properties:

1. The gene network should be embedded in the neural network.
2. The specific skin disease pattern should decide the inputs.
3. The network should have four input nodes.
4. The network should have a hidden layer with ten neurons.
5. The network should have one output node.
6. The synaptic weight for every synapse must be obtained by optimization using genetic algorithms.

2.2 Computational genetic-neuronal controller

The proposed scheme for developing an intelligent necrobot that automates the identification and treatment of skin diseases is a comprehensive approach that involves several steps shows the proposed scheme for the development of an intelligent necrobot. The first step is to capture an image of the infected skin area. This image is then processed to extract essential feature vectors that are fed into the computational genetic-neuronal controller. The next step is the optimization process, where the genetic algorithm calculates the weights of the feedforward neural network. These weights are then applied to the input layer, the hidden layer, and the link between the hidden layer and the output layer. The use of linear regression helps solve the classification problem, and the skin disease is identified with a high degree of accuracy. Once the skin disease is identified, the Necrobot generates an appropriate control action that triggers the application of medication to the affected area. This process represents a significant advancement in automated decision-making, as the computational genetic-neuronal approach is integrated into the Necrobot, making it an intelligent system capable of making informed decisions and taking appropriate action.

Fig. 14.1 provides a detailed flowchart that illustrates the entire process, highlighting the various steps involved in identifying and treating skin diseases with the intelligent necrobot. This approach is expected to revolutionize the way skin diseases are identified and treated, making the process more efficient, accurate, and accessible.

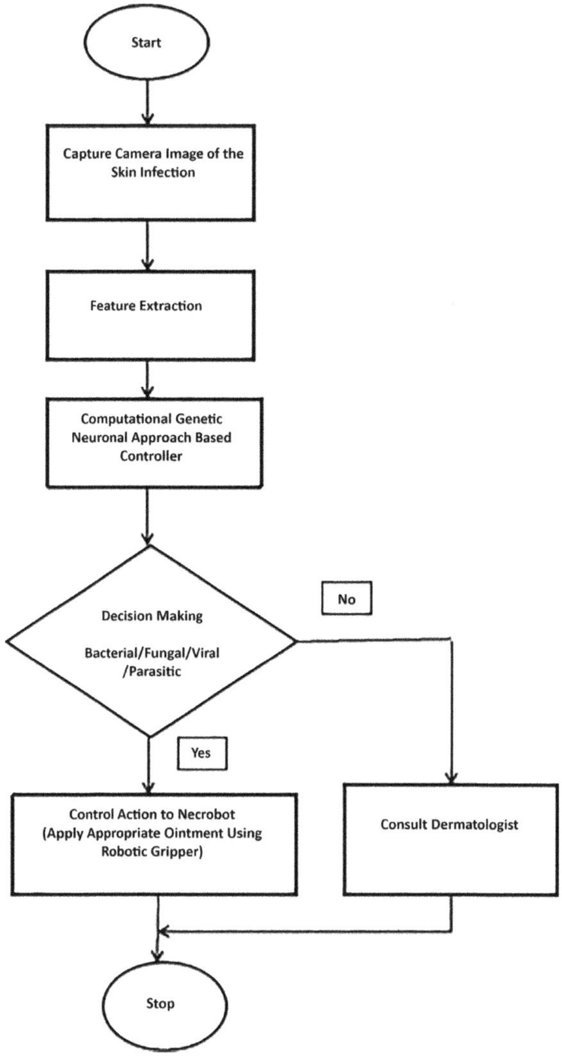

Fig. 14.1 ***The proposed methodology.*** The flow chart explains the development of intelligent necrobot for skin disease classification and cure.

3 Genetic-neuronal model for decision-making

The proposed Necrobot employs a computational genetic-neuronal model comprising a neural network embedded with genetic algorithm-based synaptic weights. The neural network consists of four input nodes, which are determined according to the specific skin disease pattern, and ten neurons in the hidden layer. The network has one output node, and the synaptic weights for each link in the network are calculated using a genetic algorithm. For a more comprehensive understanding of the proposed model, refer to Fig. 14.2: infection apply.

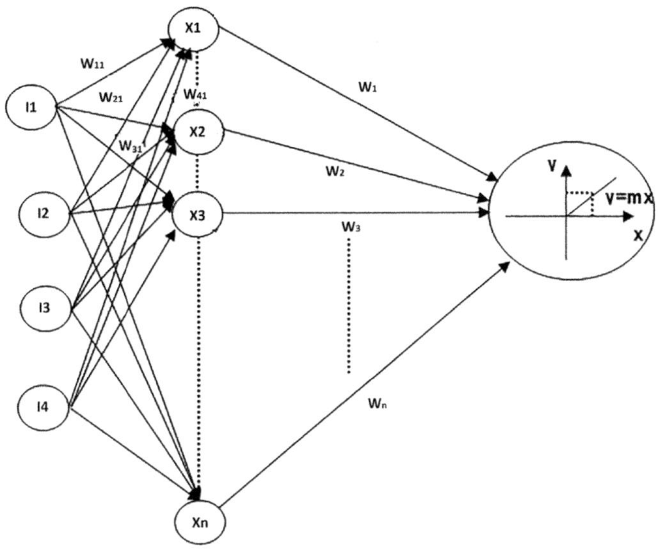

Fig. 14.2 ***The architecture of genetic-neuronal network.*** The architecture of the genetic-neuronal model for the development of a controller for the necrobot

Table 14.1 Output interpretation.

Condition	Decision criterion	Decision interpretation
Tiny pimples surrounded by red, swollen hair follicles	$0.5 \leq x < 0.6$	D_1
Swollen, red lumps that resemble flaky, scaly patches	$0.6 \leq x < 0.8$	D_2
Little pink dots	$0.8 \leq x < 0.9$	D_3
Affected patches of skin as pink or red, or brown and black	$0.9 \leq x < 1.0$	D_4

The output of the network can be interpreted as presented in Table 14.1. The decision criterion is set to identify and apply medication for a specific disease. The range of x decides the disease and its cure.

4 Result

The results of the computational genetic-neuronal model have been presented in Table 14.2. The Necrobot will execute the recommended decisions based on this data. The disease has been classified into four categories, namely bacterial, fungal, viral, and parasitic. The treatment for each type of disease will be administered accordingly.

Table 14.2 Classification of skin diseases and cure.

Condition	Pictorial view	Decision
Tiny pimples surrounded by red, swollen hair follicles	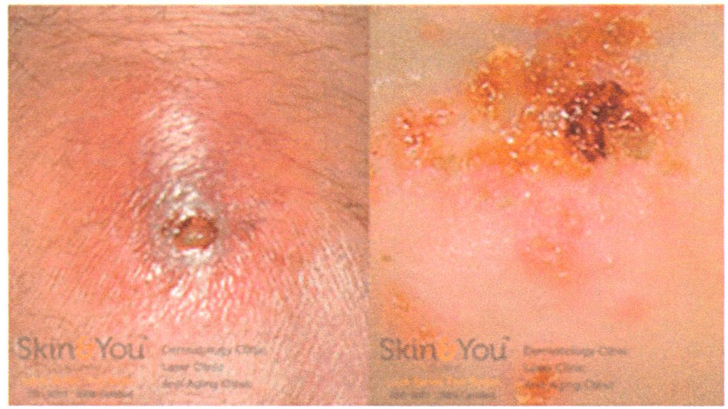 Bacterial infection from: https://www.skinandyou.in/medical-clinic/dermatology-treatment/skin-problem/bacterial-infection/	D_1 = bacterial infection: apply antibiotic ointment
Swollen, red lumps that resemble pink or red scaly patches	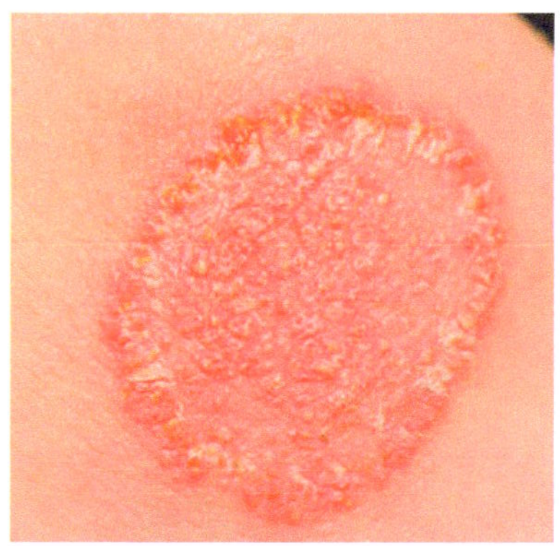 Fungal infection from: https://foradermatology.com/fungal-infections	D_2 = fungal infection, apply antifungal ointment

Typically, viral rashes include little pink dots	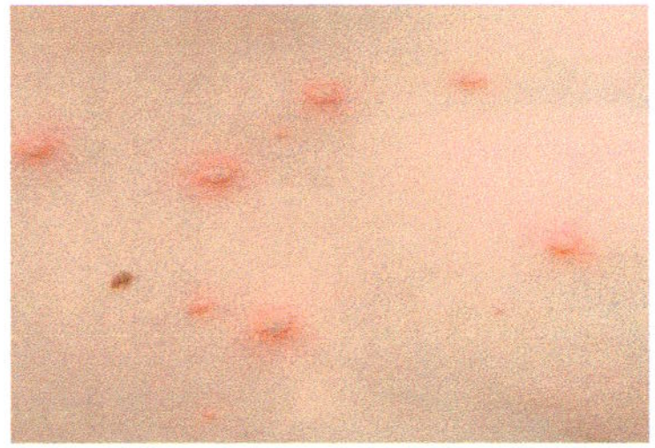 Viral infection from: https://wchh.onlinelibrary.wiley.com/doi/full/10.1002/psb.1907	D_3 =Viral infection, apply antiviral ointment
Affected patches of skin as pink or red, or brown and black	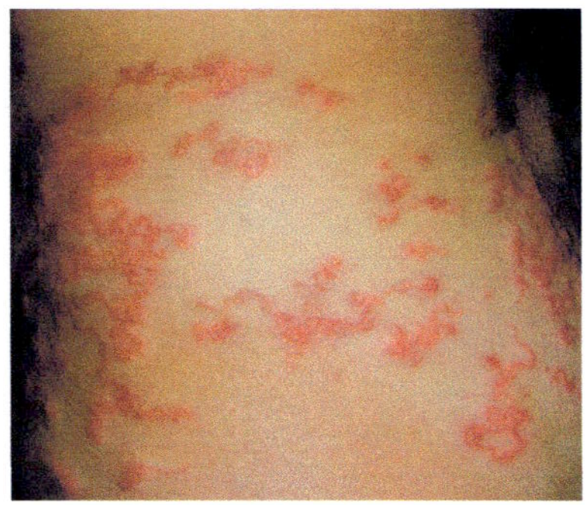 Insect bite from: https://bio.libretexts.org/Parasitic_Skin_Diseases	D_4 = apply antiparasitic lotion

5 Conclusion

This report describes a cutting-edge computational modeling necrobotics scheme for creating an intelligent necrobot that can classify and cure skin diseases. The Necrobot is equipped with an embedded genetic-neuronal controller that makes decisions similar to those of the human brain. The gene network is embedded in the neural network of the C-program, and the weights of synaptic links from the input to the hidden layer and from the hidden layer to the output are optimized using a genetic algorithm. These optimized weights help the output of the computational model converge into a particular decision index, mimicking the decision-making process of the human mind when presented with the same set of conditions.

The proposed computational model has undergone rigorous testing for skin disease classification and curing problems. The decisions made by the model have been found to be consistent with those made by a human brain when diagnosing a skin disease and suggesting a cure. The computational genetic-neuronal approach can be made even more robust by adding an evolutionary algorithm and varying degrees of fuzziness. Additionally, a hardware setup can be developed to create a prototype of an intelligent necrobot for future applications.

Overall, this innovative approach has the potential to revolutionize the field of skin disease classification and cure. The intelligent necrobot can provide faster and more accurate diagnoses and cures, leading to improved patient outcomes. The possibilities for further development in this area are vast, and the potential for future advancements is exciting.

References

Joshi, S.K., Garg, A., Joshi, G.K., 2013. Computational Genetic-Neuronal Approach for Decision Making SAPBEATS-2013.

Kambouris, M.E., Manoussopoulos, Y., Velegraki, A., Patrinos, G.P., 2023. The biote-bot hybrid. The ultimate biothreat merging nanobots, AI-enabled cybernetics and synthetic biology. Futur. Med. AI 1, 01–16.

Pasquero, C., Poletto, M., 2022. Biomorphic Intelligence. Informa UK Limited, pp. 224–238, https://doi.org/10.4324/9781003183181-21.

Pasquero, C., Poletto, M., 2023. Biodesign in the Age of Artificial Intelligence: Deep Green. Taylor and Francis, United Kingdom, pp. 1–359. Available from: http://www.tandfebooks.com/doi/book/9781000901979. https://doi.org/10.4324/9781003168614.

Ray, P.P., 2023. An introduction to necrobotics: concept, architecture, use cases, challenges, future directions. SSRN Electr. J. https://doi.org/10.2139/ssrn.4595801. 1556-5068.

Rousis, K., Eleftherakis, G., Paunovski, O., Cowling, A.J., 2012. Formal modeling of a bio-inspired paradigm capable of exhibiting emergence. In: ACM International Conference Proceeding Series Greece, pp. 223–228, https://doi.org/10.1145/2371316.2371360.

Saric, K., Redd, C., Varnfield, M., Ordwyer, J., Karunanithi, M., 2018. Increasing health care adherence through gamification, video feedback, and real-world rewards. In: 2018 Proceedings of the Annual International Conference of the IEEE Engineering in Medicine and Biology Society, EMBS. Institute of Electrical and Electronics Engineers Inc., Australia, pp. 1584–1587, https://doi.org/10.1109/EMBC.2018.8512487. 9781538636466.

Smith, G.L., Tyler, J.B., Lazarus, N., Tsang, H., Viornery, L., Shultz, J., Bergbreiter, S., 2023. Spider-inspired, fully 3D-printed micro-hydraulics for tiny, soft robotics. Adv. Funct. Mater. 33 (39). John Wiley and Sons Inc, United States. 16163028. Available from: http://onlinelibrary.wiley.com/journal/10.1002/(ISSN)1616-3028. https://doi.org/10.1002/adfm.202207435.

Vincent, J., 2022. Scientists Reanimate Dead Spiders as Robot Gripping Claws. The Verge. Archived from the original on.

Yap, T.F., Liu, Z., Rajappan, A., Shimokusu, T.J., Preston, D.J., 2022. Necrobotics: biotic materials as ready-to-use actuators. Adv. Sci. 9 (29).

Zhengwei Li, M., Saif, T.A., 2021. Mechanics of biohybrid valveless pump-bot. J. Appl. Mech. 88 (11). https://doi.org/10.1115/1.4051595. 0021-8936.

CHAPTER 15

RAEXO: A smartphone and semiactive actuation mechanism based a pilot prototype for rheumatoid arthritis rehabilitation

Snekhalatha Umapathy[a], Harsha Vardhan Reddy[b], and Murugappan M.[c]

[a]SRM Institute of Science and Technology, Biomedical Engineering, Kattankulathur, Tamil Nadu, India
[b]SRM Institute of Science and Technology, College of Engineering and Technology, Biomedical Engineering Department, Chennai, Tamil Nadu, India
[c]Kuwait College of Science and Technology, Intelligent Signal Processing (ISP) Research Lab, Department of Electronics and Communication Engineering, Kuwait, Kuwait

1 Introduction

Rheumatoid arthritis (RA) is an autoimmune bilateral symmetric inflammatory disease that impacts the synovial joints. During the initial days of RA, it affects the metacarpophalangeal (MCP) and proximal interphalangeal (PIP) joints of the hand, along with the wrist joints (Guo et al., 2018; Carbone et al., 2020). According to the recent WHO report, there are 450 million people suffering due to RA over the world, and most of their age is between 30 and 50 years (Almutairi et al., 2021). RA patients can experience symptoms including pain, swelling, stiffness, and restricted range of motion (ROM) in the affected joints. Reductions in joint strength and function can complicate the performance of daily life activities (ADL), leading to a dependence on external support and significantly affecting patients' emotional and psychological wellbeing. RA has no cure at the moment; however, advances in research and technology have brought about treatment strategies for managing the disease and preventing joint deformities (Chauhan et al., 2021).

Necrobiotics is an evolving field that combines robotics and biological organisms. It involves the parts of deceased biological organisms to develop a robotic system. This approach enables the development of robotic components by using the structural and mechanical properties of biological tissues. In this RA rehabilitation study, necrobiotics could be used as a valuable tool for enhancing semiactuation mechanisms. There are several key benefits involved in the applications of necrobiotics in the semiactive actuation mechanisms of RA rehabilitation. The key benefits are listed as follows: Enhanced biocompatibility, cost-effectiveness, sustainability, adaptability, flexibility, and therapeutic strategies. The future scope of necrobiotics in RA rehabilitation involves excellent

Necrobotics for Healthcare Applications and Management
https://doi.org/10.1016/B978-0-443-24832-0.00015-5

opportunities in the current research. Extended research and innovation in this field could lead to the development of advanced rehabilitation devices that integrate biological and technological components.

Over two-thirds of RA patients suffer from rheumatoid cachexia, which results from protein degradation and excessive metabolism. As a result, grip strength and range of motion are diminished. Muscle degradation is accelerated when physical activity is insufficient (Munsterman et al., 2012). Although pharmacological medications prescribed for RA can treat the pain, they do not improve muscle function (Roubenoff, 2003). Pedersen and Saltin (2006) demonstrated that low-intensity movement exercises increased levels of insulin-like growth factor I (IGF-I) and IGF-binding protein 6 (GFBP-6), which are responsible for the maturation of muscle cells (Pedersen and Saltin, 2006). Therefore, the effectiveness of long-term hand exercises is crucial for reversing cachexia and preventing disability. Furthermore, exercise can improve elasticity and reduce stiffness in tendons, increase ligament range of motion, and reduce joint pain. Movement rehabilitation is preferred for patients who have movement disabilities; however, they may not be able to devote sufficient time to active rehabilitation sessions due to financial, geographic, or other constraints.

Soft exoskeletons are constructed from soft materials and operated by soft actuators or tendon/cable-driven mechanisms. This contrasts with rigid exoskeletons, where motion is transmitted to the joints of the fingers or wrist through rigid mechanisms like linkages (Tiboni et al., 2022). Vélez-Guerrero et al. developed a robotic exoskeleton for upper limb rehabilitation with autonomous AI-based control (Vélez-Guerrero et al., 2021). The exoskeleton has 1-DOF that permits flexion-extension at the elbow joint. Their experimental results indicate that the device may be suitable for supporting diagnostic and rehabilitation processes related to neuromotor functions. Smartphone-based ROM tests have been developed for various joints, including the knee, shoulder, wrist, and lumbar joints (Moral-Munoz et al., 2021).

For RA patients, management and effective rehabilitation are required to improve joint function and quality of life. Conventional rehabilitation often involves physical therapy sessions, which can be time-consuming, expensive, and not easily affordable for many patients. Hence, these challenges are addressed by recent advancements in integrating modern innovations in rehabilitation technology. Once such innovative approach is the design of a smart phone and semiactive mechanism for RA rehabilitation. This prototype model provides an innovative solution by combining smart phones with an advanced actuation mechanism to enhance the personalized rehabilitation exercises for the RA patients. The smartphone has built-in sensors and connecting features for monitoring, guiding the rehabilitation exercises in real time. This enables and enhances the improved exercise performance with reduced risk of injury. At the same time, the semiactive actuation mechanism promotes muscle engagement and natural rehabilitation, enhances patients' movements, and provides targeted assistance to support the RA patients.

2 State-of-the-art works related to RA rehabilitation

In recent years, robotic technologies have consistently been used for developing assistive and rehabilitative aids. Previous studies have shown that intense rehabilitation sessions with robotic assistance can improve motor functions (Ueki et al., 2008; Kutner et al., 2010; Li et al., 2019; Wolf et al., 2006). Developing exoskeletons capable of replicating intricate anatomical structures like the human hand remains a daunting endeavor. The actuation mechanism and construction techniques required for current exoskeletons to fulfill these demands could render them costly, cumbersome, and difficult to operate independently without therapeutic aid. The most commonly used method is an underactuated multijoint serial-link structure that should be rigidly fixed to one of the finger phalanges to achieve flexion or extension. This method provides precise control over the ROM of finger joints and finger kinematics. However, misalignment of the exoskeleton joints with the finger joints is a common problem with this approach.

One of the earliest commercially available robotic hands for rehabilitation was the "Hand of Hope" (HOH). It utilized a serial-link mechanism for actuation. Although the HOH had advantages in terms of mechanical aspects, the ROM in PIP and MCP joints was limited. Furthermore, rigid serial-link actuation restricted user-specific customizations. These limitations were resolved in the ExoHand from Festo and the robotic exoskeleton developed by the Whitesides Research Group (Polygerinos et al., 2015). The Festo ExoHand used pneumatic actuators while adapting user-specific dimensioning with a three-dimensional (3D) scanner and sintering laser process. This advanced the number of DOF and the ROM. These custom-modeled silicone rubber-based soft actuators were controlled by the hydraulic system. Although the above-mentioned exoskeleton improved certain important aspects of the design, the system is bulky and expensive due to the use of a pneumatic compressor and hydraulic pump systems.

The RA Rehabilitative Device (RARD) (Matthew Chin et al., 2016) and soft robotic glove (Polygerinos et al., 2015) presented a pneumatically actuated exoskeleton design. The RARD design focused on laterally bent fingers in RA patients. The exoskeleton had a huge multiring casing and two rectangular actuating strips for actuating a single finger. Even though RARD had good ROM and subject-specific customizations, it was complex and heavier and did not allow simultaneous multifinger actuation. In contrast, the soft robotic glove used a small finger setup and increased DOF. It appeared to be lighter in weight, making it less cumbersome and more comfortable for patients to use; however, the pneumatic compressor system was heavier, weighing up to 3.8 kg, and was carried by the patient in a backpack. The soft robotic hand exoskeleton for limited hand-motor function (Rudd et al., 2019) used a tendon-driven exoskeleton with custom-built mechanical components for end-user customization. It employed both elastic and nonelastic tendons to realize finger actuation.

In Yap et al. (2015), the researchers have developed soft pneumatic actuator-based lightweight wearable gloves for rehabilitation applications. They have performed a set of experiments to examine the differences between active and passive actuation, and the developed device is highly flexible to adapt to the user's hand in different dimensions. A smartphone-based lightweight hybrid smart glove exoskeleton has been designed for finger flexion, finger abduction, and grasp quality enhancement using the tendon-drive system, pneumatic actuators, and telescopic extra thumb, respectively (Gerez et al., 2020). There are several low-cost hand exoskeletons that have been designed by different researchers using microcontrollers, servomotors, and actuators (Rudd et al., 2019;Yap et al., 2015; Gerez et al., 2020; Abdallah et al., 2017). However, very few rehabilitation devices have been developed for RA disease (Rodríguez-Sánchez-Laulhé et al., 2020; Dorado et al., 2019). Consequently, most of the devices were bulky and involved either active or passive activation mechanisms without a smartphone control application (Zhang et al., 2019; Diego et al., 2021). It highly necessitates the researchers to develop portable, low-cost, robust rehabilitation devices with smartphone control applications for rheumatic arthritis disease. Hence the proposed work focused on the development of low-cost exoskeletons with smart phone control applications for RA patients.

The main contributions of the current study are as follows:

- A developed 3D model of a hand exoskeleton that allowed RA patients to perform rehabilitation exercises using a mobile application.
- A semiactive actuation mechanism-based exoskeleton is guided by a hand motion like finger flexion and extension, wrist flexion and extension, and thumb abduction and adduction.
- Spring steel strips were employed to translate the actuation from servo motors to rotational movements of the fingers.
- This approach circumvented the misalignment problem observed in serial-link actuation systems and provided an optimal and efficient method of actuating fingers.

In addition to the traditional finger movements, it allowed wrist flexion and extension that complemented the existing DOF. The naturally acquired bending force of the spring steel combined with the actuation thrust of servos provided a sufficient force at the fingertips to allow a moderately higher ROM, which was sufficient to perform active rehabilitation sessions. The mechanical and other design aspects of the system were also taken into consideration while designing the proposed exoskeleton. Our main objective is to design and develop an exoskeleton that could drive multiple fingers and the wrist with a minimal number of actuators. The objectives include design optimization by reducing the number of components, accomplishing optimal ROM, and output fingertip force while keeping the exoskeleton movements highly agile and flexible.

The rest of the paper is organized as follows: The third section presents the proposed methodology for developing intelligent assistive and rehabilitation devices for RA; the experimental results are presented and discussed in the fourth section; and finally, the conclusion of the paper is given in the fifth section.

3 Materials and methods

The complete workflow of the proposed exoskeleton design is depicted in Fig. 15.1.

3.1 Ethical declaration

The study was approved by the ethical committee of SRM Medical College Hospital and Research Center, Chennai, India (Approval no. 2843/IEC/2021). Informed consent is obtained from the participants who participated in the experiment.

4 Design of RAEXO

We aimed to build an orthotic rehabilitation device that was simple, optimized, and highly agile, unlike most exoskeletons and orthotic arms. AutoCAD 2021 software was used to design the 3D-printed modules that could be attached and detached easily, as shown in Fig. 15.2. With the base module positioned on the dorsal side of the palm, the device was designed to fit securely and comfortably. On top of the base module should be the second module, also known as the "housing" module. It contained all the components for actuating the fingers. The third module consisted of a rectangular plastic strip supported by four cuboids with slits that held the stainless-steel strips parallel to the fingers. The thumb module had components for flexion and extension. Lastly, the radial module contained all the electronic components and was worn on the dorsal side

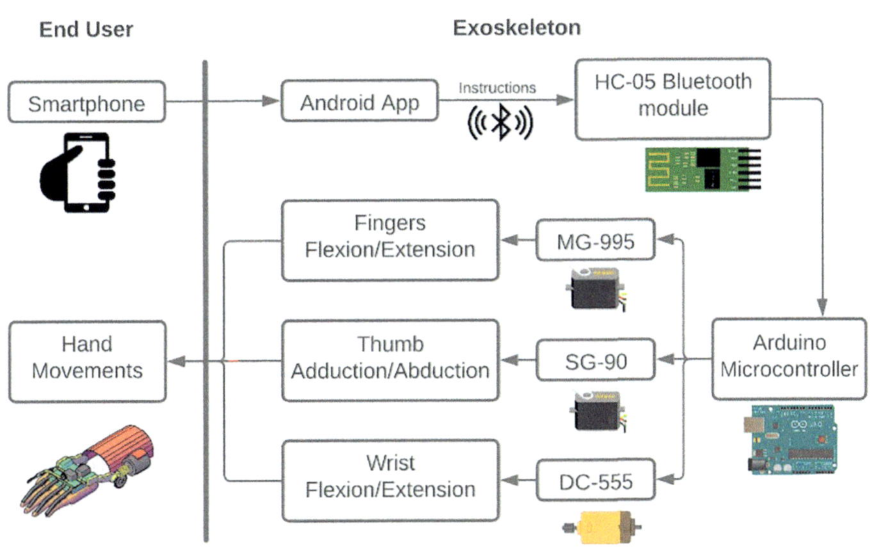

Fig. 15.1 ***Complete workflow of the proposed design.*** The entire workflow of exoskeleton design with smart phone applications for rehabilitation. (No permission required.)

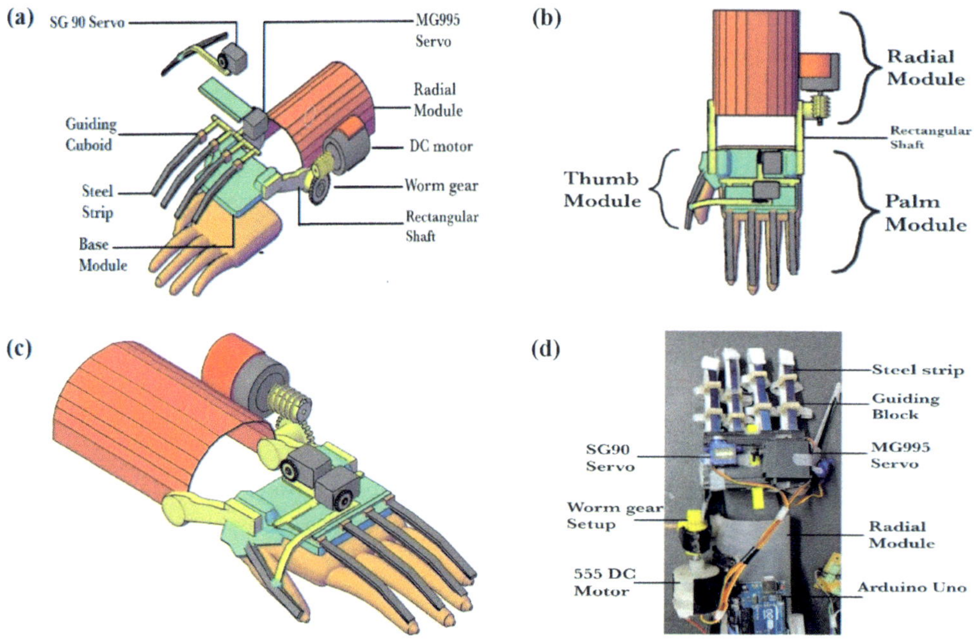

Fig. 15.2 ***Design of exoskeleton.*** (A) Exploded view of the exoskeleton, (B) top view of the design, (C) isometric lateral view displaying typical components, and (D) overview of the developed prototype. (No permission required.)

of the forearm (Fig. 15.2A). A rectangular shaft connected the palm and radial modules (Fig. 15.2B). Velcro straps (Fig. 15.2D) were used to secure the exoskeleton to the patient's hand, ensuring a firm and comfortable fit.

Flexible 3D-printing material, Flex, was used to create the exoskeleton assembly. We selected Flex because it is moderately hard, making it less brittle and more flexible than other 3D-printing materials. Stainless-steel was used to construct the skeletal structure that controlled finger extension and flexion. We used thin and wide (0.5–1 cm) high-carbon steel (1%–2%) spring strips to transfer actuation force. This spring steel had high yield strength, which made it flexible and strong, which are desirable traits for soft and adaptable actuators. A rack and pinion mechanism were attached to the dorsal side of the plastic bar attached to points on the steel strips in Fig. 15.2C. MG995 servos were used to drive the pinion. A worm and spur gear mechanism were used to achieve wrist extension and flexion. The worm gear was driven by a 12-V geared 555 DC motor that was mounted lateral to the radial module using a custom mount (Fig. 15.2).

Overall, Table 15.1 shows the three different actuation mechanisms were used to achieve six-DOF in RAEXO.

Table 15.1 Various actuation mechanisms used and associated DOFs.

Part	Actuator	Method of actuation	No. of DOF	Types of movements
Fingers	Servo motor (Rodríguez-Sánchez-Laulhé et al., 2020)	Rack and pinion	2	Flexion and extension
Wrist	DC motor (Li et al., 2019)	Worm gear pair	2	
Thumb	Servo motor (Gerez et al., 2020)	Rack and pinion	2	Adduction and abduction

No permission required.

4.1 Semiactive actuation mechanism

In the proposed design, we have adopted a semiactive actuation mechanism for controlling the exoskeleton. The actuation mechanism of the finger flexion and extension was based on the mechanical properties of the stainless-steel strips and the movement constraints of the finger joints; it was therefore described as a semiactive actuation method. This mechanism is more useful to simulate the natural movement and constraints of the human skeleton along with the mechanical force produced by the exoskeleton to achieve the required output. The flexion and extension movements were realized with the use of a worm gear pair setup. The worm gear helps to reduce the speed and increases the torque translated from the actuator. The high thrust and torque of the worm gear provided better control over the wrist movements at the angular level. Consequently, it also offers mechanical advantages, and it also contributed to optimizing the design process. Its nonintersecting axes mechanism allowed the actuator to be placed parallel to the forearm while creating movement along the radioulnar (transverse) axis of the wrist. The rotational force produced by the 555 geared DC motor (Fig. 15.2D) was translated as flexion and extension along the radioulnar axis of the wrist through the worm gear setup. A custom-designed and 3D-printed rectangular shaft was used to translate the torque of the worm gear to the palm. This moved the end effector vector of torque toward the center of the palm—that is, it pushed and pulled the palm along the dorsopalmar axis (Fig. 15.3).

The actuation mechanism of a single finger is depicted in Fig. 15.3. A stainless-steel strip was positioned along the length of the finger on the dorsal side. Wooden blocks with slits were used to keep the strips in position. When the servo rotated, it pushed the steel strip toward point D and applied force (f_a) normal to the fingertip. However, the wooden block at the fingertip with no open slit was an obstruction to the strip's movement and provided a counterforce (f_c). Since the movement was obstructed, and the force exerted by the actuator had to be directed along some other path, it was translated as the bending movement of the finger. With force (f_m) directed between points A and B, bending was created at the MCP joint with an angle (θ_m). When the MCP joint reached its bending limits, the remaining force was translated to the next diverting point between the B and C as (f_p) and bent the PIP joint with an angle (θ_p). This continued until the

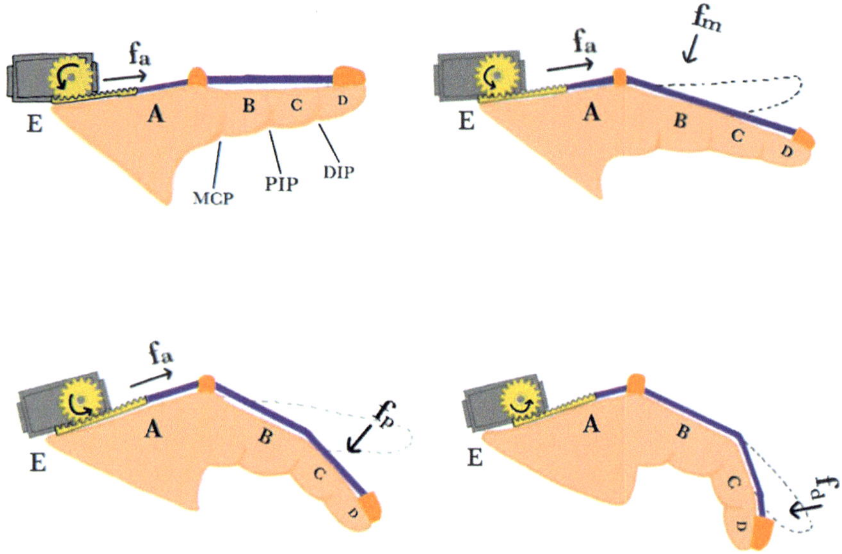

Fig. 15.3 ***Illustration of sequential forces exerted on various phalanges and finger flexion.*** Step-by-step process of sequential force applied on to various phalanges and finger flexion. (No permission required.)

tip of the steel strip (E) reached point A or the DIP joint reached its bending limits with angle (θ_d) because of the force (f_d). Adduction and abduction of the thumb were accomplished by using a rack and pinion with a flexible steel strip. A hollow slot in the palmar module of the exoskeleton makes it possible for the rack to slide. A small servomotor (SG90) was used as an actuator. The servo rotated 180 degree clockwise and pushed the strip toward the thumb. Adduction was created. As a result of counter-clockwise rotation, the rack moved into the slot, causing thumb abduction.

4.2 Embedded system

Microcontroller Arduino Uno R3 was chosen to control actuators and communicate with a mobile phone via serial. It had enough pulse width modulation (PWM) pins to drive two servo motors (MG-995 and SG-90), a direct-current (DC) motor, and a Bluetooth module (HC-05).

4.3 Android application

The most common symptoms of RA are joint stiffness and extreme pain. To avoid pain, they often avoid extreme finger kinematics. RA patients, like healthy controls, also expect to easily control any orthotic or rehabilitative equipment so that they can comfortably perform their therapy sessions and avoid joint and muscle pain. Electromyogram (EMG)-based control, attention control with brain mapping, and mirroring have been used to assist RA patients (Abdallah et al., 2017; Rodríguez-Sánchez-Laulhé et al., 2020; Dorado et al., 2019). In many cases, these systems have proven inefficient despite their high precision. Furthermore, sophisticated techniques can demand superior resources. As a result of these constraints and considerations centered on the patient, we devised an Android application for regulating exoskeleton movements. This app enabled users to tailor exoskeleton actions according to their preferences and minimized the need for expensive external control systems.

4.4 Electrical system

In the proposed investigation, an MB102 power module is incorporated into the Arduino shield to oversee voltage regulation among various modules. Moreover, for wrist movement, an L293D DC motor driver module controls the operation of the DC motor. The DC module had enough input and output ports for two 12-V DC motors. An HC-05 Bluetooth module was integrated into the shield to reduce bulk and save space. A smaller component is used to improve the power regulation and working efficiency, including a 100-nF capacitor, a 1-f/50-V capacitor, and a 5V voltage regulator.

4.5 Experimental setup

Exoskeletons are used for both therapy and assistive devices. To assess the efficiency of the exoskeleton, two types of forces called output force and grip force were utilized. A research investigation assessing the ROM and fingertip pressure applied by the prototype was carried out on RA patients at SRM Medical College Hospital and Research Center in India. Both output force and grabbing force were considered in the preliminary testing phase.

4.6 Range of motion

The ROM was computed based on the flexion angles of the finger joints. The three angles, such as θM, θP, and θD, were measured by using an MPU-6050 three-axis accelerometer. θM corresponds to the MCP, θP corresponds to PIP, and θD corresponds to DIP joints, respectively. The finger joint angles were recorded for a non-RA subject for comparison purposes. The experiment was conducted on one healthy subject (male, 26 years old) and one RA patient (male, 41 years old). We compared active and passive measurements to evaluate the exoskeleton's performance, and the healthy subject's measurements served as a baseline reference for the ROM. For each finger, one data set was collected in every session. Hence, a total of 15 data sets across the three sessions (active, passive, and for the healthy reference) were taken into consideration. The joint angles were measured simultaneously using three MPU sensors, allocating one sensor to each phalanx.

4.7 Output force

In this experimental arrangement, both a person in good health and an individual with RA were directed to bend their index finger and apply pressure to an FSR sensor situated on a flat surface in both active and passive sessions. As pressure was applied to the FSR by the subject, its resistance altered. This alteration in resistance served as an indication of the applied force. The relationship between the output voltage and FSR resistance is outlined in Eq. (15.1).

$$v_0 = v_{cc}\left[\frac{R}{R+R_f}\right] \tag{15.1}$$

The output voltage (V_o) of the FSR varied proportionally with the reciprocal of the FSR resistance (R_f). As the force exerted on the sensor intensified, its resistance declined, resulting in an elevation of the output voltage. To accurately ascertain the applied force on the sensor, we developed and incorporated a threshold-based algorithm into the Arduino Uno R3, as depicted (Fig. 15.4). The algorithm initially checked for the input

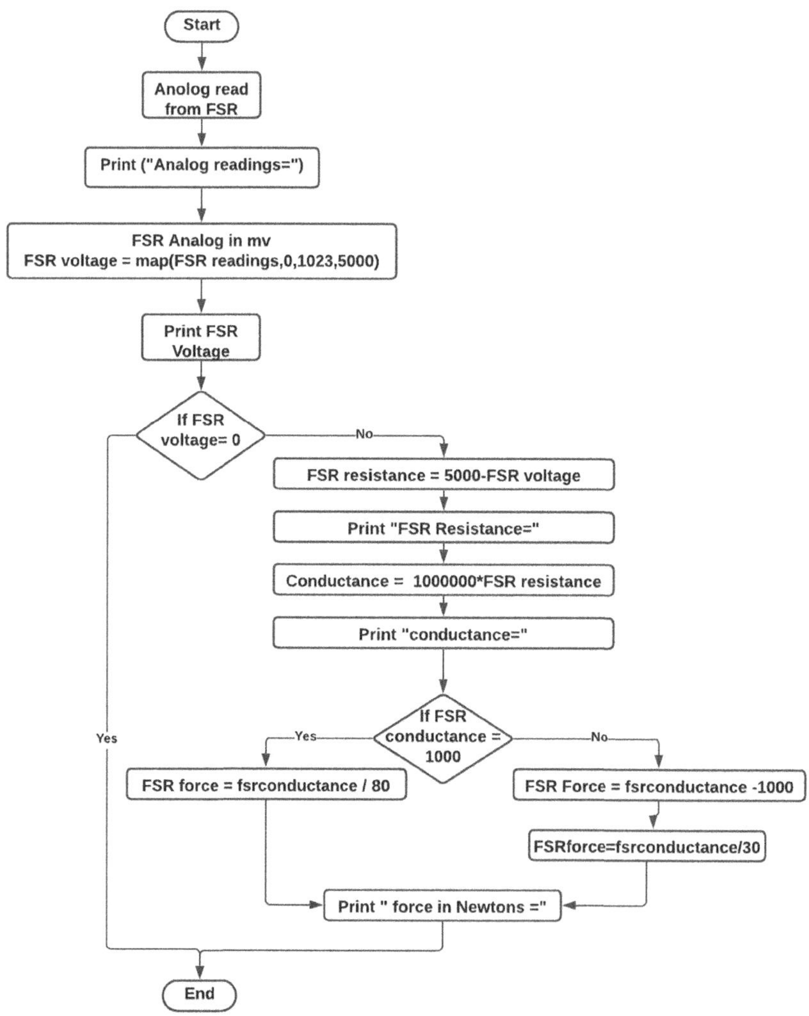

Fig. 15.4 ***Control algorithm for calculating force in Newton with FSR sensor.*** A threshold-based algorithm was developed and implemented in Arduino Uno R3 to accurately calculate the force applied to the sensor. (No permission required.)

from the FSR and then converted the analog output into voltage (using Eq. 15.1), which was presented in μV; this, in turn, was converted into conductance, which was used to calculate the force (N).

5 Results and discussion

Experiments with the proposed exoskeleton are discussed in this section. Data from the MPU-6050 sensor was analyzed and represented graphically in Fig. 15.5. Because our system only had one actuator for forefingers, all the characteristic responses had similar dynamics in finger joint-bending angles. In this study, only motion-analysis graphs were included for the MCP, PIP, and DIP joints of the index finger. Apart from the thumb, which had different joint kinematics, the same trend was seen in the analysis of all other fingers. An analysis of the wrist motion was included along with the motion analysis of the forefingers and thumb (Fig. 15.4).

Fig. 15.5 shows the output of ROM experiments with the change of pending angles in the healthy patient and active and passive sessions from the RA patient, as represented by blue, orange, and gray lines, respectively. For the MCP, PIP, and DIP of the index finger, the average joint angles of repeated sets of the experiment were: 66.09, 70.03, and 12.25 degrees for the MCP, PIP, and DIP; 69.21 degrees for wrist flexion; 67.11 degrees for wrist extension; and 27.69 degrees for thumb abduction. It was observed that the steel strips bent in a nonregular pattern between patients, but the angles measured were consistent and sufficient to aid in the rehabilitation of RA patients' fingers.

The RA patients initially applied high force during passive sessions but were unable to sustain it for a long period of time. This was due to the stiffness of the joints and the pain experienced by RA patients while bending their fingers. Using the exoskeleton, the force applied increased and decreased uniformly during active sessions. An average fingertip force of 18.53 N was measured from all five fingers. As a result, the force was higher than the 7 N maximum force that was considered ideal when performing basic gripping functions. According to the force output results, the developed exoskeleton could be used both as an assistive and a rehabilitative device (Table 15.2).

Experimental results showed that the forces exerted by the RAEXO on the fingertips and the ROM of the finger joints were not identical to those of a healthy human hand (Table 15.2). It may be because of the limited exposure and training the device has received from RA patients. Based on the experimental results, we could expect that RA patients would be able to perform like the healthy control after a sufficient number of training sessions. The ROM values of the finger joints were larger than those from the experiments with ADL (Haglo et al., 2021). Contrary to many other methods, this system did not use a particularly defined actuation kinematic, but the results were superior. Combining the natural properties of human joints with metallic substances (steel) produced the results. As a result, the proposed system was novel, highly agile, comfortable, affordable, and provided end-users with point-to-point customization.

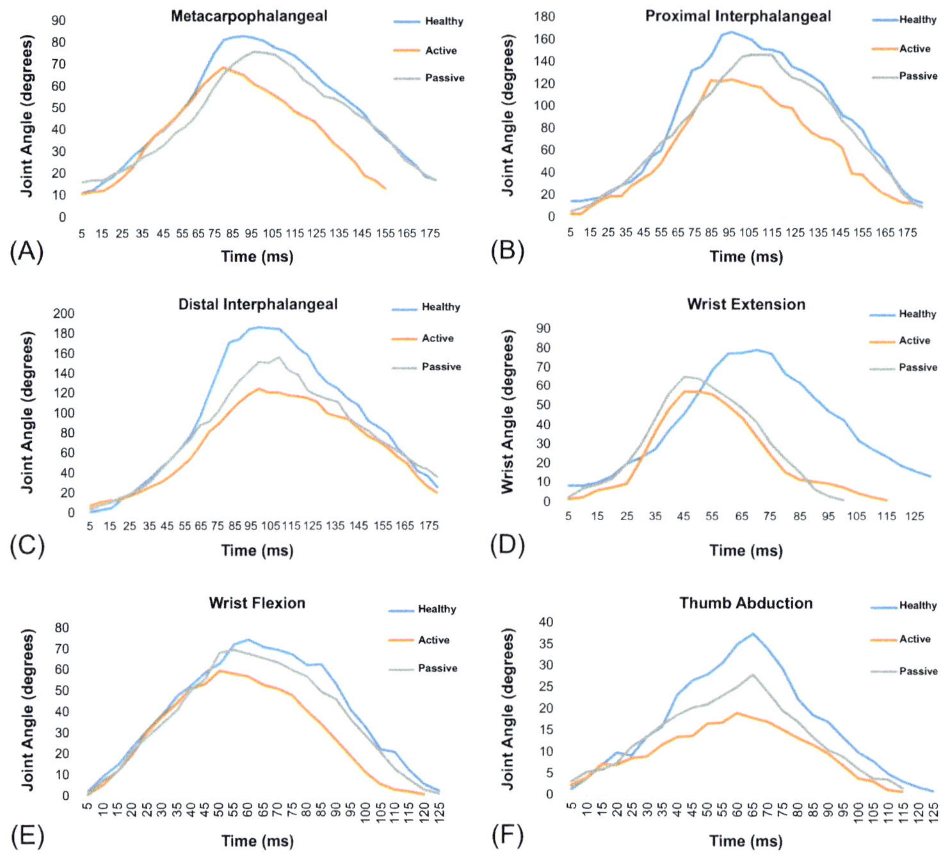

Fig. 15.5 ***Output of ROM experiments with the change of bending angles in a healthy patient, active and passive sessions of RA patient.*** Range of motion experiments carried out with the change of bending angles in a healthy patient, active and passive sessions of RA patient for various hand joints. ((A–F) No permission required.)

Table 15.2 Range of motion associated with various finger and wrist joints in degrees.

				RA Patient	
Part	**Movement**	**Joint**	**Healthy (10 trials)**	**Active (10 trials)**	**Passive (10 trials)**
Wrist	Extension	RCP	4.21–80.05	1.49–59.65	1.64–68.75
	Flexion	RCP	2.96–74.62	1.11–56.58	1.37–70.85
Thumb	Abduction	CMC	1.45–36.19	2.17–20.61	1.39–29.08
	Flexion	MCP	0.71–29.12	1.67–20.63	3.19–26.01
		IP	1.97–82.36	1.34–58.12	2.11–69.19
Forefingers	Flexion	MCP	9.31–83.15	11.31–65.18	10.69–76.09
		PIP	15.48–81.28	3.7–60.18	6.73–76.96
		DIP	1.37–18.94	1.37–6.29	0.39–12.64

No permission required.

To develop our Android app, we used App Inventor from the Massachusetts Institute of Technology (MIT). Due to the block-based programming of App Inventor, users were able to develop efficient and consistently stable Android apps with simple interfaces. Our app had a simple user interface (UI), with all four actuators having separate controls. Each wrist and forefinger had two control options (flexion and extension), and the thumb had four (flexion, extension, abduction, and adduction). Individual control buttons were designed for each option. The actuators had only one flexion mode except for the forefingers, which had two (complete and partial). There were separate buttons for each flexion mode. There is a string associated with each interfacing button. Each time the user tapped a button, the Android app transmitted the unique string via Bluetooth serial communication to the Arduino. When Arduino receives this unique string, it implements the function assigned to it in its Arduino code. The 555 DC motor rotated clockwise when the user tapped the flexion button in the wrist section. This continued until the user stopped tapping the button or tapped the Stop button, which was adjacent to the flexion button. This allowed the user to control all movements manually and to avoid pain by restricting the movement of the exoskeleton. With a very low latency of 3 ms, the Android app can control any actuator seamlessly.

Compared to existing exoskeletons such as the HOH and the Festo ExoHand, our proposed system showed improvements in many aspects. We used flexible elastic steel strips to avoid joint misalignment due to the serial-link mechanisms used in those systems. This also improved the limited ROM in MCP and PIP joints. In a similar way to Festo's ExoHand, our design provided patient-specific customization, but with an improved mechanism. Patients had to make various alterations to the Festo ExoHand before it could be used. With our system, no such modifications were necessary, as the elastic and flexible nature of the material naturally aligned itself with the finger joints. Since our prototype was developed using lightweight 3D printing material, it weighed only 730 g (including three servos, one DC motor, and all other electronic components), which is significantly less than the weights of pneumatic- and hydraulic-based bulky exoskeletons currently available on the market.

Unlike certain other systems like RARD (Matthew Chin et al., 2016), the soft robotic glove (Polygerinos et al., 2015), or hybrid systems (Ahmed et al., 2021), which relied on pneumatic systems utilizing bulky compressors for actuation, we employed a tailored Android application to regulate finger and wrist movements. As a result, the overall weight of the exoskeleton was reduced. Smartphones are common everyday items carried by most people. Patients' mobility is reduced, comfort is compromised, and the cost of the system increases when external compressors are employed. In this regard, our approach was superior, preventing the need for additional attachments such as pneumatic and hydraulic compressors. A real-time adaptation and modification of the force exerted by the steel strips were achieved by adhering to the natural kinematics of finger joints.

The overall output from the experiments was sufficient for daily rehabilitation sessions with RA patients. In comparison with the state-of-the-art designs reported in the literature, a 3D-printed exoskeleton can't deliver natural hand-like performance. The proposed system provides enough output force to create a minimal range of motion in the finger joints. In addition, the proposed system was developed as a lightweight system, allowing for point-to-point customization, and achieving acceptable performance. The system still has a lot of room for improvement. Here are the major limitations:

1. As the present system employs a simple semiactive mechanism for controlling joints, it cannot handle the forces during counteraction in the fingers. Our future works will explore counteractive threshold feedback-control mechanisms for the wrist.
2. The proposed system does not allow clinicians to visualize or investigate RA progression. Our goal is to develop an intelligent EMG-based muscle and joint health-monitoring system to visualize the progress of RA in the future.
3. The proposed system also has problems with finger joint dynamics with improved normal forces.
4. Our system has been tested with one healthy and one RA patient. In the future, we hope to collect more samples from age- and gender-matched subjects to improve the model's performance and to evaluate its practical feasibility.

6 Conclusion

The study involved the design, development, and validation of a 3D-printed exoskeleton utilized as a rehabilitative device for patients with RA. During the development of the proposed exoskeleton, the physical aspects and factors of human hands affected by RA were considered. A flexible steel strip-based exoskeleton was a safe and mechanically sound device for RA rehabilitation. According to the ROM and force exerted on the patient's fingertips, the system is feasible. The output parameters are adequate to achieve efficient actuation of the fingers and wrists. After 10 trials of experiments, it was observed that the range of flexion in forefingers of the MCP joint is high in RA patients (10.69–76.09) compared to normal subjects (9.31–83.15). As a result of using a smartphone to control the exoskeleton system, the efficiency of the process was improved, and thereby the system's overall weight was reduced. The results of the experimental setup demonstrated that the ROM and force output were higher than the ideal values needed to achieve ADL. There was also a significant difference between the values in the experiments conducted with and without an exoskeleton. The findings were consistent with the fact that the exoskeleton could be successfully utilized to perform regular exercises to counteract the progression of RA.

Future research should explore mHealth systems for a broader range of chronic diseases as a potential tool for preventing risk factors and reducing disease complications. These systems can also aid in data recording and exchange for caregivers and patients, acting as powerful complements to diagnostic processes.

References

Abdallah, I.B., Bouteraa, Y., Rekik, C., 2017. Design and development of 3D printed myoelectric robotic exoskeleton for hand rehabilitation. Int. J. Smart Sens. Intell. Syst. 10 (2), 1–26. https://doi.org/10.21307/ijssis-2017-215. 1178-5608.

Ahmed, T., Assad-Uz-Zaman, M., Islam, M.R., Gottheardt, D., McGonigle, E., Brahmi, B., Rahman, M., 2021. Flexohand: a hybrid exoskeleton-based novel hand rehabilitation device. Micromachines 12.

Almutairi, K., Nossent, J., Preen, D., Keen, H., Inderjeeth, C., 2021. The global prevalence of rheumatoid arthritis: a meta-analysis based on a systematic review. Rheumatol. Int. **41** (5), 863–877. Springer Science and Business Media Deutschland GmbH, Australia. 1437160X. Available from: http://link.springer.de/link/service/journals/00296/index.htm. https://doi.org/10.1007/s00296-020-04731-0.

Carbone, F., Bonaventura, A., Liberale, L., Paolino, S., Torre, F., Dallegri, F., Montecucco, F., Cutolo, M., 2020. Atherosclerosis in rheumatoid arthritis: promoters and opponents. Clin. Rev. Allergy Immunol. 58 (1), 1–14. https://doi.org/10.1007/s12016-018-8714-z. 1080-0549.

Chauhan, Jandu, J.S., Goyal, 2021. Arthritis. StatPearls, Treasure Island, FL.

Diego, J.-R.R., Martinez, D.W.C., Robles, G.S., Dizon, J.R.C., 2021. Development of smartphone-controlled hand and arm exoskeleton for persons with disability. Open Eng. **11** (1), 161–170. https://doi.org/10.1515/eng-2021-0016. 2391-5439.

Dorado, J., del Toro, X., Santofimia, M.J., Parreño, A., Cantarero, R., Rubio, A., Lopez, J.C., 2019. A computer-vision-based system for at-home rheumatoid arthritis rehabilitation. Int. J. Distrib. Sens. Netw. **15** (9). SAGE Publications Ltd, Spain. 15501477. Available from: http://journals.sagepub.com/loi/dsn. https://doi.org/10.1177/1550147719875649.

Gerez, L., Gao, G., Dwivedi, A., Liarokapis, M., 2020. A hybrid, wearable exoskeleton glove equipped with variable stiffness joints, abduction capabilities, and a telescopic thumb. IEEE Access **8**, 173345–173358. Institute of Electrical and Electronics Engineers Inc., New Zealand. 21693536. Available from: http://ieeexplore.ieee.org/xpl/RecentIssue.jsp?punumber=6287639. https://doi.org/10.1109/ACCESS.2020.3025273.

Guo, Q., Wang, Y., Xu, D., Nossent, J., Pavlos, N.J., Xu, J., 2018. Rheumatoid arthritis: pathological mechanisms and modern pharmacologic therapies. Bone Res. **6** (1). Sichuan University, China. 20956231. Available from: http://www.nature.com/boneres/. https://doi.org/10.1038/s41413-018-0016-9.

Haglo, H., Wang, E., Berg, O.K., Hoff, J., Helgerud, J., 2021. Smartphone-assisted high-intensity interval training in inflammatory rheumatic disease patients: randomized controlled trial. JMIR mHealth uHealth 9 (10), e28124. https://doi.org/10.2196/28124. 2291-5222.

Kutner, N.G., Zhang, R., Butler, A.J., Wolf, S.L., Alberts, J.L., 2010. Quality-of-life change associated with robotic-assisted therapy to improve hand motor function in patients with subacute stroke: a randomized clinical trial. Phys. Ther. 90 (4), 493–504. https://doi.org/10.2522/ptj.20090160. 00319023.

Li, M., He, B., Liang, Z., Zhao, C.G., Chen, J., Zhuo, Y., Xu, G., Xie, J., Althoefer, K., 2019. An attention-controlled hand exoskeleton for the rehabilitation of finger extension and flexion using a rigid-soft combined mechanism. Front. Neurorobot. **13**. Frontiers Media S.A., China. 16625218. Available from: https://www.frontiersin.org/articles/10.3389/fnbot.2019.00034/pdf. https://doi.org/10.3389/fnbot.2019.00034.

Matthew Chin, H.C., Hoon, L.J., Yeow, R.C.H., 2016. Design and evaluation of rheumatoid arthritis rehabilitative device (RARD) for laterally bent fingers. In: 2016-Proceedings of the IEEE RAS and EMBS International Conference on Biomedical Robotics and Biomechatronics. IEEE Computer Society, Singapore, pp. 839–843. http://ieeexplore.ieee.org/xpl/mostRecentIssue.jsp?punumber=5611781. https://doi.org/10.1109/BIOROB.2016.7523732. 9781509032877.

Moral-Munoz, J.A., Zhang, W., Cobo, M.J., Herrera-Viedma, E., Kaber, D.B., 2021. Smartphone-based systems for physical rehabilitation applications: a systematic review. Assist. Technol. **33** (4), 223–236. Taylor and Francis Ltd., Spain. 19493614. Available from: http://www.tandfonline.com/toc/uaty20/current. https://doi.org/10.1080/10400435.2019.1611676.

Munsterman, T., Takken, T., Wittink, H., 2012. Are persons with rheumatoid arthritis deconditioned? A review of physical activity and aerobic capacity. BMC Musculoskelet. Disord. 13. https://doi.org/10.1186/1471-2474-13-202. 14712474.

Pedersen, B.K., Saltin, B., 2006. Evidence for prescribing exercise as therapy in chronic disease. Scand. J. Med. Sci. Sports **16** (1), 3–63. Blackwell Munksgaard, Denmark. 16000838. Available from: http://onlinelibrary.wiley.com/journal/10.1111/(ISSN)1600-0838. https://doi.org/10.1111/j.1600-0838.2006.00520.x.

Polygerinos, P., Wang, Z., Galloway, K.C., Wood, R.J., Walsh, C.J., 2015. Soft robotic glove for combined assistance and at-home rehabilitation. Robot. Auton. Syst. 73, 135–143. Elsevier B.V. United States. 09218890 https://www.journals.elsevier.com/robotics-and-autonomous-systems. https://doi.org/10.1016/j.robot.2014.08.014.

Rodríguez-Sánchez-Laulhé, P., Luque-Romero, L.G., Blanquero, J., Suero-Pineda, A., Biscarri-Carbonero, Á., Barrero-Garciá, F.J., Heredia-Rizo, A.M., 2020. A mobile app using therapeutic exercise and education for self-management in patients with hand rheumatoid arthritis: a randomized controlled trial protocol. Trials 21 (1). BioMed Central Ltd, Spain. 17456215. Available from: http://www.trialsjournal.com/home/. https://doi.org/10.1186/s13063-020-04713-4.

Roubenoff, R., 2003. Exercise and inflammatory disease. Arthritis Care Res. 49 (2), 263–266. https://doi.org/10.1002/art.11008. 0893-7524.

Rudd, G., Daly, L., Jovanovic, V., Cuckov, F., 2019. A low-cost soft robotic hand exoskeleton for use in therapy of limited hand–motor function. Appl. Sci. 9 (18), 3751. https://doi.org/10.3390/app9183751. 2076-3417.

Tiboni, M., Amici, C., Gloves, S., 2022. A review on recent developments in actuation, sensing, control and applications. Actuators 11 (8), 232. https://doi.org/10.3390/act11080232. 2076-0825.

Ueki, S., Nishimoto, Y., Abe, M., Kawasaki, H., Ito, S., Ishigure, Y., Mizumoto, J., Ojika, T., 2008. Development of virtual reality exercise of hand motion assist robot for rehabilitation therapy by patient self-motion control. In: Proceedings of the 30th Annual International Conference of the IEEE Engineering in Medicine and Biology Society, EMBS'08—Personalized Healthcare through Technology. EEE Computer Society Japan, pp. 4282–4285, https://doi.org/10.1109/iembs.2008.4650156. 9781424418152.

Vélez-Guerrero, M.A., Callejas-Cuervo, M., Mazzoleni, S., 2021. Design, development, and testing of an intelligent wearable robotic exoskeleton prototype for upper limb rehabilitation. Sensors 21 (16), 5411. https://doi.org/10.3390/s21165411. 1424-8220.

Wolf, S.L., Winstein, C.J., Miller, J.P., Taub, E., Uswatte, G., Morris, D., Giuliani, C., Light, K.E., Nichols-Larsen, D., 2006. Effect of constraint-induced movement therapy on upper extremity function 3 to 9 months after stroke: the EXCITE randomized clinical trial. JAMA 296 (17), 2095–2104. American Medical Association, United States. 15383598. Available from: http://jama.ama-assn.org/cgi/reprint/296/17/2095. https://doi.org/10.1001/jama.296.17.2095.

Yap, H.K., Lim, J.H., Nasrallah, F., Goh, J.C.H., Yeow, R.C.H., 2015. A soft exoskeleton for hand assistive and rehabilitation application using pneumatic actuators with variable stiffness. In: 2015-Proceedings—IEEE International Conference on Robotics and Automation. Institute of Electrical and Electronics Engineers Inc., Singapore, pp. 4967–4972, https://doi.org/10.1109/ICRA.2015.7139889. 9781479969234 June.

Zhang, F., Lin, L., Yang, L., Yili, F., 2019. Design of an active and passive control system of hand exoskeleton for rehabilitation. Appl. Sci. 9 (11), 2291. https://doi.org/10.3390/app9112291. 2076-3417.

CHAPTER 16

Prospect of necrobotics in healthcare applications—A critical analysis from the lens of law

Puranjoy Ghosh[a] and Shrilekha Banerjee[b]
[a]KIIT (Deemed to be University), School of Law, Bhubaneswar, Odisha, India
[b]National University of Juridical Science, Law, Kolkata, West Bengal, India

1 Bioengineering and biohybridization

The character of Dr. Frankenstein created by Mary Shelley symbolizes an imaginative yet obsessive vision of science. The creation of a new lifeform by assembling living or nonliving parts was once considered madness but has inspired significant scientific pursuits that, to some degree, echo Dr. Frankenstein's manipulation of natural laws to create an enhanced being. Examples, include selecting embryos with specific traits, gene therapy, and genome sequencing, which involve locating, isolating, and manipulating genes to diverge from nature's processes of generation and life.

While *Frankenstein* did not provide details on how to accomplish genetic engineering, tissue engineering, or chemical biology, the concept of creating a new entity from assembled components resonates with the modern need for an in-depth understanding of cellular and biochemical properties. This knowledge is necessary to ensure that cells, tissues, or organs function effectively even as their natural physiology is modified. Scientific progress demands various tools, techniques, and technologies to splice, sequence, and personalize healthcare solutions. For example, stem cell research, genetic testing, and vaccine development for infectious diseases represent targeted bioengineering applications that maintain the functionality of these modified biological systems.

Bioengineering has built the foundations of human life, enabling customization of natural processes through biotechnology. In medical sciences, bioengineering has significantly advanced artificial life support systems, prosthetics, organ transplants, and devices that integrate seamlessly with human physiology. These developments have led to applications in prosthetics, organ transplants, and even the potential construction of human-like machines and devices. Biomedical engineering and biotechnology have also enabled the design of medical instruments and technologies that address public health issues while conserving environmental resources. For example, biomedical innovations

https://doi.org/10.1016/B978-0-443-24832-0.00016-7

include devices like pacemakers and cochlear implants, and bioengineering technologies have led to robotic surgery and specialized surgical tools.

These innovations often combine mechanical engineering with biotechnologies, expanding diagnostic capabilities and improving patients' quality of life. Biohybridization, which merges various biotic components through bioengineering and biotechnology, has led to the development of personalized and standardized medical products and services in health care.

Necrobotics is a promising and innovative technology that uses a natural microhydraulic gripper made from a dead spider as a component in handheld robotic actuators for handling small, lightweight, and fragile objects during surgeries. The distinctive functionality of necrobotics lies in its ability to act as a natural grasper, activated by pressurized air to perform articulated gripping, ideal for irregular shapes and able to lift objects slightly heavier than its own mass. The most remarkable feature of this gripper is its capability to maneuver light objects in tight spaces within the human body during surgery. The use of necrobotics as a natural, ready-made gripper reduces fabrication needs and conserves energy due to its neutral positioning. In addition, this biobased approach is cost-effective and environmentally friendly, as the grippers are biodegradable, thereby reducing electronic waste and environmental impact. Necrobotics thus represents a significant biotechnological achievement, offering enhanced precision in diagnostics, surgeries, and overall healthcare delivery services.

1.1 Promotion of innovative ecosystems

The United Nations sustainable development goals (SDGs) have sparked experimental initiatives across the globe to establish an inclusive global ecosystem that supports recent advancements in scientific research and development, especially within the healthcare sector. These advancements aim to improve the wellbeing and quality of life for global citizens. Sustainability requires reconfiguring existing practices in scientifically measured ways to meet both means and ends. State-of-the-art technologies, such as those blending biomedical engineering, biohybridization, and biotechnology, are central to this shift. Necrobotics, an advanced biotechnology that has achieved prominence in the healthcare sector, is an example of this integrated approach. Such innovations, developed through rigorous scientific research, must adhere to sustainability principles, ensuring that they are not harmful to public health and are environmental friendly and beneficial to society. This historic achievement of necrobotics has brought important considerations regarding intellectual property rights (IPR).

Firstly, in view of labor theory, innovative biotechnologies like necrobotics must navigate various legal frameworks, with a primary focus on ensuring safety within the healthcare system. Before commercialization, legal validation is needed to meet regulatory standards, making innovations more costly and less accessible to public health care due to the IPR framework. As per the universal IPR principles, any medical device or equipment intended for human use must comply with regulatory requirements specific to each country. In the United States, for example, the Department of Health and Human Services, through the Food and Drug and Administration, oversees compliance with the Code of Federal Regulations (21 CFR 820) and ISO 13485:2016 standards for quality systems and manufacturing practices. Each part of a device, including materials, software, and design, must meet specific requirements for public healthcare safety. Additional legal protections under the American Invents Act require proof of novelty, nonobviousness, and innovation in areas like copyright and patent law, which further contribute to time and cost expenditures before market entry. Once legal protections are obtained, inventors benefit from recognized revenue streams for a set period, reinforcing their position in the healthcare market.

Secondly, the SDGs require innovations to be ecofriendly, using natural resources in ways that minimize harm to other life forms and ecosystems. This preventive approach promotes the conservation of natural processes, biodiversity, and environmental resilience. For example, a recent bioengineering experiment at New York's Mount Siani Hospital (MinnPost, 2024) involved "organ transplantation" although xenotransplantation has been prohibited in the United States since 1999. The use of animal organs in health care highlights the complexity of bioengineering for human benefit, where innovations like necrobotics rely on resources such as dead spiders, a biotic component previously underexplored in human applications. Traditionally, insects like spiders have been valued for their roles in ecosystems rather than for direct human benefits, except as part of the natural food chain. While limited scientific attention has focused on spiders as beneficial bioresources, necrobotics' use of dead spiders as surgical grippers introduces a novel perspective. This innovation encourages further examination of the sustainable use and conservation of biotic resources, particularly insects, which aligns with SDGs 13, 14, and 15 regarding biodiversity protection.

Thirdly, the growing dominance of biomedical engineering technology could lead to oligopolistic or even monopolistic healthcare markets, amplifying inequalities in global healthcare services. Similar to the competition for green energy, tech-sovereignty is becoming a crucial factor in sustainable global development. Developed countries with technological advantages might monopolize these innovations, making them financially inaccessible to developing nations. The practice of "compulsory licensing" is often determined by sovereign jurisdiction, allowing tech-sovereign nations to capitalize on these innovations, while less affluent countries could be excluded from sustainability programs. This concentration of tech-sovereignty and bioresources may influence global healthcare dynamics and enable countries with abundant biospecies, such as spiders, to negotiate economic growth opportunities with technology-rich nations. However, monetizing natural resources in this way must be regulated to prevent environmental degradation and ensure ethical practices. If unchecked, this bioeconomic approach could result in neocolonial economic relations, where resource-rich but tech-poor countries supply biospecies for mutual economic gain, echoing past colonial resource extraction practices. An example is the ongoing elephant poaching in Botswana, despite elephants' endangered status on the International Union for Conservation of Nature (IUCN) Red List, highlighting the ethical and environmental complexities surrounding the exploitation of natural resources for technological advancement.

1.2 Biohybridization in public health care

Biohybridization is a natural part of evolution, recognized in both plants and animals, that aids survival in diverse ecological habitats. This process can be seen in migrating plant species, where new fungal pathogens emerge, and through human practices, such as domestication and crossbreeding, which introduce hybrids to enhance productivity or adaptability. Similarly, in medicinal chemistry, combining various bioactive compounds has long been used in traditional healthcare systems.

The evolution of bionic devices that integrate biology, biophysics, electronics, mechanics, and automation enables the detection of micro-signals of biological activity in patients, which is a significant contribution to health care (Yu et al., 2023). Examples of these technologies include prosthetic limbs, cochlear implants, and bionic ears. Biomimetics inspires in-depth study, replication, and application of biological systems to create innovative medical and assistive devices, as seen with robotic systems modeled after biological species, such as robotic fish (Marchese et al., 2014).

Scientific developments in public healthcare services have enabled extensive networks and biointeractions that promote sustainable use of artificial assistive products. For example, the replacement or restoration of dysfunctional human organs and tissues is achieved by integrating biocompatible, nontoxic, antibacterial, antiviral, and antimicrobial materials. These materials include nonmagnetic, nontoxic, metals, alloys, plastics, and ceramics in healthcare bioengineering and biohybridization. In addition, various medical assistive devices are created with both natural and synthetic materials for internal and external use, such as tweezers, forceps, hemostats, arterial stents, micro-electronic components, pacemakers, hearing aids, and skeletal supports. These tools enhance precision and quality in healthcare services.

Innovative biotechnological advancements in these assistive devices not only restore or replace impaired physical functions but also promote independence, integrity, and mental strength in patients (Shaha et al., 2020). Prosthetic technology has advanced significantly, introducing materials like porous and solid polydomain and monodomain biomedical devices to the healthcare sector. Examples include bioinspired pneumatic shape-morphing elastomers, dielectric elastomers for musculoskeletal conditions, artificial muscle materials, electrostatic actuators for patient mobility, liquid crystal elastomers for cell interaction and tissue growth, hydrogels for bone repair, magnetic actuators, and memory polymers. These materials have been tested and found biocompatible (Li et al., 2023).

A study by Global Market Insights Inc., United States, in 2022 reported a bionic devices market value of USD $5.3 billion, with an expected annual growth rate of more than 8.7% from 2023 to 2032, reflecting growing patient interest in technological healthcare solutions. These trends suggest an expanding healthcare market with opportunities to responsibly utilize natural resources in a competitive global marketplace.

2 Tech competition and bioeconomy

As previously mentioned, the ongoing tech cold war between European and US jurisdictions has intensified global technology competition (Godinho and Simões, 2023). This rivalry is reshaping the geopolitical and geoeconomic landscape of technological leadership, gradually diminishing institutional and economic barriers to free trade and fair competition in the healthcare sector. However, this comes with the inevitable cost of exploiting natural biological species and resources.

Over time, developing economies will increasingly join this competitive landscape, promoting homegrown research and development efforts in sustainable, high-quality healthcare technologies. Yet, immediate progress may be constrained by limited seed capital and the absence of robust long-term investment policies. This reliance on advanced technologies from wealthier economies underscores the need to geographically distribute knowledge and production capacities to ensure equitable access to high-quality healthcare worldwide, supporting the SDGs.

A strong licensing framework could enable cost-intensive, high-quality healthcare services to reach developing countries. However, the affordability of these services for the general population presents a socio-legal challenge that must be carefully examined (Haidar, 2023). To ensure that cutting-edge healthcare technologies support quality public health services on a global scale, mutually beneficial policies will be needed at institutional and legal levels. Such policies would harmonize global trade conditions for high-value, technology-based healthcare products and services, fostering sustainable development and improving living standards, especially for marginalized and vulnerable populations worldwide.

Jurisdictions with rich biodiversity and advanced biotechnologies may be uniquely positioned to leverage these innovations, potentially achieving a new form of global leadership in the biomedical field.

2.1 Necrobotics and socio-legal dilemma

Controversies in necrobotics and biotechnologies include environmental harm and ethical concerns. Reanimating dead organisms or creating life from nonliving matter challenges the natural order of mortality, raising moral questions. The pursuit of these goals not only defies natural laws but also risks promoting overly ambitious projects, such as the exponential growth of robotics. This passion for creating artificial life has resulted in the development of robots that mimic human physical appearance and abilities, posing potential threats to societal structure and natural cycles of recycling and decay.

In addition, privacy concerns emerge in such research, especially when innovative technologies are patented, which can restrict widespread benefits from these advancements. Critics refer to this process as *necrocapitalism*—the transformation of dead organisms into productive labor, which intertwines life and death to generate biogenetic capital, a concept that challenges scientific ethics. This approach commodifies biotic resources, fostering a bioeconomy (Helmreich, 2008) that could prompt a new wave of neocolonialism targeting abundant natural biospecies. The commercial use of these technologies in health care could attract significant financial investment, but it may also prompt power negotiations among wealthier jurisdictions, offering limited benefit to global populations.

Transforming nonhuman life into an economic resource, as has been done historically with metabolic, ecological, or affective labor, reflects a colonial view of nature-capital dynamics that has previously endangered countless natural species. Recent biotechnological achievements, like experimental work with both native and nonnative insect species, have shown that these creatures can influence ecosystem processes, plant nutrient cycling, and biodiversity (Lovett et al., 2016). Insects and invertebrates are vital to ecological balance, yet their use in biotechnologies for creating nanoscale devices for health care could increase demand and impact global trade, potentially leading to biological invasions.

Ongoing environmental challenges, including industrial discharge, mining, sewage, agricultural runoff, and urbanization, already strain ecosystems. Commercializing these bioengineered nanoscale devices for public health care could further endanger insect species and other biological resources. Aggressive bioengineering research and competitive information tactics may increase the demand for certain species, risking banishment or even extinction. To address these risks, integrated measures to strengthen biosecurity and monitor potential misuse are essential. Policy frameworks should include safeguards against unfair procurement practices, unscrupulous phylogenetic studies, and cost-driven exploitation of these species.

Nanoscale bioengineering, with its growing use of cell-based biology or *cellomics*, relies heavily on biocapital and computational modeling. However, the intellectual property claims surrounding these technologies mean that only a few global entities may control these high-value biotechnologies. This concentration of ownership risks antitrust issues, driving up healthcare costs. Licensing and distribution of these technologies to ensure fair market competition remains a critical legal challenge, as access may be limited to a select global population rather than the general public.

3 Environmental governance

Insects constitute a significant biomass within biotic interaction networks, influencing species coevolution, coexistence, and ecological stability over time and space (Cardoso et al., 2020). The Convention on Biological Diversity, addressing the use of harmful substances that contribute to insect extinctions and habitat loss, has long promoted the coextinction of dependent species. This trend negatively impacts the essential ecosystem services that humanity relies upon. The chronic bioaccumulation and biomagnification resulting from the use of chemical fertilizers and pesticides for increased productivity, along with industrial discharge, sewage, mining operations, and agricultural and urban runoff, have significantly impaired phylogenetic relationships and evolutionary trajectories within ecology and environmental governance.

Insects play a crucial role in ecosystems and environmental variation, primarily governed by genetic factors. The recent innovation of necrobotics has prompted researchers worldwide to experiment with insects as ecological indicators for environmental protection. Therefore, insect-specific regulations and classifications recognizing the significance of insects and their contributions to natural processes and bio-resources should be integrated into policy frameworks aimed at conserving natural habitats and the environment.

The use of biological resources for human benefit, whether through traditional methods or scientific exploration, is an age-old practice worldwide. The challenge lies in balancing the excessive commercial exploitation of bioresources with their sustainable use. It is essential to consider human intervention's impact on natural processes and the resulting adverse effects of commercial exploitation. Legal tools are crucial in promoting the sustainable use of natural biological resources without disrupting these processes. To deter the massive extraction of biological resources, including traditional knowledge, patent regimes have been introduced into legal frameworks, accompanied by penal provisions.

Insects possess innate qualities that not only allow them to germinate fungicides but also to absorb these substances throughout their life cycles. Nature itself, through these natural processes, helps maintain ecosystem harmony. However, the drive for commercial exploitation of natural habitats and biological resources threatens to outlast the resilience of climatic conditions and ecosystems, leading to the erosion of natural balancing mechanisms that cannot be easily repaired, as seen in global warming. Therefore, the approach to sustainable biological resources has gained significant importance in developmental frameworks worldwide.

4 Innovative biotechnologies and socio-legal challenges

The controversial issues surrounding biotechnologies primarily involve environmental abuse and the moral and ethical consequences of such practices. Environmental abuse, particularly in the procurement of biotic components, raises concerns. The drive to create life from the dead or to reanimate deceased organisms undermines the natural order and universal principles governing the mortality of living beings. This pursuit of unnatural goals, such as creating hypothetical creatures, not only violates the laws of nature but can also lead to harmful and ambitious outcomes, exemplified by the exponential growth of robotics. The desire to create artificial human agents has given rise to robotics that gradually replace the physical presence and capabilities of human beings, threatening societal structures and disrupting the natural recycling order.

In addition, privacy concerns linked to innovative research, along with the protection of such technologies under existing patent regimes, contribute to an exclusionary approach that limits the equitable enjoyment of these innovations.

There are various vectors within bioengineering and biotechnologies. For instance, polysaccharides found in plants, animals, and microorganisms are abundant in the biological world and possess numerous structural and functional features. Biomass-derived materials are utilized for applications such as biosensors and in the biomedical field for bone tissue engineering and artificial blood vessels, owing to their nontoxicity, renewability, and biodegradable characteristics. Necrobotics embodies the direct correlation among corporate interests, scientific endeavors (represented as bio-capital), the preservation of natural resources, and ecosystem beneficiaries.

The potential of these biotechnologies and their corresponding assistive devices and products may foster a capitalist inclination among global players to commercialize high-value devices and bioproducts, promoting the notion of "life as surplus" within global health markets. However, regenerative bioengineering faces broad challenges, including ethnographic, ethical, environmental, geopolitical, and legal issues. The potential of bio-capital species could emerge as a new currency in governance mechanisms, similar to green hydrogen in the realm of green energy.

The socioeconomic implications of these innovative biotechnologies may drive a shift towards renewable resources, fostering a new social contract and global agreement (Chatalova and Korotayev, 2022). Advanced medical technologies are welcomed, provided they do not encroach upon existing biospecies, particularly when such encroachment could harm the natural life cycles of these species and exacerbate the exploitation of biodiversity-rich natural habitats and the environmental resilience processes.

5 Conclusion

As previously mentioned, the multitude of vectors within bioengineering and biotechnologies requires comprehensive calculations to establish a robust policy framework for regulatory governance. This framework is essential for ensuring the seamless availability of these technologies and assistive products, ultimately contributing to sustainable development (Ahmed, 2023) and assistive products for achieving the central objectives of sustainable development. Comprehensive and beneficial global policy framework has become the call of the hour.

The concentration of advanced biotechnologies in healthcare services within the jurisdictions of developed economies, enriched by favorable biogeographical conditions, risks promoting an absolutist status in the realm of bioeconomy. While the spatial distribution of these technologies may harmonize power dynamics, there is an imperative to design Access-Profit Sharing Agreements that facilitate equitable access to bioresources. Failing to do so could lead to exploitative practices and undermine sustainable development efforts. Thus, formulating a comprehensive international policy framework is crucial to establishing a balanced and equitable mechanism at a global scale, aligning with the principles of sustainability.

Although the IUCN has recommended engaging indigenous and local communities in policy planning and participatory management for biodiversity conservation (2011), the existing legislative frameworks often prioritize a top-down approach. This can hinder effective environmental governance and the preservation of biodiversity.

The commercial exploitation of biotic resources not only jeopardizes the resilience of natural processes but may also incentivize the introduction of misleading units to manipulate phylogenetic studies and encroach upon vital natural habitats rich in biocapital. Therefore, it is essential to revisit and redesign domestic laws to prioritize environmental governance through protection, conservation, and sustainable procurement, ensuring biosecurity in a sustainable manner. Legislation must be firm in promoting natural processes of environmental resilience, with active engagement of indigenous and local communities to enhance effective and sustainable governance.

The IPR regime must be reformed to establish universal standards for innovative biotechnologies. This would prevent discrepancies in patent registration from exploiting healthcare markets where these technologies are intended for widespread use. The COVID-19 pandemic exemplifies the need for mutual cooperation and integration among global citizens, transcending geopolitical boundaries to ensure the sustainability of humanity. Additionally, establishing universal standards would promote accessibility, availability, affordability, and equitable opportunities for quality healthcare services, reinforcing the essence of fair competition. In exceptional cases, such as when innovative technologies have rare applications for unique diseases, patent mechanisms may be considered for free use.

Finally, it is necessary to emphasize mutual assistance, coordination, and cooperation within the policy framework to facilitate effective implementation and promote a vibrant supply chain for bioproducts derived from advanced biotechnologies like necrobotics. This includes sharing knowledge, critical technological capabilities, and providing financial assistance to developing economies in their pursuit of technological independence in healthcare services.

References

Ahmed, T., 2023. Bio-inspired artificial synapses: neuromorphic computing chip engineering with soft biomaterials. Mem. Mater. Devices Circuits Syst. 6, 100088. https://doi.org/10.1016/j.memori.2023.100088. 27730646.

Cardoso, P., Barton, P.S., Birkhofer, K., Chichorro, F., Deacon, C., Fartmann, T., Fukushima, C.S., Gaigher, R., Habel, J.C., Hallmann, C.A., Hill, M.J., Hochkirch, A., Kwak, M.L., Mammola, S., Noriega, J.A., Orfinger, A.B., Pedraza, F., Pryke, J.S., Roque, F.O., Settele, J., Simaika, J.P., Stork, N.E., Suhling, F., Vorster, C., Samways, M.J., 2020. Scientists' warning to humanity on insect extinctions. Biol. Conserv. 242. Elsevier Ltd, Finland. 00063207. Available from: https://scientistswarning.forestry.oregonstate.edu/sites/default/files/Cardoso2020_Warning.pdf. https://doi.org/10.1016/j.biocon.2020.108426.

Chatalova, L., Korotayev, A., 2022. The long cycle perspective on the emerging bio age. Futures **141**. Elsevier Ltd, Germany. 00163287. Available from: https://www.sciencedirect.com/science/article/pii/S001632872200091X?via%3Dihub. https://doi.org/10.1016/j.futures.2022.102991.

Godinho, M.M., Simões, V.C., 2023. The tech cold war: what can we learn from the most dynamic patent classes? Int. Bus. Rev. 32 (6). Elsevier Ltd, Portugal. 09695931. Available from: https://www.sciencedirect.com/science/article/pii/S0969593123000409. https://doi.org/10.1016/j.ibusrev.2023.102140.

Haidar, Z.S., 2023. Introductory Chapter: BioMimetics for HealthCare—Innovations Inspired by Nature. IntechOpen, https://doi.org/10.5772/intechopen.106328.

Helmreich, S., 2008. Species of biocapital. Sci. Cult. 17 (4), 463–478. https://doi.org/10.1080/09505430802519256. 14701189.

Li, D., Wang, Y., Huang, W., Gong, H., 2023. Biomass-derived fiber materials for biomedical applications. Front. Mater. 10. Frontiers Media S.A., China. 22968016. Available from: https://www.frontiersin.org/journals/materials/articles/10.3389/fmats.2023.1058050/full. https://doi.org/10.3389/fmats.2023.1058050.

Lovett, G.M., Weiss, M., Liebhold, A.M., Holmes, T.P., Leung, B., Lambert, K.F., Orwig, D.A., Campbell, F.T., Rosenthal, J., McCullough, D.G., Wildova, R., Ayres, M.P., Canham, C.D., Foster, D.R., La Deau, S.L., Weldy, T., 2016. Nonnative forest insects and pathogens in the United States: impacts and policy options. Ecol. Appl. 26 (5), 1437–1455. Ecological Society of America, United States. 19395582. Available from: https://esajournals.onlinelibrary.wiley.com/doi/full/10.1890/15-1176. https://doi.org/10.1890/15-1176.

Marchese, A.D., Onal, C.D., Rus, D., 2014. Autonomous soft robotic fish capable of escape maneuvers using fluidic elastomer actuators. Soft Robot. 1 (1), 75–87. Mary Ann Liebert Inc., United States. 21695180. Available from: https://www.liebertpub.com/doi/10.1089/soro.2013.0009. https://doi.org/10.1089/soro.2013.0009.

MinnPost, 2024. Hennepin County Sheriff's Office pauses DNA casework after lab contamination. https://www.minnpost.com/glean/2024/10/hennepin-county-sheriffs-office-pauses-dna-casework-after-lab-contamination/.

Shaha, R.K., Merkel, D.R., Anderson, M.P., Devereaux, E.J., Patel, R.R., Torbati, A.H., Willett, N., Yakacki, C.M., Frick, C.P., 2020. Biocompatible liquid-crystal elastomers mimic the intervertebral disc. J. Mech. Behav. Biomed. Mater. 107. Elsevier Ltd, United States. 18780180. Available from: https://www.sciencedirect.com/science/article/pii/S1751616120303118. https://doi.org/10.1016/j.jmbbm.2020.103757.

Yu, H., Li, H., Sun, X., Pan, L., 2023. Biomimetic flexible sensors and their applications in human health detection. Biomimetics 8 (3), 293. https://doi.org/10.3390/biomimetics8030293. 2313-7673.

Index

Note: Page numbers followed by *f* indicate figures and *t* indicate tables.

N

9780443248320